BIOCHEMISTRY RESEARCH TRENDS

G-QUADRUPLEX STRUCTURES, FORMATION AND ROLE IN BIOLOGY

BIOCHEMISTRY RESEARCH TRENDS

Additional books in this series can be found on Nova's website
under the Series tab.

Additional e-books in this series can be found on Nova's website
under the e-book tab.

BIOCHEMISTRY RESEARCH TRENDS

G-QUADRUPLEX STRUCTURES, FORMATION AND ROLE IN BIOLOGY

HANNAH SANTOS
EDITOR

nova publishers
New York

Library of Congress Cataloging-in-Publication Data

Names: Santos, Hannah, editor.
Title: G-quadruplex structures : formation and role in biology / editor, Hannah Santos.
Description: Hauppauge, New York : Nova Science Publishers, Inc., 2016. |
Series: Biochemistry research trends | Includes index.
Identifiers: LCCN 2016023906 (print) | LCCN 2016031885 (ebook) | ISBN 9781634855129 (hardcover) |
ISBN 9781634855501 ()
Subjects: LCSH: Quadruplex nucleic acids.
Classification: LCC QP625.Q33 G885 2016 (print) | LCC QP625.Q33 (ebook) | DDC 572.8/5--dc23
LC record available at https://lccn.loc.gov/2016023906

Published by Nova Science Publishers, Inc. † New York

CONTENTS

PREFACE

This book provides current research on G-quadruplex structures, their formation and role in biology. Chapter One summarizes the structural features of quadruplexes built by sequence extensions and site-specifically incorporated modified building blocks that have been established to induce the formation of stable, majority single G-quadruplex folding topologies, primarily from the human telomere repeat DNAs. Chapter Two discusses the G-quadruplexes and i-motifs as scaffolds for molecular engineering of DNA aptamers. Chapter Three examines the case of the natural alkaloids coptisine and chelerythrine. Chapter Four describes the G-quadruplex induced fluorescent signals that have been employed for the study of their folding and unfolding mechanism and thereby visualisation of their formation and structure. Chapter Five reviews the historical aspect of how major biological roles of G-quadruplexes have been uncovered by the use of specifically designed small molecule probes; and discusses the future applications of G-quadruplex interacting small molecules, including the development as potential drug candidates. Chapter Six focuses on the latest advances in the development of the telomerase assays and the authors further discuss the application potentiality of the assays for clinical samples.

Chapter 1 - DNA and RNA G-quadruplexes are non-B, four-stranded nucleic acid structures, which can self-assemble from guanine-rich sequences in vitro. Their structural components comprise the two or more guanine tetrads (G:G:G:G), the connecting loops between the guanine repeats, and the ends-appended sequences. The planar G-tetrads are held together by Hoogsteen-type circular double H-bonds, the tetrads stack on each other primarily via π-π interactions, and the quadruplex structures are stabilized by cations. Depending on the G-rich sequence, the length and composition of loops, type of cations, the concentration of the G-rich oligonucleotide and on the presence of cosolutes, the oligonucleotides can fold into a variety of topologies. Main topologies are the antiparallel, the (3 + 1) hybrid, and the parallel. The architectures can be built from one, two or four nucleic acid strands. Topologies can convert or be converted into each other by changing the solution conditions. The initial and also the converted folds are often polymorphic. For structural studies stable, single folds are preferentially required. Therefore, it is crucial to find the solution and sequence settings that support a predominant or a single quadruplex fold. Physiologically relevant potassium ion solutions or these with added cosolutes that mimic the crowded cellular environment are the most often used solution conditions. As for the sequences, a well-established approach is the sequence-specific mutations, for which synthetic nucleotide analogs are frequently used. This review summarizes the structural features of quadruplexes built by sequence extensions and site-specifically incorporated modified building blocks that have been established to induce the formation of stable, majority single G-quadruplex folding topologies, primarily from the human telomere repeat DNAs.

Chapter 2 - DNA aptamers are oligonucleotides with a particular three-dimensional structure that provides high affinity and high specificity to a target. Aptamers are applied as molecular recognition elements (MoRE) in nanoscience and nanotechnology as well as in theranostics as sensors, delivery systems and pharmaceuticals inhibiting target functioning; refined aptamers can be considered as functionally analogous to antibodies. Not too much is yet known about the aptamer structure–function relationship but, interestingly enough, some aptamers have a functionally significant G-quadruplex structure.

Thus, the basic thrombin-binding aptamer HD1 is a unimolecular antiparallel G-quadruplex with three short loops. Two TT-loops of the aptamer bind to exosite I on the thrombin surface. The G-quadruplex itself is a scaffold that establishes proper conformation of TT-loops: T4 and T13 are paired, while T3 and T12 are buried in hydrophobic thrombin pockets. T4-T13 pairing is absolutely required for aptamer functional activity, whereas T3 and T12 can be modified or even changed.

The scaffolding role of the G-quadruplex is highlighted by the structure–function effects of different metal cation coordination. Potassium cations provide the highest functional activity, whereas several other cations alter G-quartet fine structure and therefore change TT-loop conformation, leading to the complete loss of aptamer functional activity. On the other hand, the activity of a thrombin-binding aptamer can be increased by extending the basic G-quadruplex with either a duplex or with another G-quadruplex. Surprisingly, extra elements improve the aptamer function (i.e., by decreasing inhibition constants), although they do not provide additional interactions between the aptamer and thrombin. Most probably these extra modules support a specific G-quadruplex conformation, keeping the proper orientation of TT-loops. Therefore, a G-quadruplex scaffold could be used for modular construction of MoRE having capacity for structure–function tuning.

Another spectacular example of a scaffold for MoRE is the so-called i-motif. A DNA aptamer to hemagglutinin of influenza virus, BV42, has two continuous blocks of 5 cytosines and 22 cytosines that can fold into a unimolecular i-motif made up of four blocks of five cytosines. The structure is stable enough at slightly acidic pH; a high pK_a value of 7.1 provides the possibility for aptamer medical application under physiological conditions. Although the structure of the aptamer–hemagglutinin complex is not yet solved, the i-motif is very likely to be a scaffold for hemagglutinin-binding elements. One of the loops has GAA sequences, and two other loops have CC sequences; both kinds of sequence could be part of binding loops that have been suggested from alignment of all known aptamers to hemagglutinin. Therefore, aptamer BV42 is another example of DNA aptamers having a scaffold that supports proper conformation of binding loops.

In conclusion, both examples of DNA aptamer exhibit striking similarities with natural protein MoRE, antibodies, where DNA G-quadruplexes and i-motifs, as well as protein beta-sandwiches, are scaffolds with several loops that provide a recognition function.

Chapter 3 - Nucleic acids are highly polymorphic biomolecules: their significant structural variability is not only due to sugar-phosphate backbone flexibility, but even to the ability of purine bases to donate and/or to accept hydrogen-bonding by using different molecular sides. Together with the so-called G-quadruplex structures, abundant in many sequenced genomes, recent evidence of unusual binding motifs containing non-guanine-based quartets and other kinds of mismatches has been reported. Small molecules able to preferentially bind mismatching points, instead of the fully matched DNA stretches, are recognized to have potential as drug candidates or tools for chemical biology *in vivo*.

The two natural isoquinoline alkaloids chelerythrine and coptisine are supposed to be unable to interact through classic intercalative processes with ds-DNAs due to their elongated nature and were selected as probes of unusual binding motifs. The X-ray diffraction analyses of the CGTACG/chelerythrine, CGTACG/coptisine and CGATCG/coptisine adducts actually show chelerythrine and coptisine localized at the junction of adjacent DNA helices, in unusual intercalative binding sites containing G:G homo base mismatches, in which the guanines are hydrogen-bonded through their minor groove faces. In the CGTACG/chelerythrine and CGTACG/coptisine, the intercalation site contains six bases and is defined by two 3'-GpC dinucleotides from different chains and completed by two more guanine residues of the 5'-CpG. Also the CGATCG/coptisine complex evidences the same kind of G:G mismatch, in which two C:G:G:C platforms sandwich the alkaloid. In all the reported crystal structures, the overlapping surface between the ligands and the double stranded non-canonical binding sites, is significantly higher than what is expected for classic intercalative binding.

Chapter 4 - Fluorescently labeled oligonucleotides are the remarkable achievement for the biochemical science and become the most important probes to detect and localize specific nucleic acid targets in cells. Under certain conditions, repetitive oligonucleotide (DNA or RNA) sequences folded into a non-B form higher order secondary structure such as tetraplex. In particular G-quadruplex is a tetraplex of a single or more oligonucleotide sequence(s) formed by Hoogsteen hydrogen bonding of four guanine units. Such G-quadruplex structure present at the termini of the eukaryotic chromosome and inhibits the activity of telomerase. Moreover, they are also a potential target for novel anticancer drugs. The topology of G-quadruplexes can be parallel, antiparallel, as well as other combinations of strand orientation. The structure and folding of tetraplex have been well studied by fluorescent phenomena such as excimer, exciplex, fluorescence resonance energy transfer (FRET), fluorescence enhancement, fluorescence quenching etc. G-quartet plane is an excellent platform for stacking interaction with many small fluorophores and vice versa. Moreover, fluorophores having large π-aromatic surface and/or positively charged substituents could be an effective G-quadruplex binder. Thus, tetraplex formation leads to a readable fluorescent signal upon stacking interaction between G-quartet and the fluorophore, whether it is external or covalently attached fluorophore. This chapter describes the G-quadruplex induced fluorescent signals that have been employed for the study of their folding and unfolding mechanism and thereby visualisation of their formation and structure.

Chapter 5 - Much of what the authors have learned about G-quadruplexes stems from the research conducted by chemists, who designed specific molecules to target these structures *in vitro* and *in vivo*. Over the past decades there has been an increasing success in selectively targeting G-quadruplex DNA and RNA structures with small molecules, leading to elucidating some of their implications in biological processes *in vivo*. Biological processes, where G-quadruplexes have implications, include telomere maintenance, the DNA damage response, gene transcription and translation and genetic and epigenetic control mechanisms. This has led to the development of potential therapeutic agents, particularly as a novel anti-cancer strategy, to specifically target aspects involving these processes. This chapter deals first with the historical aspect of how major biological roles of G-quadruplexes have been uncovered by the use of specifically designed small molecule probes. This is used to highlight how fundamental research has contributed to identifying biological functions of G-quadruplexes and their potential as therapeutic targets. This will be accompanied with an overview of the major classes of small molecules in this research field. Finally, this chapter discusses the future applications of G-quadruplex interacting small molecules, including the development as potential drug candidates.

Chapter 6 - Telomeres at the ends of eukaryotic chromosomes contain accessory proteins and telomeric DNA. Telomeric DNA is composed of double-stranded tandem repeats of a G-rich motif (5'-TTAGGG-3' in humans) and a complementary C-rich motif, and a single-stranded 3' overhang of the G-rich motif, termed a G-tail. In normal somatic cells, telomeric DNA shortens with every cell division due to the end replication problem. The shortening of telomeric DNA finally leads to apoptosis. In contrast, in most tumor cells, the length of telomeric DNA is maintained by highly activated telomerase, which is a ribonucleoprotein that adds new six nucleotides, 5'-GGTTAG-3' at the 3'-end of the G-rich motif. The maintenance of telomeric DNA is a critical biological event for carcinogenesis. The specificity and universality of the telomerase activity for tumor cells demonstrates that the detection and quantification of the telomerase activity should be greatly helpful for a precise cancer diagnosis. The telomeric repeat amplification protocol (TRAP) assay, which is the most widely used telomerase assay, can detect telomerase activity with high sensitivity because it amplifies the telomerase reaction products by polymerase chain reaction (PCR). However, this assay is not appropriate for clinical applications because PCR inhibitors in clinical samples cause false negative results. Thus, researchers have proposed other telomerase assays to solve the problem. Those assays are classified into three main groups: (i) Using non enzymatic signal amplification processes. (ii) Using enzymatic signal amplification processes other than PCR. (iii) Combining processes that remove the inhibitors from the reaction mixture and enzymatic signal amplification processes including PCR. In this review, the authors focus on the latest advances in the development of the telomerase assays and further discuss the application potentiality of the assays for clinical samples.

In: G-Quadruplex Structures, Formation and Role in Biology ISBN: 978-1-63485-512-9
Editor: Hannah Santos

Chapter 1

DESIGNING G-QUADRUPLEX TOPOLOGIES BY FOCUSING ON SEQUENCE MODIFICATIONS

Janos Sagi*
Rimstone Laboratory, Carlsbad, CA, US

ABSTRACT

DNA and RNA G-quadruplexes are non-B, four-stranded nucleic acid structures, which can self-assemble from guanine-rich sequences in vitro. Their structural components comprise the two or more guanine tetrads (G:G:G:G), the connecting loops between the guanine repeats, and the ends-appended sequences. The planar G-tetrads are held together by Hoogsteen-type circular double H-bonds, the tetrads stack on each other primarily via π-π interactions, and the quadruplex structures are stabilized by cations. Depending on the G-rich sequence, the length and composition of loops, type of cations, the concentration of the G-rich oligonucleotide and on the presence of cosolutes, the oligonucleotides can fold into a variety of topologies. Main topologies are the antiparallel, the (3 + 1) hybrid, and the parallel. The architectures can be built from one, two or four nucleic acid strands. Topologies can convert or be converted into each other by changing the solution conditions. The initial and also the converted folds are often polymorphic. For structural studies stable, single folds are preferentially required. Therefore, it is crucial to find the solution and sequence settings that support a predominant or a single quadruplex fold. Physiologically relevant potassium ion solutions or these with added cosolutes that mimic the crowded cellular environment are the most often used solution conditions. As for the sequences, a well-established approach is the sequence-specific mutations, for which synthetic nucleotide analogs are frequently used. This review summarizes the structural features of quadruplexes built by sequence extensions and site-specifically incorporated modified building blocks that have been established to induce the formation of stable, majority single G-quadruplex folding topologies, primarily from the human telomere repeat DNAs.

Keywords: G-quadruplex, stability, fold, topology, telomere DNA, sequence, mutation, nucleotide analog, substitution

* jans@rimstonelab.com.

INTRODUCTION

DNA and RNA sequences can fold into canonical B-, and A-type double helices and also into noncanonical, unusual structures, such as Z-duplex, bent-, and bulged-duplexes, hairpins, triplex and quadruplex architectures, etc. Guanine (G)-rich sequences tend to fold into the non-B, four-stranded structures, called G-quadruplexes or G-tetraplexes. Depending on sequence features and solution conditions the G-rich nucleic acids can fold into multiple topologies composed of antiparallel and/or parallel-oriented strands and various numbers of molecules [1-7]. Chromosomal DNAs are abundant of G-rich sequences. Close to 700K such sequences are known presently [8] that have the potential to fold into quadruplexes. The conversion from duplex DNAs to quadruplexes are thought to be induced by superhelical stress. Selvan et al. [9] found that population of G-quadruplexes increased linearly with positive superhelicity, up to a plateau level. The molecular basis is that the flexibility of duplex DNA increases with positive superhelical stress, suggesting that G-quadruplex formation is correlated with the melting of duplex DNA. Most such sequences have been detected in gene promoters where these can act as molecular switches of gene expressions [9, 10]. The other group of G-rich sequences is in the telomere DNAs at both ends of the linear vertebrate and human chromosomes [11].

Sequences of G-rich DNA that can potentially form quadruplexes vary broadly in gene promoters [12-14], whereas, the telomere DNA is highly conserved in vertebrates and humans. Conformation and stability of structures formed by the human telomere (htel) TTAGGG repeat DNAs are the most frequently and extensively studied quadruplex models. This review mostly deals with these htel DNAs with which various defined folding topologies can be created.

Telomeres are nucleic acid–protein (Shelterin) complexes that serve as protective caps at the end of linear chromosomes, thus telomeres are critical for cell survival [15, 16]. The human telomeric DNA consists of 5-10 kilobase repeats of the hexanucleotide duplex 5'-TTAGGG-3'/5'-CCCTAA-3' that terminates in 100-200 nucleotide-long G-rich TTAGGG single-stranded overhang at the 3'-end of the duplex. The Shelterin complex includes the POT1 protein that binds the telomeric overhang [17, 18]. This length of the 3' overhang contains 16-35 TTAGGG hexameric repeats [19, 20] that could theoretically form 4-10 units of the core 21 nucleotide long (21mer) quadruplex structures of the $GGG(TTAGGG)_3$ sequence, used here later on as htel-21. These 3' G-tails can spontaneously adopt thermodynamically stable G-quadruplex structures in cationic solutions. In vivo the 3'-overhang is contained in higher order secondary structures, such as D-loop, T-loop and G-quadruplex via complex formation with the Shelterin proteins [4, 21, 22]. Each replication results in a loss of 50-200 nucleotides and telomeres also shorten as a consequence of oxidative stress, i.e., by the action of reactive oxygen species (ROS). Telomere functions are lost by critical shortening. This leads to senescence and cell death through a p53-dependent checkpoint that detects altered telomere states at the G1/S transition. In this way, telomeres serve as molecular clocks of the lifespan of human somatic cells [23]. The intact higher-order protein-telomere DNA architectures protect the ends of linear chromosomes from degradation and end-joining with other DNAs, and these compact structures make the chromosome ends also inaccessible for the cellular DNA Damage Response (repair) systems. The G-quadruplex region can obstruct the telomere extension catalyzed by the telomerase enzyme system, which extension is a major step in most human cancer cells of becoming immortal. In this way, stabilization of quadruplex structures by quadruplex binding ligands has marked the telomeric G-quadruplexes as new potential targets for anticancer strategies [6, 24].

In laboratory, the human telomere repeat sequences can be synthesized from various numbers of nucleotides. This will be indicated here as htel-21, htel-22, etc. referring to the length of the telomeric oligonucleotide. Sequence of the core 4-repeat telomere DNA sequence, htel-21 is the $GGG(TTAGGG)_3$. The extended length analogs will often be abbreviated as derivatives of the core htel-21, for instance, the 22mer $AGGG(TTAGGG)_3$ as A(htel-21).

G-quadruplexes of the telomere sequences are highly polymorphic, can adopt multiple conformations that depend on the length of the G-tract, length and composition of loops, the concentration of the oligonucleotide, the flanking nucleotides at the 5'- and 3' ends, on the chemical modifications of nucleotides, the type and concentration of the cations and the cosolvents added to the solution as well as on the binding of quadruplex-specific drugs [3, 4, 6, 25-28]. Within the three major types five subtypes of htel conformations have been defined up to now: the basket- and chair-type *antiparallels*, the two forms of the (3 + 1) *hybrids* (hybrid-1 and hyrid-2) and the *parallel*. All five topologies can assemble from a single, two, four (or even more) nucleic acid strands.

The monomolecular scaffolds, also known as intramolecular, can be described with the general formula of G_3-(N_{1-7})-G_3-(N_{1-7})-G_3-(N_{1-7})-G_{3+}, which denotes four G-tracts of at least three continuous guanine nucleotides connected by loop linkers of 1−7 residues [4,6,29]. The folding motifs of the intramolecular DNA quadruplexes are thought to be regulated by the types of loops, the syn or anti N-glycosidic conformations of guanine nucleosides and the solution conditions. These details determine the orientation of the strands and the alignment of the G-tetrads to each other [30-34]. The intramolecular antiparallel (basket-type) [35] and the parallel G-quadruplexes [36] of the A(htel-21) DNA were the first major topologies solved: the basket in Na^+-containing solution by NMR spectroscopy and the parallel by X-ray technique in K^+-crystals. The five major telomere quadruplex folds most frequently discussed in the literature are schematically illustrated in Figure 1. Three types of loops, composed of TTA nucleotides are found in the five types of quadruplex folds, the edgewise (also called lateral-type), diagonal and double chain-reversing propeller that connect the G-tetrads. The antiparallel topologies are the Na^+-stabilized basket-type and the chair-type. In the basket-type structure each strand has an antiparallel-oriented neighbor strand and the quadruplex contains both syn and anti N-glycosidic linkage-bonded 2'-deoxyguanosines (Figure 2). In the basket fold the order of loops is edgewise-diagonal-edgewise starting from the 5'-end of the sequence, for example, in the quadruplex of the A(htel-21) $AGGG(TTAGGG)_3$. The chair-type antiparallel differs from the basket-type topology in that it contains three edgewise loops (Figure 1). The two types of (3 + 1) hybrid folds were primarily observed in K^+ solutions. The (3 + 1) refers to the three parallel strands and one antiparallel oriented DNA strand in the scaffold. The two hybrids differ in the positions of the syn and anti guanosines, which determine the orientation of loops.

In hybrid-1 the order of loops is propeller-edgewise-edgewise [37-39]. In hybrid-2 topology the edgewise-edgewise-propeller is the order [39-41]. In physiologically relevant potassium solutions (100-150 mM K^+) both hybrid forms are present in various ratios in the htel-21 and A(htel-21) quadruplexes. The two hybrid forms are interconvertible due to the small energetic differences between them, 1-2 kcal/mol [42]. In the parallel topology all three

TTA loops are chain-reversal propeller types and the N-glycosidic conformation of all G-nucleotides are anti [6, 36]. To determine the exact conformation, which also determines the structural stability of the particular topology it was necessary to find ways to build single or, at least, majority single folds that are stable in various solution conditions. The focus of this review is a sum up methods currently known to build different single major topologies. For this, the natural, wild-type G-rich and the loop sequences had to be modified by site-specific incorporation of natural or synthetic building blocks, by attaching nucleotides to either terminus of the oligonucleotides, by changing the type or concentration of cations, or by adding cosolvents to the buffered aqueous solution.

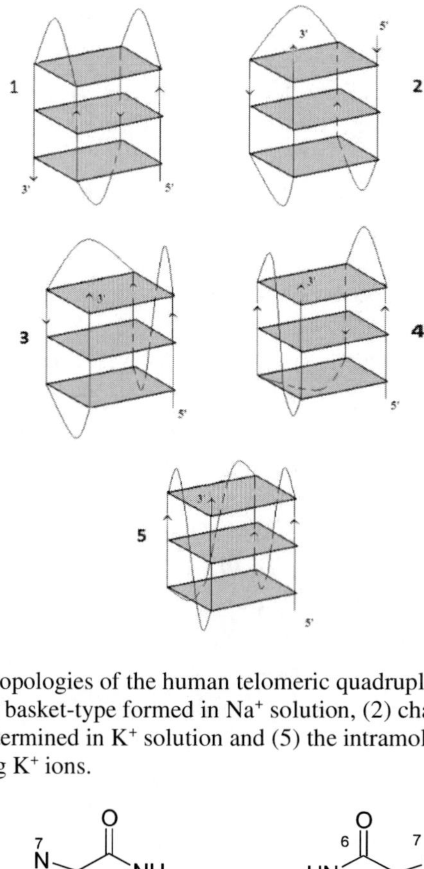

Figure 1. Schematic folding topologies of the human telomeric quadruplexes observed experimentally. Structures are (1) antiparallel basket-type formed in Na^+ solution, (2) chair-type antiparallel, (3) hybrid-1, (4) hybrid-2 topologies determined in K^+ solution and (5) the intramolecular parallel topology observed in crystal containing K^+ ions.

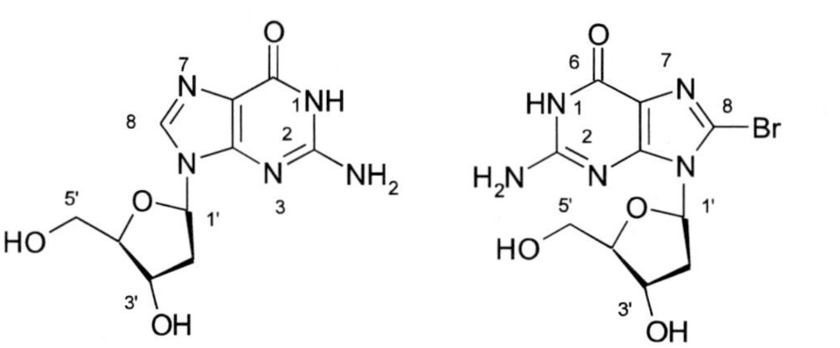

anti 2'-deoxyguanosine *syn* 8-bromo-2'-deoxyguanosine

Figure 2. The syn and anti N-glycosidic conformations of guanosines as illustrated by 8-bromo-2'-deoxyguanosines.

1. Effect of Sequence on Quadruplex Topology

1.1. G-rich Sequences That Predispose Majority Folds

Nucleotides attached either to the 5'-, 3'- or both ends of oligonucleotides, also called as the appended nucleotides can influence the assembly, conformation and stability of a quadruplex nucleic acid. The effect of chain extension is a function of location, type and sequence of the attached units.

The 22 nucleotide-long human telomere repeat deoxyoligonucleotide $AG_3(TTAG_3)_3$ or A(htel-21) forms a basket-type antiparallel scaffold in 20-150 mM buffered sodium ion solutions at neutral pH according to Wang and Patel's original NMR and CD measurements [35]. Figure 1 shows the basket form and Figure 3 illustrates its CD spectrum characterized by large positive ellipticity peak near 295 nm, a negative peak at 265 nm and a small positive maximum around 245 nm. M. Vorlíčková's group found that the one-nucleotide shorter htel-21, $G_3(TTAG_3)_3$ displayed the same spectrum as the A(htel-21) and concluded that htel-21's Na^+-solution topology was also the basket-type antiparallel [43, 44]. This was later confirmed by NMR and Raman spectroscopies [45, 46].

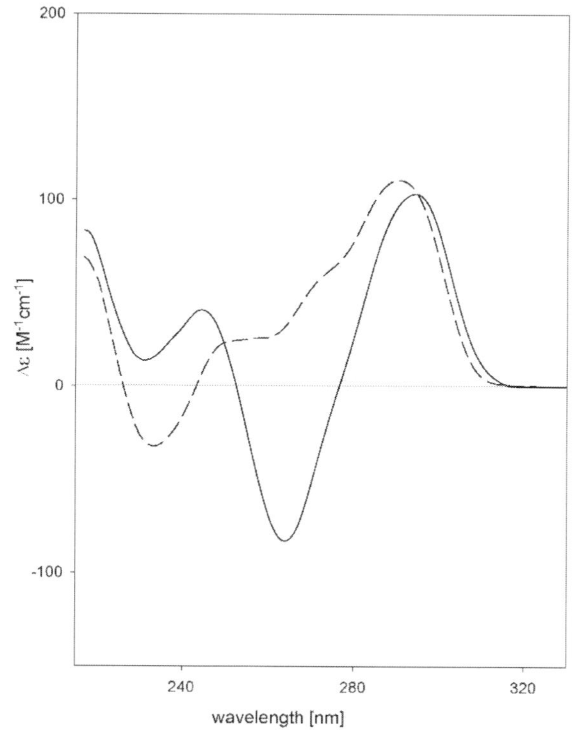

Figure 3. CD spectra of the quadruplexes adopted by the A(htel-21) oligodeoxynucleotide $AG_3(TTAG_3)_3$. Solid line illustrates the spectrum of the Na^+-stabilized antiparallel topology, the dashed-line shows the spectrum formed in K^+ ion containing solution of A(htel-21).

Until recently, the basket fold was accepted as the only topology of A(htel-21) in Na^+ solution. A recent single molecule Forster Resonance Energy Transfer microscopy (smFRET) study of Noer et al. [47] however found a dynamic equilibrium of four different states in buffered 100 mM Na^+. Although the basket fold was the most stable, there was also an unfolded and two transient conformations present. The sequence studied was the 5'- GCAGGCGTGGCACCGGTAATAGGATTAGGGTT**AGGG(TTAGGG)$_3$**-Cy3, which contained the A(htel-21) (bold, underlined) and a dye at the 3'-end. The long sequence at the 5'-end of the 22mer was in a duplex form with the complementary oligonucleotide. The two short-lived transient conformations were supposed to be a two-tetrad G-quadruplex, which was observed earlier only with the 27mer (TTAGGG)$_4$TTA htel sequence in Na^+ [48] and the chair-type antiparallel thrombin binding aptamer (TBA) [49-53] (Figure 4). Ratios of the four Na^+-forms was not specified. The transient states were not those observed as the intermediate structures during the thermal unfolding of A(htel-21), such as the triplex and the hairpin loop folds [42,48,54-58]. As we will see later in this section the appended nucleotides at either or both chain-ends can change the topology of htel quadruplexes. Therefore, results of single molecule studies with quadruplex-forming sequences that are attached to tethers may not always replicate the solution conformation of the wild-type (non-appended) quadruplexes.

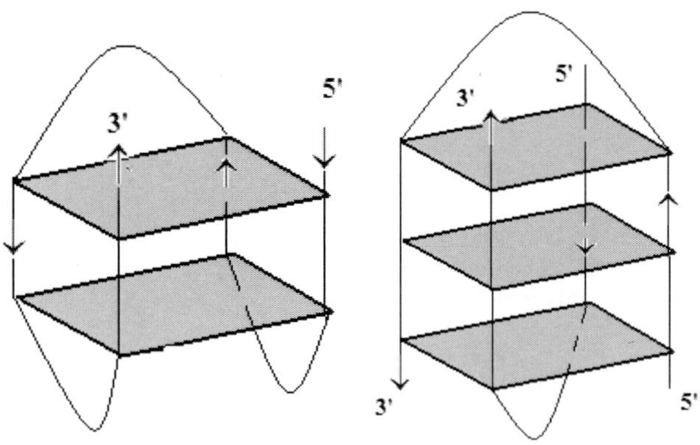

Figure 4. Schematic representation of (panel left) the two-tetrad chair-type TBA antiparallel quadruplex and (panel right) a bimolecular antiparallel quadruplex.

In 10-150 mM potassium ion containing buffers the CD spectra of htel-21 and A(htel-21) quadruplexes significantly changes. The negative peak around 260 nm of the Na^+ spectrum disappears and a strong shoulder develops around 270 nm whilst the large positive maximum close to 295 is retained (Figure 3). Position of the second positive peak also moves above 245 nm [43, 44]. These changes have been interpreted in two ways. From the non-cooperative changes in CD spectra of the Na^+-stabilized quadruplex on the effect of the of increasing K^+ ion concentrations, measured at low (3-10 μM) strand concentrations Vorlíčková concluded [43, 44] that the topology was still *mainly* the basket-type intramolecular antiparallel and the changes in spectra originated from the altered stacking interactions due to the intercalation of two K^+ ions between the three G-tetrad planes. (In the Na^+-basket fold the three smaller Na^+ ions are located in the plane formed by the four guanines of the G-tetrad [54]). This interpretation was confirmed later using different techniques [43-45, 59]. In a similar way, no change of the (3 + 1) hybrid topology of the TA(htel-21) quadruplex was observed when Na^+

ions were replaced by K^+ in spite of the extensive changes in CD spectra [46]. Conversion of the negative 260 nm peak to a strong positive shoulder at 270 nm, however, suggests an increase in parallel oriented strands, that is, formation of a mixture of antiparallel and non-antiparallel folds. The other interpretation, based on NMR results says that this spectrum reflects the presence of only the two types of hybrid conformations [37, 55]. The NMR measurements require strand concentrations in the millimolar range, whereas the micromolars are sufficient for absorption spectroscopies (UV, CD). As both the topology and the stability of htel quadruplexes depend on the concentration of the oligonucleotide [43, 60], the topology found by CD and NMR methods may not be the same. For instance, the K^+-stabilized antiparallel conformation of htel-21 observed at low, 3-5 μM strand concentrations changed toward the hybrid forms when strand concentrations were increased and around 2 mM of strands the CD spectrum unequivocally showed the characteristics of hybrid topologies with two positive peaks of similar heights at 260 nm and 290 nm (Figure 5). Above 2 mM of strands the conversion towards the parallel fold started [43, 60]. Therefore, both interpretations are true depending on which DNA concentration we are talking about. The final conclusion is, however, that the htel-21 does not have a single topology in K^+ solution at any concentration of the oligonucleotides: at low concentration the mixture contains the antiparallel as the majority fold, which may be somewhat different from the Na^+ baskets, and also contains the two hybrid forms. At high, "NMR concentrations" the two types of hybrid forms dominate or are only present.

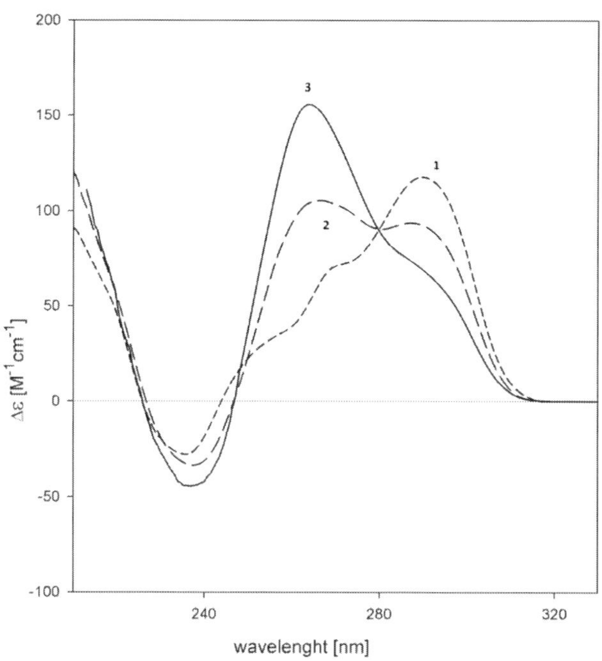

Figure 5. CD spectra of the intramolecular structures formed from the $G_3(TTAG_3)_3$ (htel-21) oligodeoxynucleotide in K^+ solution [60]. (1) the K^+-stabilized folding mixture at micromolar strand concentrations; (2) the mixture of hybrid-1 and hyrid-2 conformations at ~2 mM of strands; and (3) the conversion of folds toward the parallel topology above 2 mM of DNA strands. The shoulder near 290 nm completely disappears at 6-9 mM of strands.

The polymorphic nature of htel-21 and A(htel-21) quadruplexes observed in K^+ solutions is significantly reduced when the oligonucleotide terminal nucleotides are mutated or the chains are extended either in the 5'- or the 3'- or both directions. Diversity of folding topologies of various G-rich sequences that are not just simple appended htel structures has been reviewed by Phan and Patel earlier [12, 13]. Here we focus on the single or dominant topologies assembling from appended short htel sequences.

- The DNA quadruplex of **(htel-21)T**, i.e., the $G_3(TTAG_3)_3T$, is a 3'-end appended htel-21. This sequence in K^+ solution forms a predominant fold called hybrid-3 that contains only two G-tetrads according to Lim et al. [61]. However, Yang's group later described that this sequence forms a 2-tetrad basket-type (antiparallel) topology in K^+ solution [62]. Both were NMR spectroscopy studies.
- **TA(htel-21)** and **TTA(htel-21).** The TA(htel-21) in K^+ buffer adopted a predominant (3 + 1) hybrid-1 fold according to Luu et al. [38]. The 270 nm shoulder seen in the CD spectrum of A(htel-21) increased by the addition of T and TT to A(htel-21) and the chain extension also changed the thermodynamic stability of A(htel-21), as it was determined using DSC (Differential Scanning Calorimetry) by Hayden and Graves [63]. The DSC thermograms revealed a two component mixture during the unfolding of all three structures, which may mean the both hybrid forms were present at low temperature in K^+ solution and the predominant form was probably hybrid-1 that also had the higher stability among the two forms. The 5'-added T nucleotides reduced the stability of both components. Enthalpy of unfolding was shown to increase with the 5'-dangling nucleotides, thought to be due to enhanced interactions between the 5'-dangling nucleotide(s) and a nearby TAA loop.

Table 1. Thermostability of 5'-T-appended A(htel-21) quadruplexes determined by DSC* [63]

Sequence	T_m total	$T_{m(peak1)}$	$T_{m(peak2)}$
A(htel-21)	67.2	53.5	67.2
TA(htel-21)	62.4	52.3	66.5
TTA(htel-21)	61.3	46.2	62.4

*100 μM strand concentration in 10 mM potassium phosphate (pH 7.0), 1 mM EDTA and 100 mM KCl, Microcal VP-DSC.

The **TTA(htel-21)**, or in other formulation the **$(TTAGGG)_4$** adopts basket-type antiparallel fold in Na^+ solution [64,65]. According to McConnell and Sheardy [66] the dangling TTA sequence at the 5'-end stabilizes the hybrid-1 fold in potassium solution as compared with the quadruplex of htel-21.

- **TA(htel-21)T**, a 24mer oligonucleotide that folds into an **antiparallel** quadruplex in K^+ [29].

- **TT(htel-21)A**, another 24mer sequence that assembles into **hybrid-1** topology in K^+ solution [38].
- **TA(htel-21)TT**: This 25mer $TAG_3(TTAG_3)_3TT$ oligodeoxynucleotide was found to build a predominant hybrid-2 scaffold in K^+ solution as determined by NMR by Phan et al. [39,67]. Results of a recent investigation by C. Burrows' group [68], based on the detection of the various quadruplex folds at single-molecule level in the nanocavity of alpha-hemolysin protein's ion channel were interpreted as this htel-25, contrary to the previous NMR specifications exists of interconverting topologies of hybrid-1 and hybrid-2 as well as of a triplex intermediate in K^+ solution. In 50 mM K^+, pH 7.9, and 950 mM LiCl with 5 μM strand concentrations they found a ratio of 11:5:1 for hybrid-1, hybrid-2 and the triplex, respectively. The hybrid-1 dominated over hybrid-2 by 2.3 times, versus the NMR-found predominant hybrid-2 form [39, 67]. The reason for the variant dominant folds observed by NMR and the nanocavity techniques may originate either from the different oligonucleotide and salt concentrations used or, to some extent, the interaction of the quadruplex with the protein nanocavity. Even week interactions can catalyze the interconversions of the hybrid folds as there are only small energy differences between the two topologies [42]. However, there is no proof for this effect. On the other hand, the high salt concentration (close to 1M) used in the nanocavity experiments definitely changed the CD spectrum of htel-25, as compared to its 50 mM KCl spectrum: instead of the 270 nm shoulder a strong peak near 265 nm appeared that was almost as high as the 290 nm peak [68]. This is an indication of the presence of two forms in the high salt solution. The 5 μM strand concentrations used with the alpha-hemolysin *versus* the 5 mM strands applied with NMR [67] can also have an effect, similar to that observed with A(htel-21) [60]: This would mean that at low strand concentrations (nanocavity experiments) the htel-25 is polymorphic and the dominant fold is hybrid-1, whereas, at high strands (NMR) the polymorphic nature is strongly reduced resulting in hybrid-2 as the dominant topology.

26mers in K^+ solutions:

- **AAA(htel-21)AA**, builds a **hybrid-1** fold in K^+ [37].
- **TTA(htel-21)TT** folds into **hybrid-2** quadruplex in K^+ solution [40, 41].
- **$(TTAG_3)_4TT$** folds into **hybrid-2** topology in K^+ [27, 40, 41, 69, 70].

The 3D structure formed by the 15mer TBA DNA sequence GGTTGGTGTGGTTGG has been known as a 2-tetrad intramolecular chair-type quadruplex in both Na^+ and K^+ solutions (Figure 4) [49-53]. However, based on the result of experiments on the dependence of the CD spectra on DNA concentrations, migration in polyacrylamide gels, and the Ferguson analysis of the gel migration data Fialova et al. [71] concluded that the aptamer formed a bimolecular antiparallel tetraplex. Extension of the TBA strand at 5'end by GTAGGT hexamer or creating the 19mer analog of TBA by inserting T nucleotides into the sequence that resulted in GGTT**T**GGTTGTT**G**GTTTGG enabled the folding of this oligonucleotide into intramolecular antiparallel G-quadruplexes. Both structures ran in single bands in gel.

A special case of the effect of sequence on topology is the formation of the left-handed (Z) quadruplex, as recognized by Phan and coworkers [72] in the DNA sequence $T(GGT)_4TG(TGG)_3TGTT$, which is present in AGRO100, using CD, NMR and X-ray techniques. Its CD spectrum exhibited a profile with a negative peak at ~275 nm and a positive peak at ~250 nm, a nearly inverted shape of the spectra of the right-handed quadruplex topologies. Sequence matches of this Z-forming motif were found in various locations of the human genome [72].

Viglasky et al. [28, 73] investigated the factors influencing the thermodynamic stability of chain-extended htel quadruplexes by studying seventeen derivatives of htel-21. The appended T_n and $(TTA)_m$ that didn't form secondary structures were attached to each end of htel-21. The authors concluded that all protruding sequences stabilized the quadruplex formed and also determined the folding topology. The elevated stability was found to be enthalpy driven. The favorable enthalpy changes were not overcompensated by unfavorable entropy changes originating from structural entropy: the overhanging sequences at the 3'- and/or 5'-end caused a decrease in the degree of freedom of DNA folding. The 3'- and 5' -overhangs influenced structure and stability in different ways due to loop arrangements, i.e., the edgewise and double-chain reversal loops were present in the various topologies. Based on the CD spectra, except for the random double–extended htel-21, which formed a major parallel form, all other appended htel-21 oligonucleotides assembled into mixed topologies containing various ratios of the two hybrids and the antiparallel form in 50 mM K^+ solution.

The 100-200 nucleotide long 3'-overhang of the telomeric duplex DNA [17,18] corresponds to 16-35 TTAGGG hexameric repeats [19,20] the could form 4-10 core quadruplex units made up of 21 nucleotides, the htel-21 $GGG(TTAGGG)_3$ sequence. Considering this 4-10 units numerous experiments and theories have been published on the possible higher order structure of the repeating quadruplex units, such as the "beads-on-a-string" model, which was first formulated by M. Vorlíčková et al. [74], as well as on the formation of G-wires [74-86]. This subject as well the formation of specific topologies by ligand attachment and selection of topologies by ligands [6, 7, 87] are beyond the scope of this review.

The results described in this section demonstrate that appended nucleotides at either or both termini of G-rich oligonucleotides can greatly influence both the conformation and the stability of the quadruplex scaffold formed. In many cases this influence makes possible the assembly of quadruplexes with the desired stable majority folding topology.

1.2. Effects of Sequence and Length of the Loops on Folding Topology

Perhaps the loops are the most systematically studied structural component of quadruplexes. Both the length and composition of the loops influence the topology and stability of G-quadruplexes [3, 30, 31, 33, 88-106]. One of the first such basic studies was carried out by S. Neidle's group in 2004 [30]. They investigated the intramolecular quadruplexes assembled from two series of sequences, the 17mer $(TG_3)_4T$ and the A(htel-21), and varied the thymidine loop lengths from 1 to 7 nucleotides. The results are summarized in Table 2.

Sequence #1 folded into a parallel quadruplex characterized by a large positive ellipticity value near 265 nm and a negative peak around 245 nm. As the length of loop was increased

(#1 to #7) the thermal stability decreased continuously. Folding topology also changed from the parallel to antiparallel through a mixed one at #2. In the case of the htel-22 sequences (#8 - #15) the stabilities alternated. Changing the second TTA loop to T-to-TTT, the latter equaling to the TTA's length, increased the stability, longer T-sequences, however, decreased it although not as drastically as in the case of the $(TG_3)_4T$ series. Topology of the T_n-analogs of the wild-type htel-22 was all antiparallel (#9 - #15).

Table 2. Effect of loop length on stability and conformation* [30]

#	DNA oligonucleotide	T_m (°C)	Fold*
1.	TGGG T GGG T GGG T GGGT	>80	P
2.	TGGG T_2 GGG T_2 GGG T_2 GGT	71	P-H
3.	TGGG T_3 GGG T_3 GGG T_3 GGGT	58	AP
4.	TGGG T_4 GGG T_4 GGG T_4 GGGT	48	AP
5.	TGGG T_5 GGG T_5 GGG T_5 GGGT	34	AP
6.	TGGG T_6 GGG T_6 GGG T_6 GGGT	20	AP
7.	TGGG T_7 GGG T_7 GGG T_7 GGGT	-	AP
8.	AG_3 TTAGGG TTAGGG TTAGGGT	63	H
9.	AG_3 TTAGGG T GGG TTAGGGT	64	AP
10	AG_3 TTAGGG T_2 GGG TTAGGGT	66	AP
11	AG_3 TTAGGG T_3 GGG TTAGGGT	68	AP
12	AG_3 TTAGGG T_4 GGG TTAGGGT	61	AP
13	AG_3 TTAGGG T_5 GGG TTAGGGT	61	AP
14	AG_3 TTAGGG T_6 GGG TTAGGGT	59	AP
	Top of Form		
15	AG_3 TTAGGG T_7 GGG TTAGGGT	56	AP

*The samples were measured in 100 mM K+ solutions. P, H and AP stand for parallel, hybrid and antiparallel topologies, resp. Fold assignments are based on shapes of the CD spectra in [30].

Table 3. Effect of length of loops in the quadruplex of $(TG_3)_4T$ [92]

#	DNA oligonucleotide	T_m (°C)*	Fold**
1.	TGGG T GGG T GGG T GGGT	>90	P
2.	TGGG T GGG TTTT GGG T GGGT	87.5	P
3.	TGGG TTTT GGG T GGG T GGGT	84.6	P
4.	TGGG TTTT GGG T GGG TTTT GGGT	65.3	P
5.	TGGG TTTT GGG TTTT GGG T GGGT	67.1	P
6.	TGGG TTTT GGG TTTT GGG TTTT GGGT	56.3	AP

*Melting experiments in buffered 100 mM K^+ solution.
**CD spectra recorded in buffered 200 mM K^+ solution.

Similar results were reported by Rachwal et al. for the relationship between the loop length, stability and conformation of the $(TG_3)_4T$ model using fluorescent melting technique, CD and NMR spectroscopy (Table 3) [92].

The results suggested that if one of the three loops was a single nucleotide loop the quadruplex adopted parallel topology. Antiparallel quadruplexes formed if all three loops were longer than one nucleotide [92]. The formation parallel topology in the presence of single-nucleotide loops is probably a generally phenomenon and is applicable to all four nucleotides A, T, G and C since loop size determines the steric constraints on topology and energetics [3, 33]. The relative stability of quadruplexes provided by loops with longer than one nucleotide is also dependent on nucleotide stacking energies within loops, as has been shown by a thermodynamic profiling study [107]. With the $(TG_3)_4T$ series in Rachwal experiments [92] the quadruplex stability generally decreased with increasing loop lengths. Quadruplexes #2 and #3 with two single loops and one T_4 loop proved to be very stable and the stability didn't much depend on the position of the T_4 loop. Presence of two T_4 loops caused a 20°C drop of thermostability (#4 and #5), which was further decreased by the presence of third T_4 loop (#6).

Using the same 17mer $(TG_3)_4T$ DNA model as Hazel [30] and Rachwal [92] used, Vorlíčková et al. described different topologies for quadruplexes with single T nucleotide loop sequences. They also investigated the effect of A and C as single to triple nucleotide loops [93]. Contrary to previous results [30, 92] they found that the parallel quadruplexes of the single T (or A or C) loop sequences were not intramolecular (not built from one 17mer oligonucleotide) but from four molecules, called tetramolecular quadruplexes. Even the two-T base loop-containing sequences were found to fold into the tetramolecular parallel quadruplexes. Only the T_3 and A_{2-3} loop sequences supported the formation of intramolecular antiparallel topology. The G-rich sequences with C_2 and C_3 loops however didn't fold into quadruplexes at all in K^+ solutions (Table 4).

Table 4. Four molecular quadruplexes formed by $(TG_3)_4T$ oligodeoxynucleotides* [93]

#	DNA oligonucleotide	Fold*
1.	GGG T GGG T GGG T GGG	tetramolecular, P
2.	GGG TT GGG TT GGG TT GGG	tetramolecular, P
3.	GGG TTT GGG TTT GGG TTT GGG	bi- & intramolecular, AP
4.	GGG A GGG A GGG A GGG	tetramolecular, P
5.	GGG AA GGG AA GGG AA GGG	intramolecular, AP
6.	GGG AAA GGG AAA GGG AAA GGG	intramolecular, AP
7.	GGG C GGG C GGG C GGG	tetramolecular, P
8.	GGG CC GGG CC GGG CC GGG	hp + ds
9.	GGG CCC GGG CCC GGG CCC GGG	hp

*CD spectra were recorded in buffered 60 mM K^+ solution; P stands for parallel quadruplex; AP, antiparallel; hp, hairpin; ds, homoduplex.

Mergny and coworkers [100] carried out extensive investigations with 80 DNA sequences of the general formula $G_3T_nG_3T_nG_3T_nG_3$ with variable length T-loops to find a relationship between the sequence and the stability of quadruplex formed and also to use the data in the algorithm predicting quadruplex stability [108]. (The algorithm was recently updated [109].) The final conclusions of the results were complicated by the missing information about the folding features. However, one of the many important findings was that the thermal stability of the quadruplexes decreased linearly (R = 0.93) with the total loop length ranging from 1 to 30 nucleotides in K^+ solution.

Yadav and Sheardy [33] demonstrated with the 24mer $(TTAGGG)_4$ that changing just one base in any of three TTA loops could influence both the conformation and stability of the hybrid-1 scaffold formed in K^+. They investigated the effect of every single, double and triple sequence permutations of the loops: the TTA to TTT, TAA, and ATA for the single variations, the TTA to AAA, ATT and TAT for the double, and the TTA to AAT for the triple mutation in each loop. According to the CD spectra the quadruplex topology remained the hybrid form(s) by most sequence variations in 150 mM K^+ solution. Apparently, only the TTA to TTT single base exchange in the second and third loops (numbering from the 5'-end of the 24mer sequence), which are edgewise loops in the hybrid-1 form of the wild-type (Figure 1) resulted in the formation of antiparallel scaffolds. CD spectra of the other structures differed in peak and trough wavelength, peak or trough intensity and the height of the shoulder in the 260 - 280 nm range. The authors reasoned that both the minor and larger conformational changes arose from the stacking interactions of loop bases with themselves and/or with the tetrads bases and the preferred stacking would give rise from the lowest energy conformation of the loop. The conformational change of a loop is compensated by an overall conformational change in the quadruplex. Stacking not only influences the conformation but also the stability. It was quite remarkable that a single base change influenced the T_m, sometimes in a dramatic fashion. For example, changing TTA to TTT increased the T_m by 8°C to over 11°C, depending upon the loop location. This would also mean that the antiparallel form was more stable than the hybrid-1 of the wild-type in the above examples. Interestingly, most loop-mutated quadruplexes were more stable than the wild-type, except when all three loops were modified. The TTA to TTT, the TTA to AAA and TTA to ATT mutations destabilized the wild-type quadruplex, all in a significant way, by 7°C to 30°C.

Tippana et al. [103] published a far-reaching single-molecule FRET analysis of 20+ sequence analogs of a 34mer DNA containing a quadruplex forming sequence at the 3'-end (indicated by bold letters) including also the htel-21 sequence (not shown), in TGG CGA CGG CAG CGA GGC **GGG T GGG GA GGG T GGG.** The results confirmed the conclusions obtained earlier by UV and CD solution techniques about the effect of loop length and composition on folding conformations: One single nucleotide loop makes the sequence fold into parallel quadruplex independently of how long the other two loops are, which can contain up to nine nucleotides. Longer loops promoted antiparallel folding as well as unfolding states. As for the composition, they found that the TTT, TTA, and TAA in all three loops induced the formation of mixtures of parallel and antiparallel topologies of various ratios. The AAA loops, however, hindered the formation of quadruplexes, resulted in unfolded states. They also found that kinetics of folding also depended on loop length. The rates suggested that TTA and TTT loops stimulated the folding into parallel conformation five to six times faster than the TAA loops [103]. In solution studies with htel sequences Risitano and Fox [90] did observe formation of a quadruplex when the TTA loops were replaced with AAA, but this change caused a large decrease in quadruplex stability. Guedin et al. [98] analyzed the effect of sequences on quadruplex structure of the 3-base-long central loop while keeping the rest of the sequences unchanged. First, a series of 36 different 21mer sequences were compared that corresponded to the general formula of GGGTTTGGG**HNH**GGGTTTGGG, where H was A, C or T and N was A, C, T or G. One clear rule emerged from the experiments stating that the presence of an A at the first position of the HNH loop was detrimental to stability. In contrast, adenines had no such damaging effect when they were present at the second or third position of the loop. Cytosines were either stabilizing or destabilizing depending on their position. They also found that the correlation between T_m or ΔG^o in Na^+ and K^+ solutions was weak. To determine if these sequence effects were similar with different quadruplexes, Guedin et al. analyzed 26 additional sequences. The result confirmed the general destabilizing effect of adenine as the first base of the loop(s).

Majority of studies on the loop structure - quadruplex structure relationships were performed with model DNA sequences, including the telomere repeats. Kumar and Maiti published a comprehensive study using 36 natural gene sequences from the promoter regions of various proto-oncogenes [97]. These sequences contained diverse loop lengths and compositions. Total loop lengths varied between 5 to 18 nucleotides and composition were also different. Their investigation showed that thermodynamic stabilities of the naturally occurring quadruplexes ranged from -1.7 to -15.6 kcal/mol (ΔG^o_{37}). The results also suggested that both the length of loop and its composition affected quadruplex conformation and thermodynamics, which made it difficult to draw generalized correlations between loop length and thermodynamic stability.

Overall, the available literature suggests a strong influence of loop length and composition on quadruplex stability and folding topology. The well-confirmed rules are that a single-base loop promotes the formation of parallel quadruplexes and that quadruplexes with shorter loops display higher thermodynamic stability. As for the loop composition, a single T-to-A substitution in an intramolecular quadruplex lowers its thermodynamic stability. Though loop length has marked influence on the thermodynamic stability of quadruplexes, the variation in loop composition makes it difficult to find a significant correlation between loop length and quadruplex stability.

2. Mutations of Sequences by Natural and Synthetic Nucleotides That Stabilize Single Folds

A major instrument that shapes the conformation and stability of G-quadruplexes is the mutation, carried out either by natural nucleotides or chemically modified analogs. Building and stabilizing single or majority folding topologies in G-quadruplexes has been carried out in most cases by site-specifically introduced chemically modified nucleotide analogs. Since the conformation of the guanine base around the N-glycosidic bond is essentially related to the folding topology of G-quadruplexes, the major concept that governs these chemical modifications is to make use of the possibility to change the glycosidic torsion angles [110], that is, to incorporate modified nucleotides that favor a syn or anti guanosine conformation. For example, the syn-preferring 8-bromo-deoxyguanosine (br^8dG) (Figure 2) is incorporated in place of dG to replace an *anti* dG by a *syn* nucleotide, or the sugar modified analogs, like 2'-flour-arabinoguanosine is used to introduce a stable *anti* nucleotide to replace a *syn* dG.

2.1. Guanine Analogs in the G-tetrads

One of the earliest attempts to stabilize a topology with a non-natural building block was published by Xu et al. in 2006 [111] who site-specifically incorporated the deoxynucleotide of 8-bromoguanine (br^{8}G) into the A(htel-21) DNA sequence AGGG(TTAGGG)$_3$ and studied the conformations in K^+ solution. br^{8}G, with its preferred syn N-glycosidic conformation (Figure 2) if incorporated in place of syn-dG of a quadruplex is known to stabilize the quadruplex structure [112], similarly to 8-methylguanine (m^{8}G) [113]. The enhanced thermodynamic stability may be the driving force for the mixed folds to convert into a more stable single form. As a result of this approach, br^{8}G was later often used to stabilize a particular quadruplex conformation for various investigations [111, 114, 115]. For instance, Phan et al. [39] applied the br^{8}G modification for NMR studies with the following quadruplex-forming htel sequences:

- 23mer TA(htel-21):
TAGGG TTAGGG TTAG**<u>br^{8}G</u>**G TTAGGG - br^{8}G in position 16,

and the

- 25mer TA(htel-21)TT:
TA GGG TTAGGG TTA**<u>br^{8}G</u>**GG TTAGGG TT - br^{8}G in position 15

The 23mer TA(htel-21) mainly folds into the hybrid-1 form [38], whereas the 25mer TA(htel-21)TT adopts the hybrid-2 as the majority form in K^+ solution as observed by NMR spectroscopy [12,39,67]. Incorporation of br^{8}G into chosen (syn-dG) sequence positions, such as position G16 in the middle tetrad of the 23mer and G15 in one of the terminal tetrads of the 25mer sequence greatly stabilized the respective topology. Both dG positions were originally syn, thus br^{8}G contributed to the stability of each fold through the preferred stable syn glycosidic linkage of br^{8}G. The substitutions significantly improved the quality of the NMR spectra.

Interesting results were reported by Burrows and coworkers [68] for the incorporation of two br^{8}G nucleotides into different positions into the 25mer TA(htel-21)TT. The dual incorporations induced exclusively either the hybrid-1 or the hybrid-2 fold in the quadruplex formed from the same DNA sequence in K^+ solution, as determined by the novel method of using alpha-hemolysin protein ion channels for analysis. The syn G nucleosides are indicated in bold italics letters:

wt	TA ***G***GG TTA***GG***G TTA***G***GG TTA***G***GG TT	
in pos. 16 and 21:	TA ***G***GG TTA***G***GG TTA***G<u>br^{8}G</u>***G TTA***<u>br^{8}G</u>***GG TT	hybrid-1
in pos. 10 and 15:	TA ***G***GG TTA***G<u>br^{8}G</u>***G TTA***<u>br^{8}G</u>***GG TTA***G***GG TT	hybrid-2

The presence of br^8G enforced the syn conformation in the particular nucleotides that led to >95% hybrid-1 when br^8G was at positions 16 and 21, and to >95% hybrid-2 when br^8G occupied positions 10 and 15 of the sequence. These two modified sequences selectively eliminated one or the other type of subpopulation that might have been present in the wild-type quadruplex. These double modifications by br^8G are among the best examples for inducing a stable single fold in htel DNA quadruplexes.

With another type of sequence Brcic and Plavec [116] published an example for a single substitution: the GGGGCC repeat in the non-coding region of the C9orf72 gene, which is the cause of critical neurodegenerative disorders, forms a monomolecular mixture of G-quadruplexes in 120 mM K^+ solutions according to CD and H1-NMR spectra. A single br^8G substitution in selected positions of $(G_4C_2)_3G_4$ DNA enabled the formation of a mostly single antiparallel G-quadruplex fold with three edgewise CC loops. CD spectrum at 10 μM strand concentration of the wild-type 22mer showed two positive peaks of near equal heights around 265 nm and 290 nm, whereas CD spectrum of $(G_4C_2)_3$GG**<u>br^{8}G</u>**G had positive max near 295 nm and negative 260 nm peak. The NMR and CD spectra referred a chair-type antiparallel scaffold.

Another example for the use of double modifications is the incorporation of 8-oxoguanine (8-oxoG or O) and xanthine (X) nucleotides jointly. The two analogs of G together produced interesting folding preferences. Individually, both 8-oxoG and X are strongly destabilizing analogs of G when incorporated into the tetrad cores of the G-quadruplex due to the loss of H-bonds and also to possible steric clashes [115]. Applied together, depending on the positions of substitutions the two can also have stabilizing effects, even more, stabilizing a single fold. First, Benz and Hartig [117] incorporated two 8-oxoG and two xanthine (X) nucleotides into the 24mer htel sequence of TA(htel-21)T that resulted in X:O:X:O tetrads. The modified quadruplexes were thermally destabilized, as compared with the stability of the wild-type. Topology either remained the basket-type antiparallel or changed into parallel fold in 25 mM buffered K^+ solution due to the double substitution:

- wt TA GGG TTAGGG TTAGGG TTAGGG T

  antiparallel
- O(3):X(11):X(15):O(23) TA **<u>O</u>**GG TTAGG**<u>X</u>** TTA**<u>X</u>**GG TTAGG**<u>O</u>** T

  antiparallel
- O(3):X(9):)O(15):X(21) TA **<u>O</u>**GG TTA**<u>X</u>**GG TTA**<u>O</u>**GG TTA**<u>X</u>**GG T

  parallel

Later Cheong et al. [118] found the right positions for the O and X substitutions that didn't reduce the stability of the wild-type and also retained its topology. They inserted the G:G:X:O tetrads into the 24mer TT(htel-21)A DNA. The dual-substituted sequences all formed quadruplexes and the topologies were similar to the hybrid-1 topology of the parent quadruplex. Three of the modified quadruplexes were destabilized in 10 mm K^+ by 3°C to 12°C, two had similar stability as the wild-type had (56°C), and the G(3):O(9):X(17):G(21) tetrad in the sequence of TT**G**GG TTA**<u>O</u>**GG TTAGG**<u>X</u>** TTA**G**GGA proved to be more stable by about 2°C than the wild-type [118]. With these two analogs of G Cheong et al. could also reverse the polarity of H-bonds in a tetrad while the original folding topology of the wild-type was preserved [119].

Various single base substitutions of tetrad guanines of htel quadruplexes can stabilize a single fold. The DNA quadruplex of htel-21 $G_3(TTAG_3)_3$ that contains three G-tetrads connected via three TTA loops forms the basket-type antiparallel scaffold in Na^+ solution (Figure 1) similarly to the $AG_3(TTAG_3)_3$ that was determined by Wang and Patel in 1993 [35]. The Na^+-antiparallel fold of the wild-type htel-21 in K^+ solution converts into a mix of more thermostable, mainly K^+-stabilized antiparallel [43] and the hybrid forms [45]. The conversion can be easily followed by CD spectroscopy (Figure 3). Replacing any guanine nucleotide of the two terminal tetrads either with adenine [120], or 8-oxoG nucleotides [121] or with a tetrahydrofuranyl abasic site [122], however, hindered the conversion into the K^+-stabilized forms that is induced by the Na^+-to-K^+ exchange in the wild-type quadruplex. The modified quadruplexes remained in the original antiparallel topology also in physiologically relevant K^+ solutions.

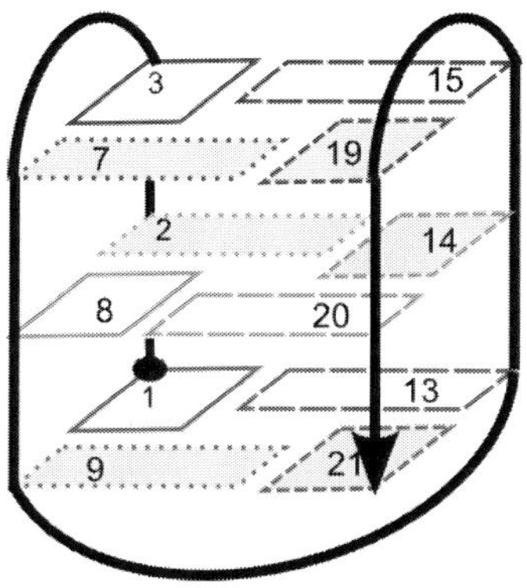

Figure 6. Antiparallel basket topology of the quadruplex of $G_3(TTAG_3)_3$ (htel-21). The rectangles represent the guanine bases, the shaded rectangles the syn deoxyguanosines [121].

Figure 6 shows the schematic antiparallel fold of the wild-type htel-21 quadruplex and Figure 7 shows an example of unchanged CD spectra on the G to 8-oxoG substitutions. The Na^+-spectra displayed two positive maxima, a large at 290 nm, a smaller near 240 nm, and a negative peak at 260 nm. The K^+ spectrum of the wild-type changed, the positive 290 nm peak widened due to the appearance of a strong shoulder near 270 nm. Shape of CD spectra of the four 8-oxoG modified structures however didn't change in K^+ independently of whether the 8-oxoG replaced a syn or anti guanosine [121]. The inhibited folding transformation by 8-oxoGs was unexpected as the modified structures were markedly destabilized as compared with the wild-type, and the reduced stability was thought to support the conformational rearrangement. The T_m value of the wild-type (75.4°C in buffered 0.1 M KCl) was reduced by 11°C, as an average, in K^+ by 8-oxoG in the four positions of the outer tetrad connected via a diagonal loop in the Na^+ form (Table 5).

Another guanine analog that inhibited a conformational transition was the O6-methylguanine (o^6G). Zhao et al. [123] incorporated o^6G into various positions of a htel sequence and, in contrast to the wild-type, the o^6G-substituted sequence analogs impeded the conversion of the G-quadruplex from the hybrid forms to parallel under molecular crowding conditions in K^+ solution.

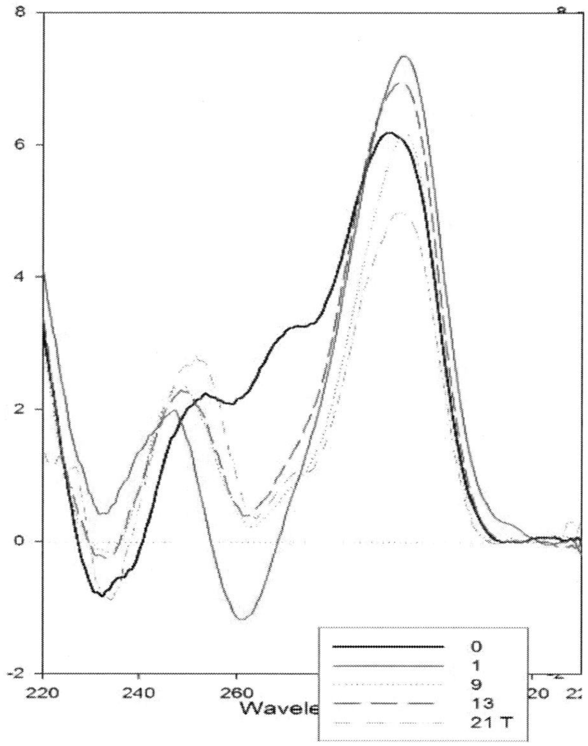

Figure 7. CD spectra of quadruplexes of the $G_3(TTAG_3)_3$ (htel-21) oligodeoxynucleotide and its 8-oxoG substituted analogs in K^+ solution [121]. The spectra were recorded in buffered 0.1 M KCl (total conc. of K^+ was 165 mM). The number 0 (thick black line) represent the spectrum of the wild-type, the 1, 9, 13 and 21(T) show the positions of guanine nucleotides substituted by 8-oxoG in the "bottom" tetrad (Figure 6). (This outer tetrad is stabilized by the diagonal loop at the 3'-end of the sequence in Na^+ solution).

Table 5. Stability of the $G_3(TTAG_3)_3$ quadruplex and its 8-oxoG analogs [121]

Abbr.	Sequence	T_m (°C)	ΔT_m (°C)
in buffered 100 M NaCl:			
wt	GGG TTAGGG TTAGGG TTAGGG	67.7	-
8-oxoG-1	**O**GG TTAGGG TTAGGG TTAGGG	64	-3.7
8-oxoG-9	GGG TTAGG**O** TTAGGG TTAGGG	64.1	-3.6
8-oxoG-13	GGG TTAGGG TTA**O**GG TTAGGG	65.3	-2.4
8-oxoG-21	GGG TTAGGG TTAGGG TTAGG**O**T	60.8	-6.9
in buffered 100 M KCl:			
wt	GGG TTAGGG TTAGGG TTAGGG	75.4	-
8-oxoG-1	**O**GG TTAGGG TTAGGG TTAGGG	63.2	-12.2
8-oxoG-9	GGG TTAGG**O** TTAGGG TTAGGG	64.2	-11.2
8-oxoG-13	GGG TTAGGG TTA**O**GG TTAGGG	64.4	-11.0
8-oxoG-21	GGG TTAGGG TTAGGG TTAGG**O**T	64.9	-10.5

*Underlined positions of the wild-type htel-21 were in syn conformation in the Na^+ basket form; O stands for 8-oxoG; total concentration of Na^+ or K^+ ions was 165 mM.

Single and double 8-bromo-, 8-methoxy-, 8-amino-substituted guanines inserted into various syn or anti positions didn't change the (3 + 1) hybrid-1 major topology of the 24mer TT(htel-21)A quadruplex in K^+ as Lech et al. determined by NMR spectroscopy with DNA concentrations of 0.2-0.7 mM [124]. CD spectrum of 2-10 μM of the wild-type and the guanine C8-modified quadruplex analogs displayed positive peaks at 290 nm, a shoulder at 270 nm. All three substituents raised the thermostability of the wild-type quadruplex, whereas 8-oxoguanine drastically decreased it. Also, its NMR spectrum was quite different from the other substituted quadruplexes; apparently it was a mixture of various folds. CD spectrum of the 8-oxoG-modified quadruplex was not published [124].

A G-to-I (inosine, i.e., hypoxanthine-2'-deoxyriboside) substitution and a deletion of an adenine nucleotide in the natural telomere sequence $(TAGGG)_4$ induced formation of single topologies (Table 6) [125].

Table 6. Folding characteristics of $(TAGGG)_4$ quadruplex analogs* [125]

Abbr.	Sequence	CD peaks (nm)*	Fold
wt	(TAGGG)2 TAGGG TAGGG	260, 295	2-tetrad AP & 3-tetrad P
I	(TAGGG)2 TAGGG TAIGG	245, -265, 295	2-tetrad AP
-A	(TAGGG)2 T GGG TAGGG	-240, 260	3-tetrad P

*CD spectra were recorded in 10 mM KCl, 40 mM LiCl and 20 mM lithium phosphate (pH 7). DNA concentration was around 5 μM. AP and P stand for antiparallel and parallel, resp.

The Giardia's 20mer telomere sequence $(TAGGG)_4$ forms two isoenergetic quadruplex topologies in K^+ solution at 25°C. One of these is, interestingly, a two-tetrad antiparallel fold, the other is a three-tetrad parallel topology. The presence of hypoxanthine in place of a guanine eliminated the three-tetrad form as the formation of a G_3I tetrad was disfavored. The presence of a single T-loop in the –A sequence supported the formation of the parallel quadruplex. Single base loops favor the formation of parallel quadruplexes, which is apparently a general rule observed with numerous other sequences (see Section 1.2).

2.2. Sugar Analogs of Tetrad Guanosines

Quadruplex topologies can also be altered by substituting the sugar moiety in a way to change the syn or anti N-glycosidic conformation of the tetrad guanosines [126-130]. Starting with the natural ribo mutations, the energetically preferred torsion angle of ribonucleosides is in the *anti*-range with North conformation sugar pucker (C3'-endo, C2'-exo). Therefore, quadruplexes built completely from ribonucleotides always have parallel conformation that contains exclusively anti nucleotides [131]. In this way, incorporation of ribonucleotides into selected positions of G-rich DNA sequences could be used to control quadruplex topology. For instance, Tang and Shafer [126] incorporated ribonucleotides into the 15mer thrombin binding aptamer (TBA) DNA by replacing the syn dG positions by riboguanosines. The TBA's two-tetrad chair-type monomolecular antiparallel fold (Figure 4) switched into a bimolecular parallel fold with simultaneous reduction of thermal stability. When the anti dG positions were substituted by ribonucleotides the original chair-type folding topology didn't change and the thermostability increased (Table 7). Increase of thermostability was probably the result of contribution of 2′-hydroxyl group to the stabilizing water structure around the quadruplex macromolecule. The full ribo TBA quadruplex was parallel but it was less stable than the deoxy quadruplex.

Table 7. Stability and folding topology of TBA quadruplexes* [126]

Sequence	T_m (°C)	CD peaks	Fold
GG TT GG TGT GG TT GG (wt)	46	240, -265, 293	antiparallel
gG TT gG TGT gG TT gG	22	-240, 265	parallel
Gg TT Gg TGT Gg TT Gg	52	240, -265, 293	antiparallel
gg uu gg ugu gg uu gg	41	-240, 265	parallel

*The underlined Gs of the wild-type sequence are the syn-Gs, and g is for riboguanosines).

Vondruskova et al. [132, 133] incorporated riboguanosines into selected G positions of the Oxytricha telomere fragment of GGGGTTTTGGGG. This deoxyoligonucleotide forms a bimolecular basket-type antiparallel quadruplex comprising four G-tetrads in both Na^+ and K^+ solutions [134, 135]. Positions G1, G9, G4, and G12 form the two terminal tetrads of the quadruplex and adopt syn, syn and anti, anti glycosidic geometries, respectively. Authors showed that the substitution of a single **g** for a **G** was able to change the quadruplex topology. With the exception of GGGGTTTTGGGg, which adopted an antiparallel structure, all the RNA/DNA hybrids formed parallel, bimolecular quadruplexes.

Peng and Damha [127] studied the effect of 2′-deoxy-2′-fluor-arabinofuranosyl nucleotides on the structure of the 15mer TBA quadruplex and other structures. The energetically preferred N-glycosidic conformation of 2′-deoxy-2′-fluor-arabinofuranosyl-guanine (abbreviated below as **F**) is anti, so if incorporated in place of a syn dG of a G-tetrad, it would affect the folding the same way as the riboguanosines did. Incorporating **F** in place of anti dG-s the analog stabilized the structure and maintained the fold of the wild-type quadruplex. With the TBA quadruplex the ΔT_m was up by 3°C by each ara-nucleotide incorporated. In cases when **F** replaced the syn dGs the T_m value of the quadruplex slightly decreased, to 45.4°C, but the CD spectrum changed from the antiparallel to parallel-type. When all nucleotides, including also the loop Ts were substituted by the fluor-ara-nucleotides the T_m increased to 54.1°C and the CD spectra portrayed parallel characteristics.

Later, Phan's group [129] investigated the effect on folding of 2'-deoxy-2'-fluor-arabinofuranosyl-guanine as well as the 2'-deoxy-2'-fluorguanosine (both abbreviated below as **F**). The DNA quadruplex models used were those arising from the 23mer TA(htel-21) that exists mainly in hybrid-1 conformation [38] and from the 25mer TA(htel-21)TT whose major form was the hybrid-2 fold in K^+ solutions, as determined by NMR spectroscopy [67]. Substitution by **F** of an anti dG nucleotide of the middle tetrad by either analog the 2'-fluor nucleotides helped increasing the dominance of single folds, *i.e.*, switching the minor conformational "contamination(s)" into the major form:

23mer hybrid-1; F-10	TA GGG TTAGFG TTAGGG TTAGGG
25mer hybrid-2; F-16	TA GGG TTAGGG TTAGFG TTAGGG TT

Not only the conformational purity but also the thermal stability increased by a single substitution in both structures, by about 3°C. When, however, all five syn dG nucleotides were replaced by **F** nucleotides the hybrid forms switched to parallel topology. The thermal stabilities also changed, increased by about 10°C as compared to the stability of the respective wild-type.

All syn dGs replaced by **F** in the:

23mer hybrid-1, F-3,9,15,16,21	TA FGG TTAFGG TTAFFG TTAFGG
25mer hybrid-2, F-3,9,10,15,21	TA FGG TTAFFG TTAFGG TTAFGG TT

The wild-type quadruplex of TAGGGTTAGGGT (htel-12) exists in a mixture of two interconverting folds in K^+ solution: a dimeric parallel and a dimeric antiparallel quadruplex (Figure 4) [136]. Martín-Pintado et al. [130] found that a single substitution of the G in position 9 of TAGGGTTA*G*GGT, which was a syn dG in the antiparallel form, by the anti-favoring analogs of 2'-deoxyguanosine, the riboguanosine (g), the arabinofuranosyl-guanine (araG), and the 2'-deoxy-2'-fluor-arabinofuranosyl-guanine [$f^{2'}$araG (**F** above)] converted the mixture of folds to the dimeric parallel G-quadruplex form. All three sugar analog nucleosides adopt the South sugar conformation (C2'-endo, C3'-exo) as determined by NMR. Thermal stabilities also changed. The $f^{2'}$araG provided the greatest thermal stabilization, 12°C. On the other hand, the 2'-deoxy-2'-fluoroguanosine substitution ($f^{2'}$dG) induced major destabilization, disruption of the wild-type structure probably due to an unfavorable electrostatic repulsion between the phosphate and the 2'-F atom. Data are collected in Table 8.

Table 8. Effect of sugar analogs of G on the stability and conformation of the htel-12 quadruplex* [130]

5'-TAGGGTTAXGGT DNA:			
X	**T_m (°C)**	**ΔT_m (°C)**	**fold**
G	41	-	dimeric mix
g	47	6	parallel
araG	50	9	parallel
$f^{2'}$araG	53	12	parallel
$f^{2'}$dG	-	-	multiple

*Determined in 15 mM potassium phosphate, 5 mM KCl, pH 7.0; Strand concentration 75 μM.

As a continuation of the work with fluoro-modified sugar derivatives Li et al. [137] incorporated 2'-deoxy-2'-fluor-ribofuranosyl-guanine, 2'-deoxy-2'-fluor-arabinofuranosyl guanine and 2′-O-4′-C-methylene-guanosine (LNA, see below) into 40 parallel and a hydrid-1 quadruplex-forming DNA sequences, the 18mer $TT(G_3T)_4$ [138] and the TT(htel-21)A [38], respectively. In the parallel fold all nucleotides are in anti glycosidic conformation, so the anti-favoring three analogs of dG if incorporated into anti dG positions strengthened both topologies and if incorporated into syn positions of TT(htel-21)A the fold of the wild-type was perturbed. The CD and NMR spectra referred to mixed hybrid conformations where the parallel strand arrangements apparently began to dominate.

LNA, the 2′-O-4′-C-methylene-linked ribonucleosides or locked nucleotide analogs are constrained in the C3′-endo sugar pucker (North) and are in the *anti*-N-glycosidic linkages [139]. LNAs have also been used to drive conformational changes in G-quadruplex DNAs [139-143]. The deoxyoligonucleotide $(G_4T_4)_3G_4$ from Oxytrichia trifalax (Oxy28) forms an intramolecular, antiparallel basket-type quadruplex with four G-tetrads. Dominick and Jarstfer [139] incorporated single locked nucleotides of G (LNA-G) and T (LNA-T) into selected positions of $(G_4T_4)_3G_4$. The bold-underlined positions of G and T were substituted, one-by-one, by an LNA nucleotide in the G**G**G**GTT**TTGGGGTT**TTGG**GGTTTTGGGG sequence. In Na^+ solution the unmodified, wild-type Oxy28 quadruplex displayed a typical antiparallel CD spectrum with a strong maximum at 295 nm and a strong minimum at 260 nm. In K^+, the Oxy28 quadruplex showed a strong maximum at 295 nm but a shallow minimum at 260 nm, still indicative of the antiparallel quadruplex. In Na^+ all LNA-containing quadruplexes showed similar spectra as the wild-type Oxy28 quadruplex did. In K^+, however, the presence and location of LNA caused distinct effects on the structure. LNA did not affect the structure if it was inserted into positions 5 and 6 (T-s in the first loop), led to a mixtures of folds in positions 4 (tetrad G), 15 or 16 (T-s in the second loop), and induced formation of intramolecular parallel folds if incorporated into positions 2, 17 and 18 (tetrad G-s) characterized by strong CD maximum at 260 nm and a minimum at 240 nm. T_m value of the unmodified quadruplexes (64°C and 85°C in Na^+ and K^+, respectively) was reduced by each single LNA, most extensively by LNA-Gs. LNA-G in position 2 caused the largest effects, -14 and -25°C, respectively. Substitution by LNA-G of the other non-terminal tetrad position in Oxy28, the G18, also reduced the T_m by 6°C and 20°C, respectively, in Na^+ and K^+.

Pradhan et al. [143] studied the effect of LNA nucleotides on the folding topology of htel-12 dimeric model quadruplex system, which in solution forms a mixture of antiparallel and parallel quadruplex topologies. LNAs incorporated into rationally selected syn or anti dG position shifted the equilibrium toward parallel G-quadruplexes. The LNA-modified parallel G-quadruplexes were thermally stabilized by about 11°C relative to the wild-type folding mixture.

Changing of the natural configuration of the sugar part of a nucleotide, the 2'-deoxy-β-D-ribofuranose to the anomeric α-D-ribofuranose also induces topology change in a G-quadruplex as the α-nucleotides are unable to adopt syn glycosidic conformation [144]. Kolganova et al. [145] incorporated α-thymidine and α-2'-dexyguanosine nucleotides into selected locations into the loops and the core parts of thirteen TBA sequences. Depending on the locations of substitution the quadruplex was either stabilized or destabilized when the α-anomeric thymidines were incorporated into the T loops, for which a few examples are shown below (Table 9), similarly when both the loops and the core were substituted. None of these substitutions changed the antiparallel topology. Replacement of the core guanines, however, either prevented assembly of a quadruplex structure or induced rearrangement of the quadruplex topology from the two-tetrad chair-type antiparallel fold to an undefined parallel architecture, as indicated by CD spectra. Replacement of all natural nucleotides with the anomeric ones also resulted in unstructured forms. The single α-dG in positions 1 and 5, which were syn positions, hindered quadruplex formation, whereas in positions 1 and 10 in one molecule, the **GG**TTGGTGT**GG**TTGG stimulated folding into a quadruplex scaffold, which, as indicated by the CD signal max at 265 nm was some sort of parallel due to the two anti dGs in place of the two syn dGs in wild type antiparallel structure.

Table 9. α-Anomeric nucleotides in TBA sequences* [145]

	Sequence	T_m	CD pos. max
	1 5 10 14		
TBA	*G*GTT*G*GTGT*G*GTT*G*G	51.0	245 & 295
	GG**T**TGGTGTGG**T**TGG	40.7	245 & 295
	GGT**T**GGTGTGGT**T**GG	60.3	245 & 295
	GG**TT**GGTGTGG**TT**GG	51.9	245 & 295
	GGTTGG**T**G**T**GGTTGG	41.7	245 & 295
	GGTTGG**TGT**GGTTGG	42.8	245 & 295
	GGT**T**GG**TGT**GGT**T**GG	54.0	245 & 295
	GGTT*G*GT**G**TGGTTGG	< 30	n/d
	GGTTGGTGT**GG**TTGG	38.9	265
	GGTT**GG**TGTGGTT**GG**	n/d	n/d
	GGTTGGTGTGGTTGG	n/d	n/d

*The syn dGs in the unmodified TBA sequence are indicated by the numbers above the Gs in italics. Underlined, bold T and G show the incorporated α-anomeric nucleotides in the sequence. Measurements were carried out in 100 mM K^+.

Another isomer-induced conformation is the L-DNA quadruplex. Instead of the natural D-nucleotides Yuan et al. [146] incorporated L-nucleotides into a G-rich sequence, which could fold into an intramolecular L-(G-quadruplex). When L-nucleotides were incorporated into a DNAzyme sequence the resulting L-DNAzyme exhibited peroxidase catalytic activity. Virgilio et al. [147,148] incorporated L-nucleotides into the 15mer TBA DNA sequences. All sequences that contained various numbers of L-nucleotides were able to fold into the antiparallel chair-type quadruplexes as the wild-type did, only thermal stability values varied by the substitutions. CD spectra were mirror images of the wild-type's spectrum.

Applicability of sugar-modified guanine nucleotides to manipulate G-quadruplex folding is well supported by the studied cited above. Most of the anti dG sugar analogs induced the conversion of a mixture of quadruplex conformers to a dominant G-quadruplex conformation.

2.3. Modification of the Sugar-Phosphate Backbone

A special group of structural alteration that can affect quadruplex folding is the isoDNA, which name refers to a modification in which the natural 3'O-OP(O^-)-5'O backbone is changed, for instance, to 2'O-OP(O⁻)-5'O backbone. In short: 2'-5'isoDNA. Experiments with TBA quadruplex showed, that most such modification in the 15mer sequence didn't change the wild-type's two-tetrad chair-type antiparallel conformation [149].

2.4. Loops with Nucleotide Analogs

Ramon Eritja and coworkers [150] demonstrated that an apparently simple modification, a mutation of the TT loop sequences of the TBA, 5'-GGTTGGTGTGGTTGG, induced basic conformational changes: The monomeric, chair-type antiparallel quadruplex (Figure 4) adapted parallel G-quadruplex that contained a novel T-tetrad (Table 10).

Table 10. Mutation in the loops of the TBA quadruplex* [150]

Abbr.	**Sequence**	**T_m**	**ΔT_m**	**Topology in K^+**	**NH_4^+**
		(°C)			
TBA	GGTTGGTGTGGTTGG	50.1	-	AP	AP
TBA2**C**[4,13]	GGT<u>**C**</u>GGTGTGGT<u>**C**</u>GG	51.5	1.4	AP	AP
TBA2**A**[4,13]	GGT<u>**A**</u>GGTGTGGT<u>**A**</u>GG	40.1	-10	AP	AP
TBA1**G**[4]	GGT<u>**G**</u>GGTGTGGTTGG	47.9	-2.2	AP	AP
TBA2G[4,13]	GGT<u>**G**</u>GGTGTGGT<u>**G**</u>GG	59.2	9.1	P (h)	mix
TBA2G[3,12]	GG<u>**G**</u>TGGTGTGG<u>**G**</u>TGG	58.5	8.4	P (h)	mix
TBA4G[3,4,12,13]	GG<u>**GG**</u>GGTGTGG<u>**GG**</u>GG	>80	-	P (h)	P

*Measured in K^+ solution; AP stands for antiparallel, P for parallel, h for hysteresis.

In the edgewise TT loops both pyrimidines, the T and C function with similar effectiveness, i.e., C can essentially substitute for T. The larger than pyrimidines purine base, A, may cause steric incompatibility and thus markedly destabilized the G-quadruplex but the quadruplex remained antiparallel. A single T-to-G mutation not, but double and tetra T-to-G mutations of loops induced basic conformational changes, these sequences folded into parallel topology accompanied by significant thermodynamic stabilization (only the T_m values are shown here). Better stacking may be the driving force for the formation of the bimolecular parallel folds instead of the antiparallel as the authors calculated from the enthalpy change values. The hysteresis (h), originating from slower folding kinetics observed during the re-annealing (cooling) of the thermally unfolded structures also refers to the formation of a quadruplex from more than one strand. T_m values increased with the DNA strand concentration, which, again, showed that the folding was not monomolecular. The extreme high thermodynamic stability of TBA4G may originate from the two continuous six guanine tracks in the bimolecular parallel fold (ΔG^o_{25} -13.5 kcal/mol versus -3.5 kcal/mol for the wild-type) where the two T-tetrads, separated by one more G-tetrad were well tolerated in this highly stabilized structure. The T-tetrads are held together by circular single O4.H3(N) H-bonds. In the presence ammonium cations the topologies didn't change with the T-to-C, T-to-A and single T-to-G loop modifications. The T_m values were lower (not shown here) but the ΔT_m-s showed similar tendency as those observed with potassium ions. Among the double and tetra T-to-G loop mutated structures however only the tetra substituted TBA4G folded into a majority parallel fold. The double mutated sequences had two maxima in their CD spectra, at 260 nm and a smaller band near 295 nm that refer to a mixture of parallel and antiparallel forms. Addition of MeOH to TBA2G [4, 13], which, in the absence of MeOH in NH_4^+, showed a mixture of parallel and antiparallel conformations, induced the formation of a parallel structure [150].

For natural base mutations in htel quadruplexes an excellent illustration is when C is substituted for the T base in the loops of htel sequences, that is, the htel telomere contains $(CTAGGG)_n$ instead of $(TTAGGG)_n$. This is actually a natural mutation in human telomeres [151]. A 22mer sequence (CTA, below) has been reported by Lim et al. [152] to fold into a chair-type antiparallel G-quadruplex in K^+ solutions (Figure 1), whereas the 22mer TTAGGG repeat folds into a mixture of antiparallel basket and hybrid folds [43,44]. The CTA quadruplex contained two full G-tetrads and a third G:C:G:C mixed base tetrad as one of the terminal tetrads. Thermal stability of the mutated quadruplex was only slightly diminished as compared with the wild-type (Table 11).

Table 11. Thermostability of quadruplexes of A(hel-21) and a T-to-C mutated analog* [152]

Abbr.	Sequence	T_m (°C)
wt	A GGG TTA GGG TTA GGG TTA GGG	64
CTA	A GGG <u>C</u>TA GGG <u>C</u>TA GGG <u>C</u>TA GGG	62

*UV melting data, measured in 100 mM K^+.

Table 12. Effect of T-to-C mutations* [155]

Abbr.	htel-22 DNA sequence	Topology	T_m (°C)	ΔT_m
wt	AGGG TTA GGG TTA GGG TTA GGG	hybrids	61.4	-
C1	AGGG <u>C</u>TA GGG <u>C</u>TA GGG <u>C</u>TA GGG	antiparallel	59.5	-1.9
CCC1	AGGG <u>CCC</u> GGG TTA GGG TTA GGG	antiparallel	60.8	-0.6
CCC3	AGGG TTA GGG TTA GGG <u>CCC</u> GGG	antiparallel	57.8	-3.6

*Measured in 50 mM K^+ solution.

Human telomeres do not only consist of TTAGG or CTAGGG repeats. Sequence analyses uncovered numerous allelic variations. These mutations were apparently restricted to loop nucleotides [153,154]. To assess the effect of mutations of loop bases on G-quadruplex folding and stability Sara Richter's group [155] synthesized 19 systematically mutated A(htel-21) sequences and by spectroscopic methods determined the conformational features and the stability of the structures formed in K^+ solution. They found that all mutated repeats studied folded into quadruplexes. The results suggested that the mutated telomeric sequences could maintain the ability to fold into G-quadruplex in vivo too. Topologies however varied according to the site(s) of mutations, in accordance with some earlier observations [30,100,152]. Among the 19 mutated A(htel-21) sequences three formed a predominantly single fold, namely the antiparallel scaffold (Table 12). Thermal stability of these folds was slightly reduced as compared with the wild-type's stability. Based on their CD spectra most mutated oligonucleotides folded into either similar mixtures of hybrid forms as the wild-type did or into different ratio of the hybrids, or into mixtures hybrid and antiparallel forms (Table 12). The wild-type's CD spectrum showed a wide peak around 290 nm, a shoulder near 270 nm and a negative peak at 240 nm, characteristics of the mixture of the K^+-antiparallel and the hybrid forms that coexist in K^+ solution [39, 43-45]. According to earlier NMR measurements [30, 100, 152] the C1 quadruplex fold contained two G:G:G:G and one G:C:G:C tetrad in an antiparallel scaffold. When the first or the third TTA tracks were replaced by a CCC triplet the antiparallel quadruplexes contained G:C:G:G and G:G:G:C tetrads. These mixed-base tetrads may have caused the reduction of thermal stability of the wild-type [155]. In these cases, even if the loop sequences were mutated the changes resulted in mutated tetrads.

We can list a polydeoxyguanosine sequence as a special case of mutations in the loop sequences. Sengar et al. [156] studied $TTG_{12\text{-}19}T$ sequences and determined by NMR spectroscopy that most sequences formed G-quadruplex structures. The $TTG_{15}T$, for example folded into a well-defined propeller-type parallel single G-quadruplex scaffold containing three G-tetrads and three single-guanine propeller loops. Authors concluded that the same structure can potentially form anywhere along long G_n stretches, making it unique for molecular recognition by other cellular molecules.

Single folds can be created not only by natural base mutations or incorporating chemically modified bases into G-rich sequences but also by base losses. Babinsky et al. [104] replaced the adenines, one-by-one, by stabilized abasic (tetrahydrofuranyl) sites in the TTA loops of the A(htel-21) DNA. Loss of adenine in any of these sequence locations didn't prevent the ability of the single-apurinic oligonucleotides to fold into intramolecular quadruplexes, as the authors proved by CD and NMR spectroscopies. While thermostability of quadruplexes formed was only slightly affected by depurination, the folding characteristics were changed. The wild-type G-quadruplex of A(htel-21) DNA in 10-160 mM K^+ solutions is fairly polymorphic that consists of a mixture of the K^+-stabilized antiparallel, the hybrid-1 and hybrid-2 structures [37-45]. The mixture is characterized by a large positive peak near 295 nm, a shoulder near 270 nm and a smaller negative peak near 245 nm in its CD spectrum [44] (Figure 3). The loss of a guanine in any G-tetrad has been shown to shift the conformational equilibrium toward the antiparallel type, as indicated by the negative CD minimum at 260 nm (Figure 3), and, concomitantly, to a significantly reduced thermodynamic stability, as compared with the wild-type [122]. In contrast, with two abasic quadruplexes (ap7 and ap19; for abbreviations and sequences see Table 13) there was no negative 260 nm peak, instead, the 270 nm shoulders noticeably increased (upper panel, Figure 8) indicating a shift toward the formation of higher population of parallel strands in the mixture, i.e., shifting the conformational equilibrium towards higher concentrations of the hybrid folds.

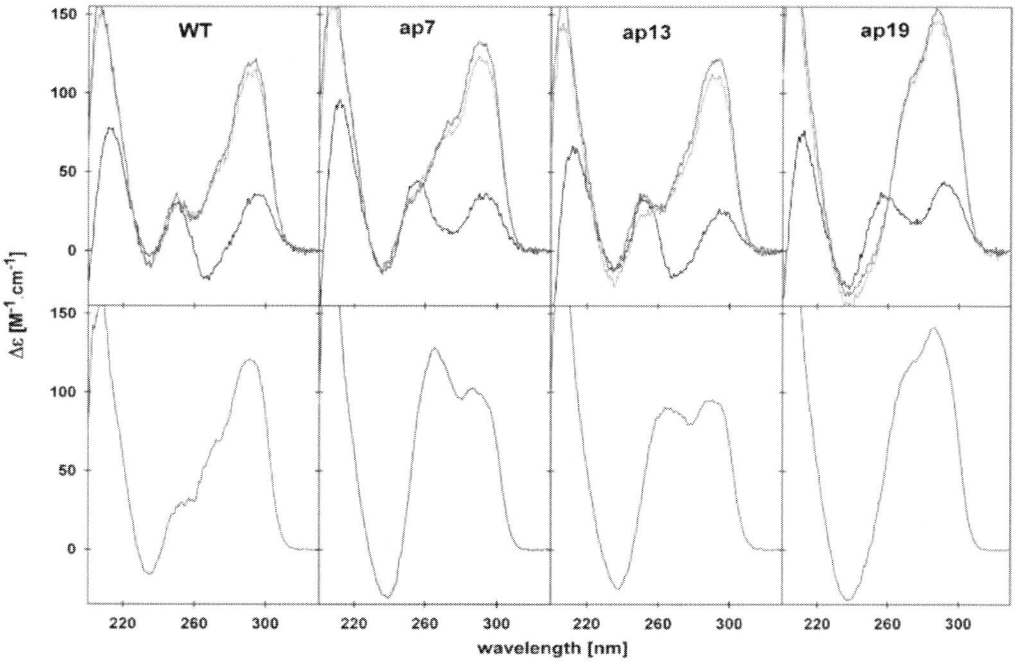

Figure 8. CD spectra of the wild-type $AG_3(TTAG_3)_3$ [A(htel-21)] quadruplex and its loop abasic site modified quadruplexes [105]. The spectra shown in the upper panel were measured at 23 μM of DNA strand concentrations in 1 mM sodium phosphate and 0.3 mM EDTA, pH 6.8 (three peaks), and in the presence of 15 mM and 165 mM K^+ ions, respectively, pH 6.9 at 23°C. The spectra displayed in the bottom panel were recorded at strand concentrations of 0.75–0.89 mM, used for NMR measurements, in 100 mM K^+, pH 7, at 23°C. WT stands for the wild-type, ap7, ap13 and ap19 for quadruplexes modified at loop positions 7, 13, and 19, respectively, by tetrahydrofuranyl abasic sites.

Folding features of wild-type htel-21 and A(htel-21) sequences has been described to depend on DNA strand concentrations above 2 mM [43,60]. As NMR measurements are generally carried out at strand concentrations about 40 times higher than the CD measurements (23 μM in the present case) Babinsky et al. [104] also recorded the CD spectra at around 0.8 mM strand concentrations (bottom panel, Figure 8) used for NMR measurements there. CD spectrum of the wild-type sample barely differed from that of 23 μM sample. However, spectrum of ap7 showed a significant increase of the 260 nm signal indicating increased presence of hybrids and/or parallel quadruplexes, as compared to the low concentration sample. The two positive bands around 260 and 295 nm of comparable heights of ap13 may originate from the increased populations of various quadruplex folds, as it was deduced from the NMR spectra. CD spectrum of ap19 did not change qualitatively by the increase of strand concentration suggesting that the AP site at the 3'-end third loop strongly stabilized a defined topology in K^+ solution even at low strand concentrations. NMR analysis also showed that topology significantly depended on the lesion's position. Though polymorphism increased when the abasic site was located in the middle loop (ap13), the abasic site in the first (ap7) and the third loops (ap19) induced single quadruplex folds. Site selective isotopic labeling demonstrated that ap7 and ap19 folded into single fold hybrid quadruplexes. Topologies were however different: ap7 folded into the hybrid-1, whereas ap19 built a hybrid-2 quadruplex. Interestingly, the abasic site in both cases was located in the propeller-like loop. In the unmodified hybrid-1 scaffold the first TTA tract forms a propeller-like loop with the A7 base exposed to the solvent, not interacting with other bases. Therefore, depurination of A7 had no dramatic effect on the stability of hybrid-1 quadruplex indicated by the T_m value of ap7 (Tables 13). In the hybrid-2 quadruplex (Figure 1) the A19 is also located in the propeller loop and does not form any stabilizing interaction with neighbors. Thus in the hybrid-2 topology too, depurination of A19 had little effect on stability. However, A13 forms an extensive interaction networks in all K^+-stabilized quadruplex folds determined so far, so depurination of A13 destabilized the quadruplex and contributed to the increase of complexity of folds. Each of the three abasic quadruplexes transformed easier than the wild-type into parallel quadruplexes in dehydrating ethanol solutions (57%). The abasic site stabilized the formation of the propeller loop connecting two parallel strands, which then facilitated the folding into propeller-type loops of the other two loops, which are specific for the parallel quadruplex topology. A triple abasic site containing sequence, the ap7,13,19 in K^+ folded into parallel quadruplex even in the absence of ethanol. This structure however didn't adopt a single quadruplex fold. Results of this study show that substitution of adenine nucleotides with abasic sites of the loop sequences of the human telomeric DNA is also a powerful tool for adjusting the folding topology of G-quadruplexes.

Table 13a. Adenine abasic site in the loops of A(htel-21) quadruplex at 23 μM strand concentrations* [104]

Abbr.	DNA sequence	T_m (°C)	ΔT_m
	1 7 13 19		
wt	AGGG TTAGGG TTAGGG TTAGGG	71.2	
ap7	AGGG TT_GGG TTAGGG TTAGGG	70.2	-1
ap13	AGGG TTAGGG TT_GGG TTAGGG	66.4	-4.8
ap19	AGGG TTAGGG TTAGGG TT_GGG	73.3	2.1

*Thermal stability determined in 165 mM K^+ solution.

Table 13b. Adenine abasic site in the loops of A(htel-21) quadruplex at around 800 μM strand concentrations* [104]

Abbr.	DNA sequence	Fold	T_m (°C)	ΔT_m
	1 7 13 19			
wt	AGGG TTAGGG TTAGGG TTAGGG	mixture	69.3	
ap7	AGGG TT_GGG TTAGGG TTAGGG	hybrid-1	71.0	1.7
ap13	AGGG TTAGGG TT_GGG TTAGGG	mixture	65.9	-3.4
ap19	AGGG TTAGGG TTAGGG TT_GGG	hybrid-2	70.7	1.4

*Thermal stability determined in 100 mM K^+ solution.

The T_m values of the wild-type and abasic A(htel-21) quadruplexes at around 0.8 mM strand and 100 mM K^+ concentrations were not much different from those of the low strand concentration values (Table 13b).

With the htel quadruplexes the 4-repeat models have been primarily studied. Phan and coworkers [157] published an interesting study concerning how the more than five repeat G-rich sequences can form a definite, single fold quadruplex topology. By using G-to-I (inosine, hypoxanthine base) substitutions in the extended, five to seven TTAGGG repeats (sequences in Table 14) they found that the extra repeats can form long loops protruding from the core due to the G-to-I substitutions, and these long-repeat-structures fold into hybrid-2 topology. The dI nucleotide hinders the formation of G_3I tetrad.

Table 14. Stability of long-repeat htel quadruplexes* [157]

Abbr.	Sequence			T_m (°C)
htel-25	TAG$_3$(TTAG$_3$)$_2$	TTA	GGG TT	62.0
htel-31	TAG$_3$(TTAG$_3$)$_2$	TTAGGGTTA	GGG TT	-
htel-31[**I**22]	TAG$_3$(TTAG$_3$)$_2$	TTAG**I**GTTA	GGG TT	49.8
htel-37[**I**22/**I**28]	TAG$_3$(TTAG$_3$)$_2$	TTAG**I**GTTAG**I**GTTA	GGG TT	47.7
htel-43[**I**22/**I**28/**I**34]	TAG$_3$(TTAG$_3$)$_2$	TTAG**I**GTTAG**I**GTTAG**I**GTTA	GGG TT	46.5
htel-43[**T**22/**T**28/**T**34]	TAG$_3$TTAG$_3$TTA$\mathbf{\underline{br^{8}G}}$GG	TTAG**T**GTTAG**T**GTTAG**T**GTTA	GGG TT	-

*T_m values were determined in in 100 mM K^+.

Risitano and Fox [31] prepared intramolecular htel quadruplexes in which the TTA loops were replaced by non-nucleosidic diol linkers, such as propanediol, octanediol or hexaethylene glycol. CD spectrum of each structure showed a positive maximum at 264 nm, negative minimum at ~244 nm, which are the characteristic of parallel quadruplexes. This is reminiscent of the effect that the single nucleotide loops do with quadruplex topology although the length of these diols can be much longer than the distance of an internucleotide linkage in the case of single nucleotide loops. Thermal stabilities were different, they increased in the order propanediol < hexaethylene glycol < octanediol. The octanediol structure displayed a biphasic melting profile suggesting that it could adopt more than one stable fold. A further special feature was that the T_m of all these non-nucleosidic structures was above 37°C in the presence of 10 mM Li^+ alone, without added K^+, in contrast to the wild-type htel quadruplex.

2.5. Topologically Constrained Quadruplexes

Interlocking G-quadruplexes can serve as a tool for assembling higher-order quadruplex structures, such as G-wires. Phan and Do [158] engineered three different interlocked dimeric G-quadruplexes in which the interlocks, like the sticky ends that connect double-stranded

DNAs were natural bases. In the dimeric quadruplexes these were guanine, thymine or adenine that participated as the fourth base of the base-deficient terminal G-triads of the neighbor quadruplex. They used the 16mer 93del GGGGTGGGAGGAGGGT DNA sequence that is an HIV-1 integrase inhibitor, which forms an extremely stable hybrid type interlocked dimeric G-quadruplex with six G-tetrads [159]. The dimer actually consisted of two parallel-stranded quadruplexes connected with opposite polarities.

Paramasivam and Doluca et al. [160-162] incorporated a twisted intercalating nucleic acid (TINA) moiety, containing phenylethynylpyren-1-yl derivatives, into a G-rich DNA sequence, which altered the folding topology of the G-quadruplex. Using CD, NMR and ESI-mass-spectra, as well as TINA excited dimer (excimer) fluorescence emission the authors proposed that insertion of the TINA monomer in the middle of a dTG_4T sequence (i.e., dTGG**X**GGT, where **X** is TINA) converted a parallel tetramolecular G-quadruplex into an assembly composed of two identical antiparallel G-quadruplex subunits stacked *via* TINA–TINA interface. Kinetic analysis showed that TINA–TINA association controlled the complex formation only in Na^+ solution but not in K^+ or in sequences with the longer G-runs, such as TGGG**X**GGGT. The results demonstrated how the tetramolecular G-quadruplexes could be used as tunable scaffolds to control the arrangement of organic chromophores.

Another example for the formation of topologically constrained G-quadruplexes was demonstrated by chain-end terminal attachments of cyclopeptide moieties to htel sequences [163-166]. Using bioconjugation techniques rigid peptide scaffolds were attached to an antiparallel DNA, and a DNA:RNA hybrid quadruplex and also to a parallel RNA (TERRA) quadruplex. The peptide moieties greatly stabilized the quadruplex architectures. Such topologically controlled G-quadruplex systems can be of great interest for the design of more selective G4 ligands.

3. Effect of Cation and Its Concentration on G-Quadruplex Folding

G-quadruplex structures are stabilized by cations, primarily by monovalent ones. Their role in stabilization and folding of quadruplexes was first reviewed comprehensively by Hud and Plavec [167] who described that the order of stabilizing the htel repeat $(TTAGGG)_4$ quadruplex was $K^+ > Na^+ \geq Rb^+ > Li^+ > Cs^+$, based on thermal stability values. Based on free energy difference values ($\Delta\Delta G^o$) the order was $K^+ > NH_4^+ > Rb^+ > Na^+ > Cs^+ > Li^+$. The free energy difference ($\Delta\Delta G^o_{37}$) between the K^+ and Na^+ ion-stabilized various wild-type htel quadruplexes has been reported to be between -1.3 to -2.4 kcal/mol [107, 121, 122, 168-173]. Thermal stability generally differs by 5 to 10°C between the two forms but much higher variances have also been published depending on the sequence. Extreme values between 1°C to 40°C have also been observed [31, 101, 120, 168, 169, 174-182]. Among intramolecular G-quadruplexes the TBA displayed one of the largest differences of the T_m values between the potassium and the sodium forms [168]. Folding of the TBA quadruplex was also found to be cation selective, and the selectivity order was $K^+ > NH_4^+ \sim Ba^+ > Cs^+ > Na^+ > Li^+$, and G-quadruplex was not detected by with Mg^{2+} and Ca^{2+} using nanopore encapsulating single-molecule method [183]. Monovalent cations have dual effects on structure. Coordinate with four O^6 atoms of guanine in the center cavity of the quadruplex fold and compensate the repulsion force of the phosphate anions within the water sphere outside of the fold. The coordination with guanine oxygen atoms is primarily responsible for the stabilization of the architecture. For the optimal stabilizations the cation must be small enough to fit in the central cavity formed by the four guanines and also should be large enough to bridge the distance between the guanine's carbonyl oxygen atoms. In this respect, the Li^+ ion is too small, the Na^+ can fit into the plane of the G-tetrad, whereas the larger K^+ ions intercalate between two G-tetrads, and this provides the highest stabilizing force. The Cs^+ ions are too large to fit between tetrads. Tl^+ is also known to support quadruplex formation. Bivalent cations have various effects on different quadruplex structures. The Sr^{2+}, Ba^{2+} and Pb^{2+} at low concentrations are stabilizing while increasing concentrations are destabilizing. Mg^{2+} and Ca^{2+} do not promote quadruplex formations by the same mechanism of action as the former mono- and bivalent cations, though they can contribute to stabilization of G-quadruplexes via general electrostatic contributions [167, 184]. Divalent cations of 1 mM concentration destabilized the antiparallel G-quadruplex of $G_4T_4G_4$. The order of destabilization was Zn^{2+} > Co^{2+} > Mn^{2+} > Mg^{2+} > Ca^{2+}. Higher concentration of certain divalent cations, such as Sr^{2+} and Ca^{2+} induced a transition from the antiparallel to parallel topology. The result suggested that divalent cations are also good tools for regulating the quadruplex structure [185-187].

Among all cations the monovalent Na^+ and K^+ are in the focus of structural studies with G-quadruplexes [188, 189]. As cells possess about ten times higher concentrations of potassium ions (100-160 mM) than sodium ions, folding topology and structural stability of quadruplexes are preferably studied in potassium containing solutions. Even the concentration of potassium ions influences folding topology of quadruplexes [32]. The main conformational effect of changing the cations is most evident for the Na^+-to-K^+ exchange with htel-21 and A(htel-21) quadruplexes: the Na^+-stabilized antiparallel basket scaffold converts into a mixture of folds, mostly into the two hybrids and the K^+-antiparallel [43, 44, 190]. Topology changes induced by other cations are also known. For instance, Pedroso et al. used Sr^{2+} to facilitate formation of parallel intermolecular DNA quadruplexes in $(TTAGGG)_{2\text{-}5}$ while longer (TTAGGG) repeats formed exclusively intramolecular structures [191]. As stated above, the K^+ ions are more stabilizing than Na^+ ions. A recent study however reports on an exception. The htel-45 $(GGGTTA)_7GGG$ sequence folds into two contiguous G-quadruplex units. Despite the lower thermal stability in Na^+ than in K^+, its Na^+-conformation proved to be more stable than the K^+-structure based on the enthalpy change values [181].

Special cases are the htel RNA quadruplexes that form parallel quadruplexes both in Na^+ [192] and K^+ solutions [193].

4. Topology Changes in Molecular Crowding – Effect of Cosolutes

Structural investigations of G-quadruplexes are generally carried out with dilute DNA solutions. These conditions, thus also the results of experiments carried out in these solutions don't have much relevance to the in vivo, cellular conditions. To understand the gene regulatory and chromosome-end protecting functions of the four-stranded nucleic acid structures as well as the ligand-targeted quadruplex structures in physiologically relevant conditions for drug development, it was necessary to clarify whether cellular conditions can influence the stability and conformation of the quadruplexes. In living cells, the biological macromolecules exist and function in a crowded intracellular environment. The intracellular space is highly packed with nucleic acids, proteins, etc. macromolecules, insoluble structural elements and small molecules constituting 30-40% (w/v) of the cellular mass, characterized by large excluded volume and low water activity [194-199]. As it was found, this crowding does affect conformation, stability, and activity of biomolecules [200-202]. To address this problem, solution conditions have been worked out that try to mimic the cellular environment [201]. Miyoshi et al. [203] demonstrated for the first time that poly(ethylene glycol) (PEG) is one of the macromolecules that can produce conditions in aqueous solutions that simulate molecular crowding in cells. They found that PEG does cause changes in properties of quadruplexes: The antiparallel topology of the G-quadruplex of $d(G_4T_4G_4)$ converted into parallel fold in 20-40% of PEG. This fold was unimolecular with four parallel strands and all G nucleotides adopted the anti glycosidic conformation that forced all three loops to form chain-reversed propeller orientation. The results suggested that both the volume excluded by PEG and the electrostatic interactions were critical for the structural conversion.

Many of the multiple quadruplex architectures formed by the various, primarily the htel repeat sequences have been studied in cosolvents-mimicked molecular crowding conditions [43, 69, 70, 204, 205]. The crowding in K^+ solutions has been imitated by various agents or cosolvents, such as EtOH (up to 60%) [43,205], polyethylene glycol (PEG) 200 (0–60%) [205-214], PEG 400 [215], PEG 1000 [216], PEG 8000 [205], PEG 35,000 [205], glycerol, ethylene glycol (0–15%) [217], histone peptide (20%) in sodium solution [209], acetonitrile (50%) [205, 218], Ficoll 400 [205, 206] and dimethyl-sulphoxide (DMSO) [205]. The important feature found by Sugimoto's group was that effect of cosolvent on nucleic acids conformation depended on their primary and secondary structures and the effect was regulated by the hydration shell around the macromolecules [206,210,219-221]. For example, while PEGs and polysaccharides, like Ficoll, decrease the thermal stability of DNA duplexes, the triplexes, G-quadruplexes, i-motifs and Holiday junctions are stabilized due to decreased water activity by the presence of the cosolutes. According to the theory and to experimental results the decreased water activity was unfavorable for the stability of Watson–Crick base pairs as its formation requires water uptake, and was favorable for the quadruplexes as their formation is accompanied by water release [221]. This theory however conflicts with the microscopic picture of the structures [222]. Sugimoto's recent results indicate that the water–water interactions around the DNA depend on the type of cosolutes and this is the interaction that contributes to the thermal stabilities of various DNA structures [221]. As for the G-quadruplexes, the common property found in K^+ solutions was that on the increase of cosolvent concentration, at low, micromolar DNA strand concentrations the quadruplex folding changed from the K^+-stabilized forms, such as the antiparallel and hybrid to the parallel fold, and the thermal and thermodynamic stabilities increased concurrently and 40% (w/v) PEG 200 was the preferred cosolute [203, 206, 207, 212, 217]. However, not all htel quadruplexes converted spontaneously into the parallel fold: Petraccone et al. [69] has shown that the hybrid-type and parallel conformations coexisted at equilibrium in $(TTAGGG)_4TT$ and $(TTAGGG)_8TT$ in the presence of PEG at physiological temperature and the degree of quadruplex conversion depended on the molecular weight of the PEG used (from PEG200 to PEG10,000). The hybrid forms remained stable when Ficoll 400 or glycerol was utilized as a crowding agent [223]. Zhoua et al. have shown that crowding agents had complex effects on the G-quadruplexes, increased the stability of antiparallel, hybrid and parallel quadruplexes in

Na^+ or K^+ solution, but in Sr^{2+}, Ba^{2+}, and Pb^{2+} the agents had no effect or destabilized the antiparallel or hybrid structures and destabilized the parallels.

Research groups first used CD spectroscopy to interpret conversion of topologies by the crowding agents [93,203,207,215]. Heddi and Phan [205] reported for the first time on an NMR-determined structure of a human telomeric G-quadruplex formed in a crowded solution. In 90 mM K^+-containing solution with 40% PEG 200, the CD and NMR profiles of four different htel G-quadruplex conformations became similar. CD spectra showed a positive peak at 260 nm and a negative peak at 240 nm, characteristic of a parallel-stranded G-quadruplex. The four DNA quadruplexes were the hybrid-1 forming TA(htel-21) [37,38,111], the hybrid-2 TA(htel-21TT [67], the hybrid-3 (two-tetrad) (htel-21)T [61] and the $AG_3(CTAG_3)_3$ sequence that in K^+ solution folded into a chair-type antiparallel-stranded G-quadruplex involving two G-tetrads and a G:C:G:C tetrad [152]. The four conformations were converted to a propeller-type parallel-stranded G-quadruplex due to water depletion and these transitions proved to be reversible.

Miller et al. [218] used 50% acetonitrile as a non-hydrogen bonding dehydrating agent and observed similar conformational transitions of the A(htel-21) quadruplex in K^+ solution as observed with PEG. They concluded that the cosolvents PEG 200 and 400 were actually dehydrating agents, not crowing agents. Their NMR analysis showed that the resulting structure of A(htel-21) contained also non-anti guanosines suggesting that the conformation present in acetonitrile is not identical to the all-parallel crystal structure, despite the supposed parallel type CD spectrum. Non-dehydrating agents, such as BSA also didn't cause full conversion to parallel, the hybrid folds remained. These results also suggested that the parallel was not the biologically relevant fold. (The sodium form of A(htel-21) showed only small changes on addition of acetonitrile.) Similarly, Hansel et al. [224] showed that conformation of htel repeats in X. laevis egg extract or in Ficoll was notably different from that observed in the presence of PEG as in X. laevis egg extract or in Ficoll the topology resembled those obtained under dilute solution conditions, whereas PEG promoted the formation of high-order parallel topologies. They concluded that PEG should not be used as a molecular crowding mimetic and suggested that proteins or egg extracts are better cosolutes to mimic cellular conditions.

Dhakal et al. [225] used laser tweezers to mechanically unfold the htel DNA quadruplex of the 27mer $(TTAGGG)_4TTA$. They found that parallel G-quadruplex coexists with non-parallel species in 1:1 ratio in 40% (w/v) dimethyl-sulphoxide (DMSO) or acetonitrile (ACN). In crowded solutions with steric cosolutes of 40% (w/v) bovine serum albumin (BSA), the parallel G-quadruplex constituted only 10% of the population. This difference unequivocally supported the notion that dehydration promotes the formation of parallel G-quadruplexes. A base modification was also reported to hinder the conversion to parallel conformation in crowding condition. Zhao and coworkers [123] found that the alkylation product O^6-methylguanine (m^6G), which is a natural base lesion, when incorporated into different positions of a htel DNA sequence impeded the conformational conversion of the G-quadruplex from the hybrid to the parallel topology under molecular crowding conditions in K^+ containing buffer.

While PEG converts the hybrid htel quadruplexes to parallel forms in K^+ solutions, it doesn't convert the Na^+-stabilized antiparallel basket structures to parallel. Based on this fact and using osmotic stress and sedimentation velocity studies Chaires and coworkers ruled out the possibility that PEG acts through a crowding mechanism, that is, by excluded volume or hydration changes resulting from osmotic stress [214]. They rather explain the conversion as a result of differential binding of PEG to the different scaffolds.

There is high osmotic pressure in cells and in solutions of molecular crowding. The pressure alone, without the presence of crowding agents can also cause structural changes in biomolecules, such as all forms of DNA. Takahashia and Sugimoto [226] investigated the effect of high pressure, 400 MPa, on the folding of duplex, i-motif and G-quadruplex DNAs. While the i-motif tetraplex was stabilized by high pressure, the G-quadruplex was somewhat destabilized. Interestingly, the effects were similar to the effect of 40% ethylene-glycol. According to the *P–T* diagrams the duplex was more stable than the G-quadruplex and i-motif structures under low pressure, but the i-motif dominated the structural composition under high pressure. Under crowding conditions, the *P–T* diagrams indicated that the duplex does not form under high pressure, and i-motif and G-quadruplex structures did form. The results suggest that pressure impacts the function of nucleic acids by stabilizing the non-canonical structures.

5. Effect of High DNA Concentrations on the Folding Topology of Quadruplexes

Mimicking cellular molecular crowding conditions are generally studied by adding cosolutes to dilute G-quadruplex solutions of 3-50 μM stand concentrations (Section 4). It is known that higher DNA concentrations (and also high salt concentrations) can cause conformational transition of G-quadruplexes from unimolecular to bimolecular structures [159, 227-231]. Renciuk [43], Abu-Ghazalah [232], Palacky [45], Kejnovská [60], and Li [233] et al., however, demonstrated more striking topological changes in htel quadruplexes on the effect of increasing stand concentrations from the micromolars to millimolars. Raising the DNA concentrations, that is, when the quadruplex itself was the crowding agent, produced similar folding changes than various cosolvents induced in htel quadruplexes of micromolar strand concentrations: At high DNA concentrations the parallel folding topologies arose. Renciuk [43], Palacky [45] and Keynovska [60] used the (htel-21), Abu-Ghazalah [232] the A(htel-21), the hybrid-1 quadruplex forming AAA(htel-21)AA [37] and hybrid-2 forming TTA(htel-21)TT [40,41] DNA sequences, and Li [233] investigated 19 loop-modified sequence analogs of A(htel-21). Although TTA(htel-21)TT and AAA(htel-21)AA required high salt (1 M NaCl) to transform, the authors suggested that the conversion of monomolecular quadruplexes into multimolecular, self-associated parallel topologies is a general phenomenon.

In Kejovska's study [60] the strand concentration of the (htel-21) deoxyoligonucleotide was increased from ~3 μM to ~9 mM in 165 mM K^+ solution (Table 15). Thermostability of the quadruplexes did not change up to ~ 2 mM of strands indicating the presence of intramolecular structures. Above 2 mM the T_m values increased linearly with the DNA concentrations implying that unfolding of quadruplexes was not monomolecular any more. The results were explained by self-association of G-quadruplexes above ~2 mM of strands. Abu-Ghazalah et al. [232] have also found that above 2 mM of strands of the A(htel-21) quadruplexes in 0.1 M and 1 M NaCl formed higher-order, multimolecular parallel structures, which they also called as aggregated complexes. At low strand concentrations in [60] multiple quadruplex folds were detected with htel-21, the K^+-stabilized antiparallel and the hybrid folds with the K^+-antiparallel dominating. At ~2 mM of strands only the hybrids were present, and in 6 to 9 mM only parallel quadruplexes were observed by CD spectroscopy (Figure 5, Table 15). Close to 9 mM the concentration of the DNA phosphate anions approached the concentration of the K^+ counter ions in the solution. Volume exclusion was assumed to promote the topological changes of quadruplexes toward the parallel, and the decreased screening of anions probably contributed to the reduced free energy changes in the crowded DNA solutions, while thermal stabilities increased all over (Table 15) [60].

Table 15. Thermodynamic and CD spectral data for the formation of $G_3(TTAG_3)_3$ quadruplexes* [60]

Strand Conc. (mM)	T_m (°C)	ΔT_m (°C)	ΔG^o_{37} (kcal/mol)	CD $\Delta\varepsilon$ at 265 nm	CD $\Delta\varepsilon$ at 290 nm
0.0036	75.9	0	-6.51	65.9	118.2
0.033	76.6	0.7	-6.48	69.1	117.6
0.424	76.0	0.1	-6.05	56.4	117.7
1.72	76.1	0.2	-6.06	78.0	99.5
3.73	77.0	1.1	-5.8	84.1	107.7
4.25	79.9	4.0	-5.54	102.9	104.1
6.35	82.1	6.2	-5.22	105.1	92.4
8.79	86.5	10.6	-5.03	155.1	68.8

CONCLUSION

The TTAGGG repeats constituting the human telomeric DNA 3'-overhang sequence can form a variety of G-quadruplex conformations in physiologically relevant potassium ion containing solutions in vitro. In structural studies stable single folding arrangements of quadruplexes are preferred or required. Here, we summarized numerous methods reported in the literature for creating a single or a predominant and stable folding topology, mainly in the human telomeric DNA quadruplexes. The focuses were on the extension and mutations of the sequence repeats using natural or synthetic nucleotide analogs in both the G-rich parts, forming the core of a quadruplex and the connecting loops of the potentially quadruplex forming oligodeoxynucleotides. In addition to the clear and encouraging result reached so far, several ambiguities remained to be solved. For instance, whether the single-molecular analyses can accurately replicate the results of solutions studies, or whether the topologies observed are the same when those are determined by different techniques. This problem comes up when the topologies detected by low concentration requiring absorption spectroscopic studies are compared with the high DNA concentration demanding NMR results.

As for the significance of the main DNA model examined here: The G-quadruplexes assembling from the 21 nucleotide long human telomere fragment $G_3(TTAG_3)_3$ and its longer sequence analogs have been considered as just model structures since such quadruplexes exist in tightly-bond protein complexes (Shelterin) at the ends of linear chromosomes. Until recently no such free quadruplexes were assumed to be present in cells. Graves et al. [234] however demonstrated the presence of circulating telomeric DNAs in the serum of cancer patients released from cancer cells in response to chemotherapy-induced cell death. As this DNA was readily detectable it was assumed that the fragments folded into G-quadruplexes that are more resistant to nucleases than the unfolded telomeric oligodeoxynucleotides. The in vivo studies also showed that these structures were very effective in the induction of TLR9-mediated cancer cell invasion. These results can change the view of the human telomere quadruplexes from simple model structures to real players in vivo and to their role as diagnostics.

REFERENCES

[1] Kypr, J., Fialová, M., Chládková, J., Tůmová, M. and Vorlíčková, M. (2001). Conserved guanine-guanine stacking in tetraplex and duplex DNA. *Eur. Biophys. J., 30*, 555-558.

[2] Kypr, J. and Vorlíčková, M. (2002). Circular dichroism spectroscopy reveals invariant conformation of guanine runs in DNA. *Biopolymers (Biospectroscopy)*, *67*, 275-277.

[3] Burge, S., Parkinson, G.N., Hazel, P., Todd, A.K. and Neidle, S. (2006). Quadruplex DNA: sequence, topology and structure. *Nucleic Acids Res.*, *34*, 5402–5415.

[4] Neidle, S. and Balasubramanian, S. (Eds.) (2006). *Quadruplex nucleic acids.* Cambridge; RSC Publishing.

[5] Collie, G.W. and Parkinson, G.N. (2011). The application of DNA and RNA G-quadruplexes to therapeutic medicines. *Chem. Soc. Rev., 40*, 5867-5892.

[6] Neidle, S. Therapeutic applications of quadruplex nucleic acids, Academic Press; 2012.

[7] Neidle, S. (2016). Quadruplex Nucleic Acids as Novel Therapeutic Targets. J. Med. Chem. Article ASAP; DOI: 10.1021/acs.jmedchem.5b01835.

[8] Chambers, V.S., Marsico, G., Boutell, J.M., Di Antonio, M., Smith,G.P. and Balasubramanian, S. (2015). High-throughput sequencing of DNA G-quadruplex structures in the human genome. *Nat.Biotechnol.*, *33*, 877–881.

[9] Selvam, S., Koirala, D., Yu, Z. and Mao, H. (2014). Quantification of Topological Coupling between DNA Superhelicity and G-quadruplex Formation. *J. Am. Chem. Soc., 136*, 13967−13970.

[10] Rhodes, D. and Lipps, H.J. (2015). G-quadruplexes and their regulatory roles in biology. *Nucleic Acids Res.*, *43*, 8627–8637.

[11] Wellinger, R.J. and Sen, D. (1997). The DNA structures at the ends of eukaryotic chromosomes. *Eur. J. Cancer, 33*, 735–749.

[12] Phan, A.T., Kuryavyi, V., Luu, K.N. and Patel, D.J. Structural diversity of G-quadruplex scaffolds. In Neidle, S. and Balasubramanian, S. (Eds.) (2006), pp 81-99. Quadruplex nucleic acids. Cambridge; RSC Publishing.

[13] Patel. D.J., Phan, A.T. and Kuryavyi, V. (2007). Human telomere, oncogenic promoter and 5'-UTR G-quadruplexes: diverse higher order DNA and RNA targets for cancer therapeutics. *Nucleic Acids Res.*, *35*, 7429–7455.

[14] Onel, B., Carver, M., Wu, G., Timonina, D., Kalarn, S., Larriva, M. and Yang, D. (2016). A New G-Quadruplex with Hairpin Loop Immediately Upstream of the Human BCL2 P1 Promoter Modulates Transcription. *J. Am. Chem. Soc., 138*, 2563–2570.

[15] Blackburn, E.H. (2000). Telomere states and cell fates. *Nature, 408*, 53–56.

[16] De Cian, A., Lacroix, L., Douarre, C., Temime-Smaali, N., Trentesaux, C., Riou, J.F. and Mergny, J.L. (2008). Targeting telomeres and telomerase. *Biochimie, 90*, 131–155.

[17] Blackburn, E.H. and Gall, J.G. (1978). A tandemly repeated sequence at the termini of extrachromosomal ribosomal RNA genes in Tetrahymena. *J. Mol. Biol., 120*, 33–53.

[18] Moyzis, R.K., Buckingham, J.M., Cram, L.S., Dani, M., Deaven, L.L., Jones, M.D., Meyne, J., Ratliff, R.L. and Wu, J.R. (1988). A highly conserved repetitive DNA sequence, (TTAGGG)n, present at the telomeres of human chromosomes. *Proc. Natl Acad. Sci. USA, 85*, 6622–6626.

[19] Makarov, V.L., Hirose, Y. and Langmore, J.P. (1997). Long G tails at both ends of human chromosomes suggest a C strand degradation mechanism for telomere shortening. *Cell, 88*, 657–666.

[20] Wright, W.E., Tesmer, V.M., Huffman, K.E., Levene, S.D. and Shay, J.W. (1997). Normal human chromosomes have long G-rich telomeric overhangs at one end. *Genes Dev., 11*, 2801–2809.

[21] Paeschke, K., Simonsson, T., Postberg, J., Rhodes, D. and Lipps, H.J. (2005) Telomere end-binding proteins control the formation of G-quadruplex DNA structures in vivo. *Nat. Struct. Biol., 12*, 847–854.

[22] Chang, C.-C., Chu, J.-F., Kao, F.-J., Chiu, Y.-C., Lou, P.-J., Chen, H.-C. and Chang, T.-C. (2006). Verification of antiparallel G-quadruplex structure in human telomeres by using two-photon excitation fluorescence lifetime imaging microscopy of the 3,6-Bis(1-methyl-4-vinylpyridinium)carbazole diiodide molecule. *Anal. Chem., 78*, 2810–2815.

[23] Harley, C.B. (1991). Telomere loss: mitotic clock or genetic time bomb? *Mutation Res. 256*, 271–282.

[24] Jullien, L., Mestre, M., Roux, P. and Gire, V. (2013). Eroded human telomeres are more prone to remain uncapped and to trigger a G2 checkpoint response. *Nucleic Acids Res., 41*, 900–911.

[25] Rachwal, P.A., Brown, T. and Fox, K.R. (2007). Effect of G-Tract Length on the Topology and Stability of Intramolecular DNA Quadruplexes. *Biochemistry, 46,* 3036 - 3044.

[26] Gaynutdinov, T.I., Neumann, R.D. and Panyutin, I.G. (2008). Structural polymorphism of intramolecular quadruplex of human telomeric DNA: effect of cations, quadruplex-binding drugs and flanking sequences. *Nucleic Acids Res., 36*, 4079–4087.

[27] Dai, J., Carver, M. and Yang, D. (2008). Polymorphism of human telomeric quadruplex structures. *Biochimie, 90,* 1172-1183.

[28] Vıglasky, V., Bauer, L. and Tluckova, K. (2010). Structural Features of Intra- and Intermolecular G-Quadruplexes Derived from Telomeric Repeats. *Biochemistry* 49, 2110–2120.

[29] Phan, A.T. (2010). Human telomeric G-quadruplex: structures of DNA and RNA sequences. *FEBS J., 277*, 1107–1117.

[30] Hazel, P., Huppert, J., Balasubramanian, S. and Neidle, S. (2004). Loop-length-dependent folding of G-quadruplexes. *J. Am. Chem.Soc., 126*, 16405–16415.

[31] Risitano, A. and Fox, K.R. (2004). Influence of loop size on the stability of intramolecular DNA quadruplexes. *Nucleic Acids Res., 32,* 2598–2606.

[32] Rujan, I.N., Meleney, J.C. and Bolton, P.H. (2005). Vertebrate telomere repeat DNAs favor external loop propeller quadruplex structures in the presence of high concentrations of potassium. *Nucleic Acids Res., 33,* 2022–2031.

[33] Yadav, D. and Sheardy, R.D. (2012). A single base permutation in any loop of a folded intramolecular quadruplex influences its structure and stability. *J. Biophys. Chem., 3,* 341-347.

[34] Gray, D.M., Wen, J.-D., Gray, C.W., Repges, R., Repges, C., Raabe, G. and Fleischhauer, J. (2008). Measured and calculated CD spectra of G-quartets stacked with the same or opposite polarities. *Chirality, 20,* 431-440.

[35] Wang,Y. and Patel, D.J. (1993). Solution structure of the human telomeric repeat d[AG3(T2AG3)3] G-tetraplex. *Structure, 1,* 263–282.

[36] Parkinson, G.N., Lee, M.P.H. and Neidle, S. (2002). Crystal structure of parallel quadruplexes from human telomeric DNA. *Nature 417,* 876–880.

[37] Ambrus, A., Chen, D., Dai, J., Bialis, T., Jones, R.A. and Yang, D. (2006). Human telomeric sequence forms a hybrid-type intramolecular G-quadruplex structure with mixed parallel/antiparallel strands in potassium solution. *Nucleic Acids Res., 34,* 2723–2735.

[38] Luu, K.N., Phan, A.T., Kuryavyi, V., Lacroix, L. and Patel, D.J. (2006). Structure of the human telomere in K^+ solution: an intramolecular (3 + 1) G-quadruplex scaffold. *J. Amer. Chem. Soc., 128,* 9963–9970.

[39] Phan, A.T., Kuryavyi, V., Luu, K.N. and Patel, D.J. (2007) Structure of two intramolecular G-quadruplexes formed by natural human telomere sequences in K^+ solution. *Nucleic Acids Res., 35,* 6517–6525.

[40] Dai, J., Carver, M., Punchihewa, C., Jones, R.A. and Yang, D. (2007). Structure of the hybrid-2 type intramolecular human telomeric G-quadruplex in K^+ solution: insights into structure polymorphism of the human telomeric sequence. *Nucleic. Acids Res., 35,* 4927–4940.

[41] Dai, J., Punchihewa, C., Ambrus, A., Chen, D., Jones, R.A. and Yang, D. (2007). Structure of the intramolecular human telomeric G-quadruplex in potassium solution: a novel adenine triple formation. *Nucleic Acids Res., 35,* 2440–2450.

[42] Gray, R.D., Buscaglia, R. and Chaires, J.B. (2012). Populated intermediates in the thermal unfolding of the human telomeric quadruplex. *J. Am. Chem. Soc., 134,* 16834−16844.

[43] Renciuk, D., Kejnovská, I., Skolakova, P., Bednarova, K., Motlova, J. and Vorlíčková, M. (2009). Arrangements of human telomere quadruplex in physiologically relevant K^+ solutions. *Nucleic Acids Res., 37,* 6625-6634.

[44] Vorlíčková, M., Kejnovská, I., Sagi, J., Renciuk, D., Bednarova, K., Motlova, J. and Kypr, J. (2012). Circular dichroism and guanine quadruplexes. *Methods, 57,* 64–75.

[45] Palacký, J. Vorlíčková, M., Kejnovská, I. and Mojzeš, P. (2013). Polymorphism of human telomeric quadruplex structure controlled by DNA concentration: a Raman study. *Nucleic Acids Res., 41,* 1005-1016.

[46] Wang, Z.-F., Li, M.-H., Hsu, S.-T.D. and Chang, T.-C. (2014). Structural basis of sodium-potassium exchange of a human telomeric DNA quadruplex without topological conversion. *Nucleic Acids Res., 42*, 4723-4733.

[47] Noer, S.L., Preus, S., Gudnason, D., Aznauryan, M., Mergny, J.-L. and Birkedal, V. (2016). Folding dynamics and conformational heterogeneity of human telomeric G-quadruplex structures in Na^+ solutions by single molecule FRET microscopy. *Nucleic Acids Res., 44,* 464–471.

[48] Lim, K.W., Ng, V.C.M., Martın-Pintado, N., Heddi, B. and Phan, A.T. (2013). Structure of the human telomere in Na^+ solution: an antiparallel (2 + 2) G-quadruplex scaffold reveals additional diversity. *Nucleic Acids Res., 41,* 10556–10562.

[49] Macaya, R.F., Schultze, P., Smith, F.W., Roe, J.A. and Feigon, J. (1993). Thrombin binding DNA aptamer forms a unimolecular quadruplex structure in solution. *Proc. Natl. Acad. Sci. USA, 90*, 3745-3749.

[50] Wang, K.Y., McCurdy, S., Shea, R.G., Swaminathan, S. and Bolton, P.H. (1993). A DNA aptamer which binds to and inhibits thrombin exhibits a new structural motif for DNA. *Biochemistry, 32,* 1899-1904.

[51] Schultze, P., Macaya, R.F. and Feigon, J. (1994). Three-dimensional solution structure of the thrombin-binding DNA aptamer d(GGTTGGTGTGGTTGG). *J. Mol. Biol., 235,* 1532-1547.

[52] Marathias, V.M. and Bolton, P.H. (2000). Structures of the potassium-saturated, 2:1, and intermediate, 1:1, forms of a quadruplex DNA. *Nucleic Acids Res., 28,* 1969-1977.

[53] Mao, X. and Gmeiner, W.H. (2005). NMR study of the folding-unfolding mechanism for the thrombin-binding DNA aptamer d(GGTTGGTGTGGTTGG). *Biophys. Chem., 113,* 155-160.

[54] Gray, R.D. and Chaires, J.B. (2008). Kinetics and mechanism of K^+- and Na^+-induced folding of models of human telomeric DNA into G-quadruplex structures. *Nucleic Acids Res., 36*, 4191–4203.

[55] Mashimo, T., Yagi, H., Sannohe, Y., Rajendran, A. and Sugiyama, H. (2010). Folding pathways of human telomeric type-1 and type-2 G-quadruplex structures. *J. Am. Chem. Soc., 132*, 14910–14918.

[56] Koirala, D., Mashimo, T., Sannohe, Y., Yu, Z., Mao, H. and Sugiyama, H. (2012). Intramolecular folding in three tandem guanine repeats of human telomeric DNA. *Chem. Commun., 48,* 2006–2008.

[57] Li, W., Hou, X.M., Wang, P.Y., Xi, X.G. and Li, M. (2013). Direct measurement of sequential folding pathway and energy landscape of human telomeric G-quadruplex structures. *J. Am. Chem. Soc., 135*, 6423–6426.

[58] Gray, R.D., Trent, J.O. and Chaires, J.B. (2014). Folding and unfolding pathways of the human telomeric G-quadruplex. *J. Mol. Biol., 426*, 1629–1650.

[59] Hightower, J.B., Olmos, D.R. and Walmsley, J.A. (2009). Supramolecular structure and polymorphism of alkali metal salts of guanosine 5'-monophosphate: SEM and NMR Study, *J. Phys. Chem. B, 113,* 12214-12219.

[60] Kejnovská, I., Vorlíčková, M., Brázdová, M. and Sagi, J. (2014). Stability of human telomere quadruplexes at high DNA concentrations. *Biopolymers, 101*, 428-438.

[61] Lim, K.W., Amrane, S., Bouaziz, S., Xu, W., Mu, Y., Patel, D.J., Luu, K.N. and Phan, A.T. (2009). Structure of the human telomere in K^+ solution: A stable basket-type G-quadruplex with only two G-tetrad layers. *J. Am. Chem. Soc.*, *131,* 4301–4309.

[62] Zhang, Z. Dai, J., Veliath, E., Jones, R.A. and Yang, D. (2010). Structure of a two-G-tetrad intramolecular G-quadruplex formed by a variant human telomeric sequence in K^+ solution: insights into the interconversion of human telomeric G-quadruplex structures. *Nucleic Acids Res., 38*, 1009–1021.

[63] Hayden, K.L. and Graves, D.E. (2014). Addition of Bases to the 5'-end of Human Telomeric DNA: Influences on Thermal Stability and Energetics of Unfolding. *Molecules, 19*, 2286-2298.

[64] Balagurumoorthy, P. and Brahmachari, S.K. (1994). Structure and stability of human telomeric sequence. *J. Biol. Chem.*, *269*, 21858–21869.

[65] Rezler, E.M., Seenisamy, J., Bashyam, S., Kim, M.Y., White, E., Wilson, W.D. and Hurley, L.H. (2005). Telomestatin and diseleno sapphyrin bind selectively to two different forms of the human telomeric G-quadruplex structure. *J. Am. Chem. Soc.*, *127*, 9439–9447.

[66] McConnell, K. and Sheardy, R.D. (2014). DNA G-quadruplexes with dangling ends. Abstracts of Papers, 248th ACS National Meeting and Exposition, San Francisco, CA, United States, August 10-14, 2014.

[67] Phan, A.T., Luu, K.N. and Patel, D.J. (2006). Different loop arrangements of intramolecular human telomeric (3 + 1) G-quadruplexes in K^+ solution. *Nucleic Acids Res., 34*, 5715−5719.

[68] An, N., Fleming, A.M., Burrows, C.J. (2013). Interactions of the Human Telomere Sequence with the Nanocavity of the α-Hemolysin Ion Channel Reveal Structure-Dependent Electrical Signatures for Hybrid Folds. *J. Am. Chem. Soc.*, *135*, 8562−8570.

[69] Petraccone, L., Malafronte, A., Amato, J. and Giancola, C. (2012). G-quadruplexes from human telomeric DNA: How many conformations in PEG containing solutions? *J. Phys. Chem. B.*, *116*, 2294−2305.

[70] Petraccone, L., Pagano, B. and Giancola, C. (2012). Studying the effect of crowding and dehydration on DNA G-quadruplexes. *Methods, 57*,76-83.

[71] Fialová, M., Kypr, J. and Vorlíčková, M. (2006). The thrombin binding aptamer GGTTGGTGTGGTTGG forms a bimolecular guanine tetraplex. Biochem. Biophys. Res. Commun., *344*, 50–54.

[72] Chung, W.J., Heddi, B., Schmitt, E., Lim, K.W., Mechulam, Y. and Phan, A.T. (2015). Structure of a left-handed DNA G-quadruplex. *Proc. Natl. Acad. Sci. USA, 112*, 2729–2733.

[73] Viglasky, V., Bauer, L., Tluckova, K. and Javorsky, P. (2010). Evaluation of Human Telomeric G-Quadruplexes: The Influence of Overhanging Sequences on Quadruplex Stability and Folding. *J. Nucleic Acids*, ID 820356.

[74] Vorlíčková, M., Chladkova, J., Kejnovská, I., Fialová, M. and Kypr, J. (2005). Guanine tetraplex topology of human telomere DNA is governed by the number of (TTAGGG) repeats. *Nucleic Acids Res.*, *33*, 5851–5860.

[75] Yu, H.-Q., Miyoshi, D. and Sugimoto, N. (2006). Characterization of Structure and Stability of Long Telomeric DNA G-Quadruplexes. *J. Am. Chem. Soc., 128*, 15461-15468.

[76] Chang, C.-C., Chien, C.-W., Lin, Y.-H., Kang, C.-C. and Chang, T.C. (2007). Investigation of spectral conversion of d(TTAGGG)4 and d(TTAGGG)13 upon potassium titration by a G-quadruplex recognizer BMVC molecule. *Nucleic Acids Res., 35,* 2846–2860.

[77] Smargiasso, N., Rosu, F., Hsia, W., Colson, P., Baker, E.S., Bowers, M.T., De Pauw, E. and Gabelica, V. (2008). G-Quadruplex DNA Assemblies: Loop Length, Cation Identity, and Multimer Formation. *J. Am. Chem. Soc.*, *130*, 10208–10216.

[78] Collie, G.W., Parkinson, G.N., Neidle, S., Rosu, F., De Pauw, E. and Gabelica, V. (2010). Electrospray Mass Spectrometry of Telomeric RNA (TERRA) Reveals the Formation of Stable Multimeric G-Quadruplex Structures. *J. Am. Chem. Soc.*, *132*, 9328–9334.

[79] Petraccone, L., Spink, C., Trent, J.O., Garbett, N.C., Mekmaysy, C.S., Giancola, C. and Chaires, J.B. (2011). Structure and Stability of Higher-Order Human Telomeric Quadruplexes. *J. Am. Chem. Soc., 133*, 20951–20961.

[80] Wang, H., Nora, G.J., Ghodke, h. and Opresko, P.L. (2011). Single Molecule Studies of Physiologically Relevant Telomeric Tails Reveal POT1 Mechanism for Promoting G-quadruplex Unfolding. *J. Biol. Chem. 286*, 7479–7489.

[81] Payet, L. and Huppert, J.L. (2012). Stability and Structure of Long Intramolecular G-Quadruplexes. *Biochemistry, 51*, 3154−3161.

[82] Amrane, S., Adrian, M., Heddi, B., Serero, A., Nicolas, A., Mergny, J.L. and Phan, A.T. (2012). Formation of Pearl-Necklace Monomorphic G-Quadruplexes in the Human CEB25 Minisatellite. *J. Am. Chem. Soc.*, *134*, 5807−5816.

[83] Yu, H., Gu, X., Nakano, S., Miyoshi, D. and Sugimoto, N. (2012). Beads-on-a-string structure of long telomeric DNAs under molecular crowding conditions. *J. Am. Chem. Soc.*, *134*, 20060-20069.

[84] Zhou, J., Rosu, F., Amrane, S., Korkut, D.N., Gabelica, V. and Mergny, J.L. (2014). Assembly of chemically modified G-rich sequences into tetramolecular DNA G-quadruplexes and higher order structures. *Methods, 67,* 159–168.

[85] Tothova, P., Krafcíkova, P. and Víglasky, V. (2014). Formation of Highly Ordered Multimers in G-Quadruplexes. *Biochemistry, 53*, 7013−7027.

[86] Punnoose, J.A., Cui, Y., Koirala, D., Yangyuoru, P.M., Ghimire, C., Shrestha, P. and Mao, H. (2014). Interaction of G-Quadruplexes in the Full-Length 3′ Human Telomeric Overhang. *J. Am. Chem. Soc.*, *136*, 18062−18069.

[87] Marchand,A., Granzhan, A., Iida, K., Tsushima, Y., Ma, Y., Nagasawa, K., Teulade-Fichou, M.P. and Gabelica, V. (2015). Ligand-Induced Conformational Changes with Cation Ejection upon Binding to Human Telomeric DNA G-Quadruplexes. *J. Am. Chem. Soc.*, *137*, 750−756.

[88] Guo, Q., Lu, M. and Kallenbach, N.R. (1993). Effect of thymine tract length on the structure and stability of model telomeric sequences. *Biochemistry, 32*, 3596–3603.

[89] Smirnov, I. and Shafer, R.H. (2000). Effect of loop sequence and size on DNA aptamer stability. *Biochemistry, 39*, 1462-1468.

[90] Risitano, A. and Fox, K.R. (2003). Stability of intramolecular DNA quadruplexes: comparison with DNA duplexes. *Biochemistry*, *42*, 6507–6513.

[91] Cevec, M. and Plavec, J. (2005). Role of loop residues and cations on the formation and stability of dimeric DNA G-quadruplexes. *Biochemistry, 44*, 15238–15246.

[92] Rachwal, P.A., Findlow, I.S., Werner, J.M., Brown, T. and Fox, K.R. (2007). Intramolecular DNA quadruplexes with different arrangements of short and long loops. *Nucleic Acids Res.*, *35*, 4214–4222.

[93] Vorlíčková, M., Bednarova, K., Kejnovská, I. and Kypr, J. (2007). Intramolecular and intermolecular guanine quadruplexes of DNA in aqueous salt and ethanol solutions. *Biopolymers, 86*, 1–10.

[94] Rachwal, P.A., Brown, T. and Fox, K.R. (2007). Sequence effects of single base loops in intramolecular quadruplex DNA. *FEBS Lett.*, *44*, 1657–1660.

[95] Bugaut, A. and Balasubramanian, S. (2008). A sequence-independent study of the influence of short loop lengths on the stability and topology of intramolecular DNA G-quadruplexes. *Biochemistry, 47*, 689–697.

[96] Guedin, A., De Cian, A., Gros, J., Lacroix, L. and Mergny, J.-L. (2008). Sequence effects in single-base loops for quadruplexes. *Biochimie, 90*, 686–696.

[97] Kumar, N. and Maiti, S. (2008). A thermodynamic overview of naturally occurring intramolecular DNA quadruplexes. *Nucleic Acids Res.* 36, 5610–5622.

[98] Guedin, A., Alberti, P. and Mergny, J.L. (2009). Stability of intramolecular quadruplexes: sequence effects in the central loop. *Nucleic Acids Res.* 37, 5559–5567.

[99] Arora, A. and Maiti, S. (2009). Stability and molecular recognition of quadruplexes with different loop length in the absence and presence of molecular crowding agents. *J. Phys. Chem. B.*, *113*, 8784–8792.

[100] Guédin, A., Gros, J., Alberti, P. and Mergny, J.L. (2010). How long is too long? Effects of loop size on G-quadruplex stability. *Nucleic Acids Res.*, *38*, 7858-7868.

[101] Aggrawal, M., Joo, H., Liu, W., Tsai, J. and Xue, L. (2012). 8-Oxo-7,8-dihydrodeoxyadenosine: the first example of a native DNA lesion that stabilizes human telomeric G-quadruplex DNA. *Biochem. Biophys. Res. Commun.*, *421*, 671–677.

[102] Kwok, C.K., Sherlock, M.E. and Bevilacqua, P.C. (2013). Effect of loop sequence and loop length on the intrinsic fluorescence of G-quadruplexes. *Biochemistry, 52*, 3019–3021.

[103] Tippana, R., Xiao, W. and Myong, S. (2014). G-quadruplex conformation and dynamics are determined by loop length and sequence. *Nucleic Acids Res.*, *42*, 8106–8114.

[104] Babinsky, M., Fiala, R., Kejnovská, I., Bednařova, K., Marek, R., Sagi, J., Sklenař, V. and Vorlíčková, M. (2014). Loss of loop adenines alters human telomere d[AG3(TTAG3)3] quadruplex folding. *Nucleic Acids Res.*, *42,* 14031-14041.

[105] Konvalinova, H., Dvorakova, Z., Renciuk, D., Bednarova, K., Kejnovská, I., Trantírek, L., Vorlíčková, M. and Sagi, J. (2015). Diverse effects of naturally occurring base lesions on the structure and stability of the human telomere DNA quadruplex. *Biochimie, 118* (2015) 15-25.

[106] Virgilio, A., Esposito, V., Mayol, L., Giancola, C., Petraccone, L. and Galeone, A. (2015). The oxidative damage to the human telomere: effects of 5-hydroxymethyl-2'-deoxyuridine on telomeric G-quadruplex structures. *Org. Biomol. Chem.*, *13*, 7421-7429.

[107] Olsen, C.M., Gmeiner, W.H. and Marky, L.A. (2006). Unfolding of G-Quadruplexes: Energetic, and Ion and Water Contributions of G-Quartet Stacking. *J. Phys. Chem. B, 110*, 6962-6969.

[108] Stegle, O., Payet, L., Mergny, J.L. MacKay, D.J. and Leon, J.H. (2009). Predicting and understanding the stability of G-quadruplexes. *Bioinformatics, 25*, i374–i382.
[109] Bedrat, A., Lacroix, L. and Mergny, J.L. (2016). Re-evaluation of G-quadruplex propensity with G4Hunter. *Nucleic Acids Res., 44*, 1746–1759.
[110] Dias, E., Battiste, J.L. and Williamson, J.R. (1994). Chemical probe for glycosidic conformation in telomeric DNAs. *J. Am. Chem. Soc., 116*, 4479–4480.
[111] Xu, Y., Noguchi, Y. and Sugiyama, H. (2006). The new models of the human telomere d[AGGG(TTAGGG)3] in K^+ solution. *Bioorg. Med. Chem., 14*, 5584–5591.
[112] Xu, Y., Suzuki, Y., Lonnberg, T. and Komiyama, M. (2009). Human telomeric DNA sequence-specific cleaving by G-quadruplex formation. *J. Am. Chem. Soc., 131*, 2871-2874.
[113] Xu, Y. and Sugiyama, H. (2006). Formation of the G-quadruplex and i-motif structures in retinoblastoma susceptibility genes (Rb). *Nucleic Acids Res., 34*, 949–954.
[114] Okamoto, K., Sannohe, Y., Mashimo, T., Sugiyama, H. and Terazima, M. (2008). G-quadruplex structures of human telomere DNA examined by single molecule FRET and BrG-substitution. *Bioorg. Med. Chem., 16*, 6873–6879.
[115] Sagi, J. (2014). G-quadruplexes incorporating modified constituents: a review. *J. Biomolec. Struct. and Dynamics, 32*, 477-511.
[116] Brcic, J. and Plavec, J. (2015). Solution structure of a DNA quadruplex containing ALS and FTD related GGGGCC repeat stabilized by 8-bromodeoxyguanosine substitution. *Nucleic Acids Res., 43*, 8590–8600.
[117] Benz, A. and Hartig, J.S. (2008). Redesigned tetrads with altered hydrogen bonding patterns enable programming of quadruplex topologies. *Chem. Commun., 34,* 4010-4012.
[118] Cheong, V.V., Heddi, B., Lech, C.J. and Phan, A.T. (2015). Xanthine and 8-oxoguanine in G-quadruplexes: formation of a G·G·X·O tetrad. *Nucleic Acids Res., 43*, 10506–10514.
[119] Cheong, V.V., Lech, C.J., Heddi, B. and Phan, A.T. (2016). Inverting the G-Tetrad Polarity of a G-Quadruplex by Using Xanthine and 8-Oxoguanine. *Angew. Chem. Int. Ed., 55*, 160 –163.
[120] Tomasko, M., Vorlíčková, M. and Sagi, J. (2009). Substitution of adenine for guanine in the quadruplex-forming human telomere DNA sequence G3(T2AG3)3. *Biochimie, 91*, 171-179.
[121] Vorlíčková, M., Tomasko, M., Sagi, A.J., Bednarova, K., Sagi, J. (2012). 8-Oxoguanine in a quadruplex of the human telomere DNA sequence. *FEBS J., 279*, 29-39.
[122] Školáková, P., Bednarova, K., Vorlíčková, M. and Sagi, J. (2010). Quadruplexes of human telomere dG3(TTAG3)3 sequences containing guanine abasic sites. *Biochem. Biophys. Res. Commun., 399*, 203-208.
[123] Zhao, A., Zhao, C., Tateishi-Karimata, H., Ren, J., Sugimoto, N. and Qu, X. (2016). Incorporation of O6-methylguanine restricts conformational conversion of human telomere G-quadruplex under molecular crowding. *Chem. Commun., 52*, 1903-1906.
[124] Lech, C.J., Lim, J.K.C., Lim, J.M.W., Amrane, S., Heddi, B. and Phan, A.T. (2011). Effects of Site-Specific Guanine C8-Modifications on an Intramolecular DNA G-Quadruplex. *Biophysical J., 101*, 1987–1998.

[125] Hu, L., Lim, K.W., Bouaziz, S. and Phan, A.T. (2009). Giardia Telomeric Sequence d(TAGGG)4 Forms Two Intramolecular G-Quadruplexes in K^+ Solution: Effect of Loop Length and Sequence on the Folding Topology. *J. Am. Chem. Soc.*, *131*, 16824-16831.

[126] Tang, C-F. and Shafer, R.H. (2006). Engineering the quadruplex fold: Nucleoside conformation determines both folding topology and molecularity in guanine quadruplexes. *J. Am. Chem. Soc.*, *128*, 5966–5973.

[127] Peng, Ch.G. and Damha, M.J. (2007). G-quadruplex induced stabilization by 2′-deoxy-2′-fluoro-D-arabinonucleic acids (2′F-ANA). *Nucleic Acids Res.*, *35*, 4977–4988.

[128] Webba da Silva, M., Trajkovski, M., Sannohe, Y., Maani Hessari, N., Sugiyama, H. and Plavec, J. (2009). Design of a G-Quadruplex Topology through Glycosidic Bond Angles. *Angew. Chem., Int. Ed.*, *48*, 9167-70.

[129] Lech, C.J., Li, Z., Heddi, B. and Phan, A.T. (2012). 2'-F-ANA-guanosine and 2'-F-guanosine as powerful tools for structural manipulation of G-quadruplexes. *Chem. Commun.*, *48*, 11425–11427.

[130] Martín-Pintado, N., Yahyaee-Anzahaee, M., Deleavey, G.F., Portella, G., Orozco, M., Damha, M.J. and Gonzalez, C. (2013). Dramatic Effect of Furanose C2′ Substitution on Structure and Stability: Directing the Folding of the Human Telomeric Quadruplex with a Single Fluorine Atom. *J. Am. Chem. Soc.*, *135*, 5344−5347.

[131] Collie, G.W., Haider, S.M., Neidle, S. and Parkinson, G.N. (2010). A crystallographic and modelling study of a human telomeric RNA (TERRA) quadruplex. *Nucleic Acids Res.*, *38*, 5569–5580.

[132] Vondruskova, J., Kypr, J., Kejnovská, I., Fialová, M. and Vorlíčková, M. (2008). Guanine quadruplex formation by RNA/DNA hybrid analogs of Oxytricha telomere G4T4G4 fragment. *Biopolymers, 89*, 797-806.

[133] Vondruskova, J., Kypr, J., Kejnovská, I., Fialová, M. and Vorlíčková, M. (2008). Role of loops in the guanine quadruplex formation by DNA/RNA hybrid analogs of G4T4G4. Int. J. Biol. Macromol., *43*, 463-467.

[134] Schultze, P., Hud, N.V., Smith, F.W. and Feigon, J. (1999). The effect of sodium, potassium and ammonium ions on the conformation of the dimeric quadruplex formed by the Oxytricha nova telomere repeat oligonucleotide d(G4T4G4). *Nucleic Acids Res.*, *27*, 3018–3028.

[135] Haider, S.M., Parkinson, G.N. and Neidle, S. (2003). Structure of a G-quadruplex–Ligand Complex. *J. Mol. Biol.*, *326*, 117–125.

[136] Phan, A.T. and Patel, D.J. (2003). Two-Repeat Human Telomeric d(TAGGGTTAGGGT) Sequence Forms Interconverting Parallel and Antiparallel G-Quadruplexes in Solution: Distinct Topologies, Thermodynamic Properties, and Folding/Unfolding Kinetics. *J. Am. Chem. Soc.*, *125*, 15021-27.

[137] Li, Z., Lech, C.J. and Phan, A.T. (2014). Sugar-modified G-quadruplexes: effects of LNA-, 2'F-RNA– and 2'F-ANA-guanosine chemistries on G-quadruplex structure and stability. *Nucleic Acids Res.*, *42*, 4068–4079.

[138] Do, N.Q. and Phan, A.T. (2012). Monomer-dimer equilibrium for the 5'-5' stacking of propeller-type parallel-stranded G-quadruplexes: NMR structural study. *Chem. Eur. J.*, *18*, 14752–14759.

[139] Dominick, P.K. and Jarstfer, M.B. (2004). A conformationally constrained nucleotide analogue controls the folding topology of a DNA quadruplex. *J. Am. Chem. Soc.*, *126*, 5050-5051.

[140] Randazzo, A., Esposito, V., Ohlenschlager, O., Ramachandran, R. and Mayol L. (2004). NMR solution structure of a parallel LNA Quadruplex. *Nucleic Acids Res.*, *32*, 3083-3092.

[141] Randazzo. A., Esposito, V., Ohlenschlager, O., Ramachandran, R., Virgilio, A. and Mayol, L. (2005). Structural studies on LNA quadruplexes. *Nucl. Nuclt. Nucleic Acids.* *24*, 795-800.

[142] Nielsen, J.T., Arar, K. and Petersen, M. (2006). NMR solution structures of LNA (locked nucleic acid) modified quadruplexes. *Nucleic Acids Res., 34*, 2006–2014.

[143] Pradhan, D., Hansen, L.H., Vester, B. and Petersen, M. (2011). Selection of G-Quadruplex Folding Topology with LNA-Modified Human Telomeric Sequences in K^+ Solution. *Chem.–Eur. J., 17*, 2405–2413.

[144] Latha, Y.S. and Yathindra, N. (1992). Stereochemical studies on nucleic acid analogues. I. Conformations of alpha-nucleosides and alpha-nucleotides: interconversions of sugar puckers via O4'-exo. *Biopolymers, 32*, 249-269.

[145] Kolganova, N.A., Varizhuk, A.V., Novikov, R.A., Florentiev, V.L., Pozmogova, G.E., Borisova, O.F., Shchyolkina, A.K., Smirnov, I.P., Kaluzhny, D.N. and Timofeev, E.N. (2014). Anomeric DNA quadruplexes. *Artificial DNA: PNA and XNA, 4*, e28422.

[146] Yuan, L., Tian, T., Chen, Y., Zhang, Z. and Zhou, X. (2013). An L-DNA G-Quadruplex: Application for Peroxidase DNAzyme. *Nucl. Nuclt. Nucleic Acids, 32*, 589-598.

[147] Virgilio, A., Varra, M., Scuotto, M., Capuozzo A., Irace, C., Mayol, L., Esposito, V. and Galeone, A. (2014). Expanding the potential of G-quadruplex structures: Formation of heterochiral TBA analogue. *ChemBioChem.*, *15*, 652-655.

[148] Virgilio, A., Mayol, L., Varra, M., Galeone, A. and Esposito, V. (2015). Unusual Chair-Like G-Quadruplex Structures: Heterochiral TBA Analogues Containing Inversion of Polarity Sites. J. Chem. ID 473051.

[149] Aher, M.N., Erande, N.D., Fernandes, M. Kumar, V.A. (2015). Unimolecular antiparallel G-quadruplex folding topology of 2'–5'-isoTBA sequences remains unaltered by loop composition. *Org. Biomol. Chem.*, *13*, 11696-11703.

[150] Avino, A., Portella, G., Ferreira, R., Gargallo, R., Mazzini, S., Gabelica, V., Orozco, M. and Eritja, R. (2014). Specific loop modifications of the thrombin-binding aptamer trigger the formation of parallel structures. *FEBS J.*, *281*, 1085–1099.

[151] Mendez-Bermudez, A., Hills, M., Pickett, H.A., Phan, A.T., Mergny, J.L., Riou, J.F. and Royle, N.J. (2009). Human telomeres that contain (CTAGGG)n repeats show replication dependent instability in somatic cells and the male germline. *Nucleic Acids Res.*, *37*, 6225–6238.

[152] Lim, K.W., Alberti, P., Guedin, A., Lacroix, L., Riou, J.-F., Royle, N.J., Mergny, J.-L. and Phan, A.T. (2009). Sequence variant (CTAGGG)(n) in the human telomere favors a G-quadruplex structure containing a G:C:G:C tetrad. *Nucleic Acids Res.*, *37*, 6239–6248.

[153] Baird, D.M., Jeffreys, A.J. and Royle, N.J. (1995). Mechanisms underlying telomere repeat turnover, revealed by hypervariable variant repeat distribution patterns in the human Xp/Yp telomere. *EMBO J., 14*, 5433-5443.

[154] Coleman, J., Baird, D.M. and Royle, N.J. (1999). The plasticity of human telomeres demonstrated by a hypervariable telomere repeat array that is located on some copies of 16p and 16q. *Hum. Mol. Genet., 8,* 1637-1646.

[155] Sattin, G., Artese, A., Nadai, M., Costa, G., Parrotta, L., Alcaro, S., Palumbo, M. and Richter, S.N. (2013). Conformation and Stability of Intramolecular Telomeric G-Quadruplexes: Sequence Effects in the Loops. *PLOS ONE, 8*, e84113.

[156] Sengar, A., Heddi, B. and Phan, A.T. (2014). Formation of G-Quadruplexes in Poly-G Sequences: Structure of a Propeller-Type Parallel-Stranded G-Quadruplex Formed by a G15 Stretch. *Biochemistry, 53*, 7718-7723.

[157] Yue, D.J.E., Lim, K.W. and Phan, A.T. (2011). Formation of (3 + 1) G-Quadruplexes with a Long Loop by Human Telomeric DNA Spanning Five or More Repeats. *J. Am. Chem. Soc., 133*, 11462–11465.

[158] Phan, A.T. and Do, N.Q. (2013). Engineering of interlocked DNA G-quadruplexes as a robust scaffold. *Nucleic Acids Res.* 41, 2683–2688.

[159] Phan, A.T., Kuryavyi, V., Ma, J.-B., Faure, A., Andreola, M.-L. and Patel, D.J. (2005). An interlocked dimeric parallel-stranded DNA quadruplex: a potent inhibitor of HIV-1 integrase. *Proc. Natl. Acad. Sci. USA, 102,* 634–639.

[160] Paramasivam, M., Cogoi, S., Filichev, V.V., Bomholt, N., Pedersen, E.B. and Xodo, L.E. (2008). Purine twisted-intercalating nucleic acids: a new class of anti-gene molecules resistant to potassium-induced aggregation. *Nucleic Acids Res., 36*, 3494-3507.

[161] Doluca, O., Withers, J.M. and Filichev, V.V. (2013). Molecular Engineering of Guanine-Rich Sequences: Z-DNA, DNA Triplexes, and G-Quadruplexes. *Chem. Rev., 113*, 3044−3083.

[162] Doluca, O., Withers, J., Loo, T.S., Edwards, P., Gonzalez, C. and Filichev, V. (2015). Interdependence of pyrene interactions and tetramolecular G4-DNA assembly. *Org. Biomol. Chem., 13,* 3742-3748.

[163] Murat, P., Cressend, D., Spinelli, N., Van der Heyden, A., Labbe, P., Dumy, P. and Defrancq, E. (2008). A Novel Conformationally Constrained Parallel G Quadruplex. *ChemBioChem, 9,* 2588-2591.

[164] Murat, P., Gennaro, B., Garcia, J., Spinelli, N., Dumy, P. and Defrancq, E. (2011). The Use of a Peptidic Scaffold for the Formation of Stable Guanine Tetrads: Control of a H-bonded Pattern in Water. *Chem. Eur. J., 17*, 5791-5795.

[165] Bonnet, R., Dejeu, J., Lavergne, T. and Defrancq, E. (2014). Topologically Controlled DNA and RNA G-Quadruplexes. *The 5th International Meeting on Quadruplex Nucleic Acids*, Bordeaux, France.

[166] Bonnet, R., Lavergne, T., Gennaro, B., Spinelli, N. and Defrancq, E. (2015). Construction of anti-parallel G-quadruplexes through sequential templated click. *Chem. Commun., 51*, 4850-4853.

[167] Hud, N. and Plavec, J. The role of cations in determining quadruplex structure and stability. In Neidle, S. and Balasubramanian, S. (Eds.) (2006), pp. 100-130. Quadruplex nucleic acids. Cambridge; RSC Publishing.

[168] Mergny, J.L., Phan, A.T. and Lacroix, L. (1998). Following G-quartet formation by UV-spectroscopy. *FEBS Lett.*, *435*, 74–78.

[169] Sacca, B., Lacroix, L. and Mergny, J.L. (2005). The effect of chemical modifications on the thermal stability of different G-quadruplex-forming oligonucleotides. *Nucleic Acids Res.*, *33*, 1182–1192.

[170] Lane, A.N., Chaires, J.B., Gray, R.D. and Trent, J.O. (2008). Stability and kinetics of G-quadruplex structures. *Nucleic Acids Res.*, *36*, 5482-5515.

[171] Sagi, J., Renciuk, D., Tomasko, M. and Vorlíčková, M. (2010). Quadruplexes of human telomere DNA analogs designed to contain G:A:G:A, G:G:A:A, and A:A:A:A tetrads. *Biopolymers*, 93, 880-886.

[172] Chaires, J.B. (2010). Human telomeric G-quadruplex: thermodynamic and kinetic studies of telomeric quadruplex stability. *FEBS J.*, *277*, 1098–1106.

[173] Fan, H.Y., Shek, Y.L., Amiri, A., Dubins, D.N., Heerklotz, H., Macgregor Jr., R.B. and Chalikian, T.V. (2011). *J. Am. Chem. Soc.*, *133*, 4518–4526.

[174] Hardin, C.C., Watson, T., Corregan, M. and Bailey, C. (1992). Cation-dependent transition between the quadruplex and Watson-Crick hairpin forms of d(CGCG3GCG). *Biochemistry, 31*, 833–841.

[175] Deng, J., Xiong, Y. and Sundaralingam, M. (2001). X-ray analysis of an RNA tetraplex (UGGGGU)4 with divalent Sr^{2+} ions at subatomic resolution (0.61 Å). *Proc. Natl. Acad. Sci. USA, 98*, 13665–13670.

[176] Caceres, C., Wright, G., Gouyette, C., Parkinson, G. and Subirana, J.A. (2004). A thymine tetrad in d(TGGGGT) quadruplexes stabilized with Tl^{+}/Na^{+} ions. *Nucleic Acids Res., 32*, 1097-1102.

[177] Risitano, A. and Fox, K.R. (2005). Inosine substitutions demonstrate that intramolecular DNA quadruplexes adopt different conformations in the presence of sodium and potassium. *Bioorg. Med. Chem. Lett.*, *15*, 2047–2050.

[178] Gros, J., Rosu, F., Amrane, S., De Cian, A., Gabelica, V., Lacroix, L. and Mergny, J.L. (2007). Guanines are a quartet's best friend: impact of base substitutions on the kinetics and stability of tetramolecular quadruplexes. *Nucleic Acids Res.*, *35*, 3064–3075.

[179] Tran, P.L., Mergny, J.L. and Alberti, P. (2011). Stability of telomeric G-quadruplexes. *Nucleic Acids Res.*, *39*, 3282–3294.

[180] Marusic, M. and Plavec, J. (2015). The effect of DNA sequence directionality on G-Quadruplex folding. *Angew. Chem. Int. Ed.*, *54,* 11716–11719.

[181] Saintome, C., Amrane, S., Mergny, J.L. and Alberti, P. (2016). The exception that confirms the rule: a higher-order telomeric G-quadruplex structure more stable in sodium than in potassium. *Nucleic Acids Research*, Advance access; doi:10.1093/nar/gkw003.

[182] Largy, E., Marchand, A., Amrane, S., Gabelica, V. and Mergny, J.-L. (2016). Quadruplex Turncoats: Cation-Dependent Folding and Stability of Quadruplex-DNA Double Switches. *J. Am. Chem. Soc.*, *138*, 2780–2792.

[183] Shim, J.W., Tan, Q. and Gu, L-Q. (2009). Single-molecule detection of folding and unfolding of the G-quadruplex aptamer in a nanopore nanocavity. *Nucleic Acids Res., 37*, 972–982.

[184] Campbell, N.H. and Neidle, S. (2012). G-quadruplexes and metal ions. In: Sigel, A., Sigel, H. and Sigel, R.K. (eds). Interplay between Metal Ions and Nucleic Acids. Metal Ions in Life Sciences. Springer, Vol. 10, pp. 119–134.

[185] Miyoshi, D., Nakao, A., Toda, T. and Sugimoto, N. (2001). Effect of divalent cations on antiparallel G-quartet structure of d(G4T4G4). *FEBS Lett.*, *496*, 128-133.

[186] Miyoshi, D., Nakao, A. and Sugimoto, N. (2003). Structural transition from antiparallel to parallel G-quadruplex of d(G4T4G4) induced by Ca^{2+}. *Nucleic Acids Res.*, *31*, 1156-1163.

[187] Miyoshi, D., Matsumura, S., Li, W. and Sugimoto, N. (2003). Structural Polymorphism of Telomeric DNA Regulated by pH and Divalent Cation. *Nucl. Nuclt. and Nucleic Acids, 22,* 203 – 221.

[188] Sen, D. and Gilbert, W. (1990). A sodium-potassium switch in the formation of four-stranded G4-DNA. *Nature, 344*, 410–414.

[189] Smestad, J. and Maher, L.J., III. (2013). Ion-dependent conformational switching by a DNA aptamer that induces remyelination in a mouse model of multiple sclerosis. *Nucleic Acids Res.*, *41*, 1329–1342.

[190] Gray, R.D., Petraccone, L, Trent, J.O. and Chaires, J.B. (2010). Characterization of a K^+-Induced Conformational Switch in a Human Telomeric DNA Oligonucleotide Using 2-Aminopurine Fluorescence. *Biochemistry, 49,* 179–194.

[191] Pedroso, I.M., Duarte, L.F., Yanez, G., Baker, A.M. and Fletcher, T.M. (2007). Induction of parallel human telomeric G-quadruplex structures by Sr^{2+}. *Biochem. Biophys. Res. Commun.*, *358*, 298–303.

[192] Xu, Y., Kaminaga, K. and Komiyama, M. (2008). G-Quadruplex Formation by Human Telomeric Repeats-Containing RNA in Na^+ Solution. *J. Am. Chem. Soc.*, *130*, 11179–11184.

[193] Martadinata, H. and Phan, A.T. (2009). Structure of Propeller-Type Parallel-Stranded RNA.

[194] G-Quadruplexes, Formed by Human Telomeric RNA Sequences in K^+ Solution. *J. Am. Chem. Soc.*, *131*, 2570–2578.

[195] Minton, A.P. and Wilf, J. (1981). Effect of macromolecular crowding upon the structure and function of an enzyme: glyceraldehyde-3-phosphate dehydrogenase. *Biochemistry, 20*, 4821–4826.

[196] Zimmerman, S.B. and Minton, A.P. (1993). Macromolecular crowding: biochemical, biophysical, and physiological consequences. *Annu. Rev.Biophys. Biomol. Struct.*, *22*, 27–65.

[197] Minton, A.P. (1998). Molecular crowding: analysis of effects of high concentrations of inert cosolutes on biochemical equilibria and rates in terms of volume exclusion. *Methods Enzymol.*, *295*, 127–149.

[198] Luby-Phelps, K. (1999). Cytoarchitecture and physical properties of cytoplasm: volume, viscosity, diffusion, intracellular surface area. *Int. Rev. Cytol.*, *192*, 189–221.

[199] Minton, A.P. (2001). The influence of macromolecular crowding and macromolecular confinement on biochemical reactions in physiological media. *J. Biol. Chem.*, *276*, 10577–10580.

[200] Ellis, R.J. (2001). Macromolecular crowding: an important but neglected aspect of the intracellular environment. *Curr. Opin. Struct. Biol.*, *11*, 114–119.

[201] Zimmerman, S.B. and Trach, S.O. (1991). Estimation of macromolecule concentrations and excluded volume effects for the cytoplasm of Escherichia coli. *J. Mol. Biol.*, *222*, 599–620.

[202] Ellis, R.J. and Minton, A.P. (2003). Cell biology: Join the crowd. *Nature, 425*, 27–28.

[203] Zhou, H.X., Rivas, G. and Minton, A.P. (2008). Macromolecular Crowding and Confinement: Biochemical, Biophysical, and Potential Physiological Consequences. *Annu. Rev. Biophys., 37*, 375–397.

[204] Miyoshi, D., Nakao, A. and Sugimoto, N. (2002). Molecular crowding regulates the structural switch of the DNA G-quadruplex. *Biochemistry, 41,* 15017–15024.

[205] Vorlíčková, M., Bednarova, K. and Kypr, J. (2006). Ethanol is a better inducer of DNA guanine tetraplexes than potassium cations. *Biopolymers, 82*, 253–260.

[206] Heddi, B. and Phan, A.T. (2011). Structure of human telomeric DNA in crowded solution. *J. Am. Chem. Soc., 133*, 9824–9833.

[207] Miyoshi, D. Karimata, H. and Sugimoto, N. (2006). Hydration Regulates Thermodynamics of G-Quadruplex Formation under Molecular Crowding Conditions. *J. Am. Chem. Soc., 128*, 7957–7963.

[208] Xue, Y., Kan, Z., Wang, Q., Yao, Y., Liu, J., Hao, Y. and Tan, Z. (2007). Human telomeric DNA forms parallel G-quadruplex in K^+ solution under molecular crowding condition. *J. Am. Chem. Soc., 129*, 11185–11191.

[209] Kan, Z., Lin, Y., Wang, F., Zhuang, X., Zhao, Y., Pang, D., Hao, Y. and Tan, Z. (2007). G-quadruplex formation in human telomeric (TTAGGG)4 sequence with complementary strand in close vicinity under molecularly crowded condition. *Nucleic Acids Res., 35*, 3646–3653.

[210] Miyoshi, D., Nakamura, K., Muhuli, S., Tateishi-Karimata, H. and Sugimoto, N. (2008). Thermodynamics of DNA structures under molecular crowding conditions with neutral and positive charged cosolutes. *Nucleic Acids Symp. Series, 52*, 413-414.

[211] Miyoshi, D., Nakamura, K., Tateishi-Karimata, H., Ohmichi, T. and Sugimoto, N. (2009). Hydration of Watson-Crick base pairs and dehydration of Hoogsteen base pairs inducing structural polymorphism under molecular crowding condition. *J. Am. Chem. Soc., 131*, 3522-3531.

[212] Zheng, K., Chen, Z., Hao, Y. and Tan, Z. (2010). Molecular crowding creates an essential environment for the formation of stable G-quadruplexes in long double-stranded DNA. *Nucleic Acids Res., 38*, 327-338.

[213] Fujimoto, T., Nakano, S., Miyoshi, D. and Sugimoto, N. (2011). The effects of molecular crowding on the structure and stability of G-quadruplexes with an abasic site. *J. Nucleic Acids*, ID 857149.

[214] Lannan, F.M., Mamajanov, I. and Hud, N.V. (2012). Human Telomere Sequence DNA in Water-Free and High-Viscosity Solvents: G-Quadruplex Folding Governed by Kramers Rate Theory. *J. Am. Chem. Soc., 134*, 15324–15330.

[215] Buscaglia, R., Miller, M.C., Dean, W.L., Gray, R.D., Lane, A.N., Trent, J.O. and Chaires, J.B. (2013). Polyethylene glycol binding alters human telomere G-quadruplex structure by conformational selection. *Nucleic Acids Res., 41*, 7934–7946.

[216] Li, J., Correia, J.J., Wang, L., Trent, J.O. and Chaires, J.B. (2005). Not so crystal clear: the structure of the human telomere G-quadruplex in solution differs from that present in a crystal. *Nucleic Acids Res., 33*, 4649-4659.

[217] Kilburn, D., Roh, J.H., Guo, L., Briber, R.M., Woodson, S.A. and Jenkins, T.C. (2010). Molecular Crowding Stabilizes Folded RNA Structure by the Excluded Volume Effect. *J. Am. Chem. Soc., 132*, 8690-8696.

[218] Kumar, N. and Maiti, S. (2008). Role of molecular crowding in perturbing quadruplex-Watson Crick duplex equilibrium. *Nucleic Acids Symp. Series, 52*, 157-158.

[219] Miller, M.C., Buscaglia, R., Chaires, J.B., Lane, A.N. and Trent, J.O. (2010). Hydration Is a Major Determinant of the G-Quadruplex Stability and Conformation of the Human Telomere 3′ Sequence of d(AG3(TTAG3)3). *J. Am. Chem. Soc.*, *132*, 17105−17107.

[220] Miyoshi, D., Karimata, H. and Sugimoto, N. (2007). Hydration regulates the thermodynamic stability of DNA structures under molecular crowding conditions. *Nucl. Nuclt. Nucleic Acids, 26,* 589-95.

[221] Miyoshi, D. and Sugimoto, N. (2008). Molecular crowding effects on structure and stability of DNA. *Biochimie, 90*, 1040–1051.

[222] Nakano, M., Tateishi-Karimata, H., Tanaka, S., Tama, F., Miyashita, O., Nakano, S. and Sugimoto, N. (2015). Thermodynamic properties of water molecules in the presence of cosolute depend on DNA structure: a study using grid inhomogeneous solvation theory. *Nucleic Acids Res.*, *43*, 10114-10125.

[223] Son, I., Shek, Y.L., Dubins, D.N. and Chalikian, T.V. (2014). Hydration changes accompanying helix-to-coil DNA transitions. *J. Am. Chem. Soc.*, *136*, 4040–4047.

[224] Zhoua, J., Tateishi-Karimatac, H., Mergny, J.-L., Chenga, M., Fenga, Z., Miyoshic, D., Sugimoto, N. and Li, C. (2016). Reevaluation of the stability of G-quadruplex structures under crowding conditions. *Biochimie, 121,* 204–208.

[225] Hansel, R., Lohr, F., Foldynova-Trantirkova, S.; Bamberg, E.; Trantirek, L. and Dotsch, V. (2011). The parallel G-quadruplex structure of vertebrate telomeric repeat sequences is not the preferred folding topology under physiological conditions. *Nucleic Acids Res., 39*, 5768-5775.

[226] Dhakal, S., Cui, Y., Koirala, D., Ghimire, C., Kushwaha, S., Yu, Z., Yangyuoru, P.M. and Mao, H. (2013). Structural and mechanical properties of individual human telomeric G-quadruplexes in molecularly crowded solutions. *Nucleic Acids Res., 41*, 3915–3923.

[227] Takahashia, S. and Sugimoto, N. (2015). Pressure-dependent formation of i-motif and G-quadruplex DNA structures. *Phys. Chem. Chem. Phys.*, *17*, 31004-31010.

[228] Do, N.Q., Lim, K.W., Teo, M.H., Heddi, B. and Phan, A.T. (2011). Stacking of G-quadruplexes: NMR structure of a G-rich oligonucleotide with potential anti-HIV and anticancer activity. *Nucleic Acids Res.*, *39*, 9448-9457.

[229] Trajkovski, M., da Silva, M.W. and Plavec, J. (2012). Unique Structural Features of Interconverting Monomeric and Dimeric G-Quadruplexes Adopted by a Sequence from the Intron of the N-myc Gene. *J. Am. Chem. Soc., 134*, 4132-4141.

[230] Kuryavyi, V., Cahoon, L.A., Seifert, H.S. and Patel, D.J. (2012). RecA-Binding pilE G4 Sequence Essential for Pilin Antigenic Variation Forms Monomeric and 5' End-Stacked Dimeric Parallel G-Quadruplexes. *Structure, 20*, 2090-2102.

[231] Wei, D., Todd, A.K., Zloh,M., Gunaratnam, M., Parkinson, G.N. and Neidle, S. (2013). Crystal Structure of a Promoter Sequence in the B-raf Gene Reveals an Intertwined Dimer Quadruplex. *J. Am. Chem. Soc., 135*, 19319-19329.

[232] Adrian, M., Ang, D.J., Lech, C.J., Heddi, B., Nicolas, A. and Phan, A.T. (2014). Structure and conformational Dynamics of a Stacked Dimeric G-Quadruplex Formed by the Human CEB1 Minisatellite. *J. Am. Chem. Soc., 136*, 6297−6305.

[233] Abu-Ghazalah, R.M., Rutledge, S., Lau, L.W.Y., Dubins, D.N., Macgregor, R.B. Jr. and Helmy, A.S. (2012). Concentration-dependent structural transitions of human telomeric DNA sequences. *Biochemistry, 51*, 7357-7366.

[234] Li, Y.Y., Abu-Ghazalah, R., Zamiri, B. and Macgregor. R.B., Jr. (2016). Concentration-dependent conformational changes in GQ-forming ODNs. *Biophys. Chem., 211*, 70-75.

[235] Graves, D.E., Harris, K., Selander, K., Hayden, K. and Tucker, B. (2015). Nucleic acids: The target or the ligand. Abstracts, 67th Southeast/71st Southwest Joint Regional Meeting of the American Chemical Society, Memphis, TN, United States, November 4-7 (2015), SERMACS-SWRM-898.

In: G-Quadruplex Structures, Formation and Role in Biology ISBN: 978-1-63485-512-9
Editor: Hannah Santos

Chapter 2

G-QUADRUPLEXES AND I-MOTIFS AS SCAFFOLDS FOR MOLECULAR ENGINEERING OF DNA APTAMERS

Elena Zavyalova* and Alexey Kopylov
Chemistry Department, Lomonosov Moscow State University,
Moscow, Russian Federation

ABSTRACT

DNA aptamers are oligonucleotides with a particular three-dimensional structure that provides high affinity and high specificity to a target. Aptamers are applied as molecular recognition elements (MoRE) in nanoscience and nanotechnology as well as in theranostics as sensors, delivery systems and pharmaceuticals inhibiting target functioning; refined aptamers can be considered as functionally analogous to antibodies. Not too much is yet known about the aptamer structure–function relationship but, interestingly enough, some aptamers have a functionally significant G-quadruplex structure.

Thus, the basic thrombin-binding aptamer HD1 is a unimolecular antiparallel G-quadruplex with three short loops. Two TT-loops of the aptamer bind to exosite I on the thrombin surface. The G-quadruplex itself is a scaffold that establishes proper conformation of TT-loops: T4 and T13 are paired, while T3 and T12 are buried in hydrophobic thrombin pockets. T4-T13 pairing is absolutely required for aptamer functional activity, whereas T3 and T12 can be modified or even changed.

The scaffolding role of the G-quadruplex is highlighted by the structure–function effects of different metal cation coordination. Potassium cations provide the highest functional activity, whereas several other cations alter G-quartet fine structure and therefore change TT-loop conformation, leading to the complete loss of aptamer functional activity. On the other hand, the activity of a thrombin-binding aptamer can be increased by extending the basic G-quadruplex with either a duplex or with another G-quadruplex. Surprisingly, extra elements improve the aptamer function (i.e., by decreasing inhibition constants), although they do not provide additional interactions between the aptamer and thrombin. Most probably these extra modules support a specific

* Corresponding author: Zavyalova Elena, Chemistry Department, Lomonosov Moscow State University, Leninskie gory 1, Moscow, 119991 Russian Federation, tel: +74959393149, fax: +74959393181, email: zlenka2006@ gmail.com.

G-quadruplex conformation, keeping the proper orientation of TT-loops. Therefore, a G-quadruplex scaffold could be used for modular construction of MoRE having capacity for structure–function tuning.

Another spectacular example of a scaffold for MoRE is the so-called i-motif. A DNA aptamer to hemagglutinin of influenza virus, BV42, has two continuous blocks of 5 cytosines and 22 cytosines that can fold into a unimolecular i-motif made up of four blocks of five cytosines. The structure is stable enough at slightly acidic pH; a high pK_a value of 7.1 provides the possibility for aptamer medical application under physiological conditions. Although the structure of the aptamer–hemagglutinin complex is not yet solved, the i-motif is very likely to be a scaffold for hemagglutinin-binding elements. One of the loops has GAA sequences, and two other loops have CC sequences; both kinds of sequence could be part of binding loops that have been suggested from alignment of all known aptamers to hemagglutinin. Therefore, aptamer BV42 is another example of DNA aptamers having a scaffold that supports proper conformation of binding loops.

In conclusion, both examples of DNA aptamer exhibit striking similarities with natural protein MoRE, antibodies, where DNA G-quadruplexes and i-motifs, as well as protein beta-sandwiches, are scaffolds with several loops that provide a recognition function.

Aptamers as a Class of Molecular Recognition Elements

A paradigm of nucleic acid aptamers is based on the possibility of finding an artificial synthetic oligonucleotide that binds to a given target with high affinity and high specificity. This possibility is not obvious in the case of targets that do not bind nucleic acids in nature. Nevertheless, hundreds of successful cases of aptamer selection have been reported to date. There are natural examples of RNA aptamers in some viruses which bind host proteins that are essential for virus genome replication and host antiviral defense [1, 2]. These aptamers are a result of natural selection driven by evolutionary mechanisms.

A specific technique has been developed to select both DNA and RNA aptamers *in vitro* to a given target. SELEX (systematic evolution of ligands by exponential enrichment) is the selection of aptamers from a huge oligonucleotide library. Selection is driven by high affinity of an aptamer to a target, i.e., it complies an extraction of the most kinetically stable complexes of oligonucleotides with a target that result from competitive binding of all possible oligonucleotides to the target. To make the selection process successful, it is repeated in several selection cycles.

To generate a huge library, oligonucleotides with partially randomized sequences are synthesized, having two defined terminal regions for PCR primers. The theoretical capacity of the library is 4^N, where N is the number of randomized nucleotides. Under real experimental conditions with limited oligonucleotide amounts, the library is under-represented to a considerable extent but even if the best binder is absent, SELEX will select related sequences that could be further improved with rational design.

Several selection approaches have been successfully applied to generate aptamers [3-6]. The use of capillary electrophoresis provides high separation efficiency of aptamer–target complexes that decreases the number of selection repeats down to 1-4 cycles [5, 7]. Microfluidics [8], surface plasmon resonance [9] and magnetic beads [5] are other useful examples of selection approaches. Whole cells, virions, proteins and organic molecules have been used as targets for aptamer selection [7, 10-12].

Nucleic acid aptamers could also be called "chemical antibodies," because they have high affinity and high specificity to the target just like protein antibodies. Aptamers have several advantages in terms of biotechnology. For example, aptamer selection proceeds for several days *in vitro*, whereas selection of monoclonal antibodies requires much more time and proceeds *in vivo*. Aptamers can be developed for a wide variety of targets, whereas antibodies can not be developed for highly toxic substances, highly conserved proteins or substances which are low immunogenic or unstable *in vivo* [3, 4, 11-13].

Nucleic acid aptamers can be readily modified chemically, which improves their pharmacokinetic and pharmacodynamic properties. Moreover, modifications can considerably enhance the aptamer affinity to the target. For example, aptamer derivatives, SOMAmers, combine the advantages of classical aptamers and the diversity of amino acid side chains of proteins. SOMAmers have heterocyclic bases that are extended with additional hydrophobic groups (e.g., amino acid analogues), thus creating a surface with a hybrid nature for better target binding [14-16]. A SOMAmer-based high-content biomarker discovery platform can measure dozens, hundreds and even a thousand protein analytes in one sample with high sensitivity [14, 16]. Another example was reported by Hirao et al. [17, 18]: unnatural heterocyclic base pairs were introduced into the DNA aptamer, substantially increasing its thermal stability and nuclease resistance.

The specific 3D structure of an aptamer is simpler than those of antibodies, so it can be easily assembled and renatured. This attribute permits large variation in purification protocols, as well as expanding storage conditions. Aptamers have rational antidotes; a complementary oligonucleotide readily transforms the unique aptamer 3D structure into double-stranded DNA with no affinity to the target. Non-modified nucleic acid aptamers have extremely low toxicity being, in addition, fully biodegradable, yielding natural non-toxic metabolites. All these properties suggest aptamers as next generation therapeutics with an extremely high efficacy/risk ratio [10, 11, 13, 19, 20].

Since the development of SELEX in 1990, thousands of aptamers to hundreds of targets have been reported. The most attractive therapeutic targets are viral proteins (HIV-1, hepatitis C, influenza, etc.), parasites (*Trypanosoma*, *Plasmodium*, etc.), prions, angiogenesis-related proteins (VEGF, angiopoietin-2, etc.), inflammation-related proteins (platelet-derived growth factor, P- and L-selectins, nuclear factor κ-B, complement C5, etc.), proliferation-related proteins (transcription factor E2F, prostate-specific membrane antigen, tenascin-C, etc.), antibodies and immune system proteins (antibody MA20 to anti-insulin receptor, monoclonal antibody to acetylcholine receptor, immunoglobulin E, interferon-γ, etc.), as well as hemostasis proteins (thrombin, factors VIIa and IXa, von Willebrand factor, etc.) [4, 11-13, 21, 22].

Aptamers are of great pharmaceutical promise; dozens of aptamers have already entered clinical trials yielding only one marketed aptamer so far, pegaptanib. Implementation of this class of novel drugs has to proceed with caution. Several aspects have a high impact on the advance of aptamers, but most of them seem to be a consequence of a lack of preclinical and clinical data, and they have to be overcome in the nearest future [23, 24].

Similar to antibodies, aptamers can be widely used in diagnostics. This field is more predictable than therapeutic application; for example, diagnostics can utilize aptamers with any chemical modifications. Fluorescent labels can be easily attached to oligonucleotides to improve the detection limit. "Aptasensors" have been demonstrated to detect and estimate the content of amino acids, a huge variety of proteins, viruses, bacteria, etc. [3, 4, 11-13].

Generally, aptamers are promising molecular recognition elements (MoRE). High affinity and high specificity ensure successful usage in diagnostics. The availability of rational antidotes encourages their therapeutic implications. Besides practical applications, aptamers are interesting for the fundamental sciences as rather small artificial structured oligomers with high affinity and specificity to a target. Further, we will focus on the details of DNA aptamer recognition of protein targets as well as on the peculiarities of DNA aptamer–protein interactions. We start by reviewing the well-studied thrombin DNA aptamers with a G-quadruplex core structure, and continue with recently found hemagglutinin aptamers with an i-motif core structure.

HD1, A KEY DNA APTAMER TO THROMBIN

DNA aptamers to thrombin are among the leaders in terms of comprehension of structure–function relationships. The first and most studied aptamer is a 15-mer with the sequence GGTTGGTGTGGTTGG, HD1 (also called ARC183, 15-TBA, GS522, G15D and HTQ) [25, 26]. It was proposed by Bock et al. [26] as a consensus sequence of 32 selected aptamers (Table 1). Analysis of all selected aptamers revealed that the pattern GGNTGGN_{2-5}GGNTGG is represented by 87% of the sequences. The frequency of occurrence of T in the third or 12th positions is about 69%. The frequency of appearance of three nucleotides in the middle of the sequence is about 66%. Notably, there are only five cases of a TGT-sequence in the middle of the aptamer, that is 16% of all aptamers; the second frequent variant is a GA-sequence (three cases, 9%). Only one aptamer from the family has the exact sequence of HD1. These data confirm that SELEX yields a family of related sequences with varied affinity to the target; these aptamers will be improved during further research.

Based on the data in Table 1, 10 nucleotides are highly conserved, namely G1, G2, T4, G5, G6, G10, G11, T13, G14 and G15. These data can be fully understood from detailed structural analysis of aptamer HD1. According to the NMR data, eight guanines form two G-quartets of antiparallel G-quadruplex, whereas T4 and T13 make a TT-pair that is in a stacking with the bottom G-quartet (Figure 1). Therefore, all highly conserved nucleotides are involved in the hydrogen bonding creating the core structure of the aptamer [27-29]. If not otherwise specified, in this review "aptamer HD1" means just the G-quadruplex-based structure that is stabilized by coordinating one potassium cation in the center of the molecule.

T3 and T12 appear with 69% frequency; these nucleotides are unstacked in the HD1 structure [27-29], and are buried in the hydrophobic pockets of thrombin exosite I when HD1 is complexed with thrombin [30, 31]. The TGT-loop is variable in the HD1 family; it does not interact with thrombin, but is partially stacked with the top G-quartet [30, 31]. Interestingly, T7 is unstacked in NMR structures of aptamer HD1 and in one X-ray structure of HD1–thrombin complex [27-30]; on the contrary, T9 is unstacked in the recent X-ray structure of HD1–thrombin complex [31]. These data indicate that stacking of G8 and T9 with the top of the G-quartet is more thermodynamically favorable, but alternative stacking is also possible if it is driven by dense crystal packaging.

Table 1. Alignment of the aptamers to thrombin selected by Bock et al. [26]

Aptamer number	Numbering according to HD1 sequence												
	1	2	3	4	5	6	7-9	10	11	12	13	14	15
1	**G**	**G**	T	**T**	**G**	**G**	GTC	**G**	**G**	T	**T**	**G**	**G**
2	**G**	**G**	A	**T**	**G**	**G**	TTT	**G**	**G**	T	**T**	**G**	**G**
3	**G**	**G**	T	**T**	**G**	**G**	GA	**G**	**G**	G	**T**	**G**	**G**
4	**G**	**G**	T	**T**	**G**	**G**	CGA	**G**	**G**	A	**T**	**G**	**G**
5	**G**	**G**	T	**T**	**G**	**G**	GTAG	T	**G**	T	**T**	**G**	**G**
6	**G**	**G**	T	**T**	**G**	**G**	GCT	**G**	**G**	T	**T**	**G**	**G**
7	**G**	**G**	T	**T**	**G**	**G**	GA	**G**	**G**	T	**T**	**G**	**G**
8	**G**	**G**	T	**T**	**G**	**G**	GTC	**G**	**G**	T	**T**	**G**	**G**
9	**G**	**G**	A	**T**	**G**	**G**	TGT	**G**	**G**	T	**T**	**G**	**G**
10	**G**	**G**	T	**T**	**G**	**G**	CAG	**G**	**G**	A	**T**	**G**	**G**
11	**G**	**G**	A	**T**	**G**	**G**	TGA	**G**	**G**	T	**T**	**G**	**G**
12	**G**	**G**	G	**T**	**G**	**G**	TTA	**G**	**G**	T	**T**	**G**	**G**
13	**G**	**G**	G	**T**	**G**	**G**	TTA	**G**	**G**	T	**T**	**G**	**G**
14	**G**	**G**	T	**T**	**G**	**G**	GTTG	**G**	**G**	A	**T**	**G**	**G**
15	**G**	**G**	T	**T**	**G**	**G**	TGT	**G**	**G**	T	**T**	**G**	**G**
16	**G**	**G**	T	**T**	**G**	**G**	TGT	**G**	**G**	G	**T**	**G**	**G**
17	**G**	**G**	G	**T**	**G**	**G**	ATA	**G**	**G**	T	**T**	**G**	**G**
18	**G**	T	G	**T**	**G**	**G**	TAGTT	T	**G**	T	**T**	**G**	**G**
19	**G**	**G**	T	**T**	**G**	**G**	TTACT	**G**	**G**	T	**T**	**G**	**G**
20	**G**	**G**	T	**T**	**G**	**G**	TCT	**G**	**G**	G	**T**	**G**	**G**
21	**G**	**G**	T	**T**	**G**	**G**	GTT	**G**	**G**	G	**T**	**G**	**G**
22	**G**	**G**	T	**T**	**G**	**G**	CCA	**G**	**G**	T	**T**	**G**	**G**
23	T	A	G	C	**G**	**G**	CAGT	**G**	**G**	T	**T**	**G**	**G**
24	**G**	**G**	G	**T**	**G**	**G**	GGA	**G**	**G**	T	**T**	**G**	**G**
25	**G**	**G**	T	**T**	**G**	**G**	TTT	**G**	**G**	G	**T**	**G**	**G**
26	**G**	**G**	T	**T**	**G**	**G**	TTAG	**G**	**G**	T	**T**	**G**	**G**
27	**G**	**G**	A	**T**	**G**	C	GGT	**G**	**G**	T	**T**	**G**	**G**
28	**G**	**G**	T	**T**	**G**	**G**	TTAT	**G**	**G**	T	**T**	**G**	**G**
29	**G**	**G**	T	**T**	**G**	**G**	TGT	**G**	**G**	T	**T**	**G**	**G**
30	**G**	**G**	T	**T**	**G**	**G**	TGT	**G**	**G**	G	**T**	**G**	**G**
31	**G**	**G**	T	**T**	**G**	**G**	GA	**G**	**G**	T	**T**	**G**	**G**
32	**G**	**G**	T	**T**	**G**	**G**	TGGGT	**G**	**G**	A	**T**	**G**	**G**
Consensus	**G**	**G**	T	**T**	**G**	**G**	NNN	**G**	**G**	T	**T**	**G**	**G**
Abundance, %	97	94	69	97	100	97	66	94	100	69	100	100	100

The structure of HD1 is slightly altered in the complex with thrombin: the main difference is within the TT-pair, which is broken in the complex. In uncomplexed HD1, the TT-pair has two symmetrical hydrogen bonds N3–H–O2; the N3–O2 distances are 2.7–2.9 Å (pdb id 148D). In both known HD1–thrombin complexes, T4 and T13 are separated; the N3–O2 distances are about 6 Å. These thymines form new hydrogen bonds with arginines from thrombin exosite I, as well as with a water molecule that is placed just between T4 and T13. The stacking between these thymines and the bottom G-quartet is retained.

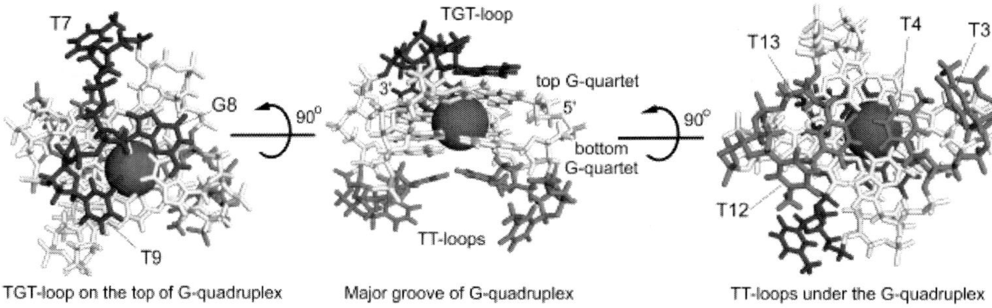

Figure 1. Structure of HD1 stabilized with a potassium cation (NMR data, pdb id 148D). Views from three sides: the TGT-loop, the major groove and the TT-loops. G-quartets are shown in white, TGT-loop in black and TT-loops in grey.

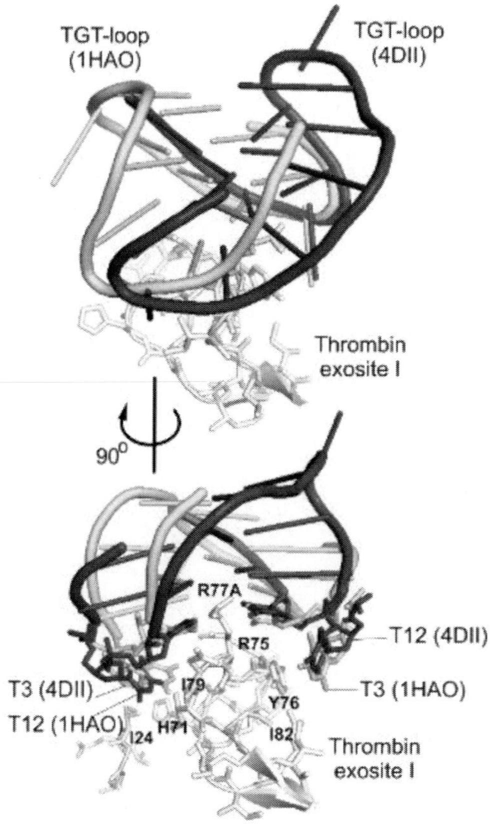

Figure 2. Two modes of HD1–thrombin complexes based on X-ray structures with pdb id 1HAO (in grey) and 4DII (in black). Thrombin exosite I is shown in white. Views from the minor (top) and major (bottom) aptamer grooves. The nucleotides and amino acids that provide interactions are labeled.

Symmetrical TT-loops allow formation of two different complexes of HD1 with thrombin (Figure 2). In both types of complex, TT-loops interact with thrombin exosite I that is a structured positively charged loop surrounded with two hydrophobic pockets. However, the aptamer in the structure with pdb id 1HAO is turned by 180° compared to the aptamer in the structure with pdb id 4DII; the overall topologies of HD1 in both complexes are very similar, except for the differences in TGT-loop stacking mentioned above. Whether these two complexes are thermodynamically identical or whether crystal forces compensate for a thermodynamically less favorable conformation are questions still waiting for answers.

Thrombin complexes with other G-quadruplex aptamers demonstrate the same ambiguity. mTBA is HD1 with reversed 5'-5'-chain polarity between T3 and T4. In mTBA–thrombin complex, the aptamer orientation is the same as in the structure with pdb id 1HAO, and T7 is unstacked in both cases [32]. Aptamer RE31 is HD1 with an additional duplex region. In RE31–thrombin complex, the aptamer orientation is the same as in the structure with pdb id 4DII that is contrary to mTBA, but T7 is unstacked just as in mTBA and the majority of HD1 structures [33]. These data further confirm the existence of several similar conformations of aptamer HD1 that have similar structural properties and are interchangeable.

Altering the Sequence of HD1

The structural data reveal some conformational polymorphisms of aptamer HD1 and its complex with thrombin. Extensive study of HD1 and its analogues with modified sequences has revealed more peculiarities of the aptamer.

The existence of a G-quadruplex can be easily monitored with circular dichroism (CD) spectroscopy, as antiparallel G-quadruplexes have a characteristic CD spectrum. Thermal denaturation of the structure can be easily studied with CD and UV spectroscopy by monitoring changes of spectrum intensity at 295 nm [34, 35].

Aptamer functional activity to block thrombin exosite I can be estimated by hydrolysis of protein substrates of thrombin, as hydrolysis of low molecular weight substrates is generally unaffected with the aptamers. Thrombin exosite I is distant from the active site of the enzyme; exosite I binds protein substrates and arranges them properly for further hydrolysis. For the aptamers, apparent inhibition constants can be obtained with a turbidimetric assay that is based on the monitoring of fibrin associates formed during fibrinogen hydrolysis with thrombin [36, 37].

Millimolar concentrations of potassium cations or equimolar quantities of thrombin are required to assemble the HD1 structure. Depending on the conditions, typical HD1–G-quadruplex melting temperatures are in the range 46-51°C [34, 35, 38], whereas the apparent inhibition constant is about 12-16 nM [34, 39, 40].

Altering Guanines of G-Quartets

Eight guanines of two G-quartets are required for G-quadruplex stability and aptamer inhibitory activity. The replacement of any G-quadruplex guanine with 7-deazaguanine, inosine or an abasic site results in the total loss of aptamer activity. Replacement with 8-methylguanine decreases the inhibitory activity moderately, by 1-8 times [41].

On the contrary, Bini et al. [42] proposed several aptamers with one or two substituted guanines of the G-quadruplex using a computational approach. The authors marked out the "best" aptamer G5T_G14A (ggttTgtgtggttAg), the "medium" aptamer T3C_G11A (ggCtggt gtgAttgg) and the "worst" aptamer T4A_G11A_G15C (ggtAggtgtgAttgC) with substitutions.

Bini et al. estimated that the best aptamer possessed more affinity to thrombin than HD1 aptamer. In our comparative study, both medium and worst aptamers were extremely weak thrombin inhibitors with inhibition constants above 300 nM, whereas the best aptamer had an inhibition constant of 70 ± 8 nM (unpublished data) that is 5.7-fold higher than the constant for HD1. This unusual aptamer is likely to have some alternative, non-G-quadruplex, structure with properly structured TT-loops.

Altering the TGT-Loop

The TGT-loop does not participate in thrombin binding, but it can affect the conformation of the whole G-quadruplex, having some impact on structure stability and inhibitory activity. Data on functional activity and thermal stability of aptamers with altered TGT-loops are summarized in Table 2.

Substitutions in the TGT-loop just slightly increase aK_i; the values do not exceed two-fold aK_i of HD1. A minimal change is observed for the substitution of T7, which is commonly unstacked in HD1. The maximum change is observed for substitution of T9, giving a 1.7-1.8-fold decrease of functional activity (Table 2). Both G8 and T9 positions are stacked with the G-quadruplex: abasic sites in these positions decrease inhibitory activity 6.0-7.5-fold, whereas an abasic site in T7 only slightly affects aptamer activity [41]. Finally, TGT-loop alterations decrease aptamer stability, decreasing melting temperatures by 2-10°C [34, 35] which correlates with the decrease in inhibitory activity.

Table 2. Apparent inhibition constants (aK_i) and melting temperatures (T_m) of the G-quadruplex of the aptamer HD1 with an altered TGT-loop

ID	Sequence	aK_i, nM	T_m, °C	
			25 mM K^+	120 mM K^+
HD1	ggttggtgtggttgg	12.2	46.4#	51*
T7A	ggttgg**A**gtggttgg	19.4		41.9*
T7C	ggttgg**C**gtggttgg	14.1		
T7G	ggttgg**G**gtggttgg	17.5		
G8A	ggttggt**A**tggttgg	17.2	40.2#	
G8C	ggttggt**C**tggttgg	20.8		
G8T	ggttggt**T**tggttgg	15.9	42.2#	
T9A	ggttggtg**A**ggttgg	22.2		44.2*
T9C	ggttggtg**C**ggttgg	22.2		
T9G	ggttggtg**G**ggttgg	20.8		

Inhibition constants were determined by turbidimetric assay (unpublished data). # data from Smirnov and Shafer [35]; * data from Nagatoishi et al. [34].

Table 3. Apparent inhibition constants (aK_i) and melting temperatures of G-quadruplex (T_m) for aptamer HD1 with altered TT-loops

ID	Sequence	aK_i, nM	T_m, °C		
			25 mM K^+	120 mM K^+	2 × excess of thrombin
HD1	ggttggtgtggttgg	12.2	46.4[#]	51[*]	51.3[*]
T3A	gg**A**tggtgtggttgg	8.5		43.3[*]	48.1[*]
T4A	ggt**A**ggtgtggttgg	>200		51.1[*]	36[*]
T12A	ggttggtgtgg**A**tgg	10.8		43.5[*]	48[*]
T13A	ggttggtgtggt**A**gg	>200		51.1[*]	37.7[*]
T3C	gg**C**tggtgtggttgg	46	46.8[#]		
T4C	ggt**C**ggtgtggttgg	>200	40.9[#]		
T12C	ggttggtgtgg**C**tgg	104	47.1[#]		
T13C	ggttggtgtggt**C**gg	>200	39.5[#]		
T4I	ggt**I**ggtgtggttgg	>200			
T13I	ggttggtgtggt**I**gg	>200			

Inhibition constants were determined by turbidimetric assay (unpublished data). [#] data from Smirnov and Shafer [35]; [*] data from Nagatoishi et al. [34].

Several studies are available on modifications with non-natural nucleotides. These modifications in the TGT-loop affect both the stability and inhibitory activity of the aptamer; the result depends on several factors such as the decrease of structure tension and the possibility to form additional interactions with the G-quartet [43-45].

Altering TT-Loops

HD1 interacts with thrombin exosite I with two TT-loops [30, 31]; that is why alterations of these loops are a reasonable way to affect affinity and selectivity of the aptamer. In HD1, T4 and T13 residues form the TT-pair, whereas T3 and T12 residues are unstacked. In HD1–thrombin complex, T4 and T13 are unpaired and interact with the enzyme forming polar contacts, whereas T3 and T12 are in close contact with thrombin hydrophobic pockets, forming mainly hydrophobic contacts [27-31].

The existence of a TT-pair in the aptamer is crucial for thrombin inhibition; substitutions of T4 or T13 with adenine, cytosine or inosine lead to the loss of inhibitory activity (Table 3) [34, 46]. The same has been found for abasic modifications [41]. Contrary to the functional activity, T–T pairing is not crucial for G-quadruplex stability (Table 3); the alternative A–T and I–T pairs can be formed in principle, but they considerably distort the aptamer conformation.

Substitutions of T3 and T12 residues could be manifold, as these residues form hydrophobic contacts with thrombin, having no rigorous geometry. Alterations of T3 or T12 with adenine slightly decreases aK_i (Table 3), whereas substitutions with cytosine significantly increase the constant, 3.8-8.5-fold (Table 3). Alteration of the T3 residue leads to a higher aK_i than the substitution of the T12 residue. The inhibitory activity correlates with larger hydrophobicity of adenine compared to cytosine. Addition of non-natural hydrophobic groups to T3 or T12 greatly increases inhibitory activity [44, 45], whereas abasic sites decrease inhibitory activity three-fold [41].

Summarizing, the TT-pair seems to be obligatory for proper orientation of T3 and T12; these residues can be successfully altered with more hydrophobic nucleotides.

ALTERING THE CONFORMATION OF HD1 WITH CATIONS

The G-quadruplex structure of HD1 is stabilized with a potassium cation that is coordinated by oxygen atoms of guanines in the center of the G-quadruplex. Several other cations could be coordinated in a similar manner. The structures of HD1 with Sr^{2+} [47], Pb^{2+} [48, 49] and NH_4^+ [50] have been characterized with NMR spectroscopy, and the structure of thrombin–HD1–Na^+ has been resolved with X-ray analysis [31]. Some other mono- and divalent cations have been studied in terms of thermal stability of aptamer–cation complexes; the data are summarized in Table 4. The most stable complexes are formed with Sr^{2+}, Ba^{2+}, Pb^{2+} and K^+. The data agree well with the thermal stability of other G-quadruplexes with these cations. Ionic radius in the range 1.3-1.4 Å is the optimum size for a cation to be coordinated within G-quadruplex; other cations are either too small or too big [51].

Besides coordination within the G-quadruplex, some cations can be coordinated with phosphates in the minor grooves. Two Mn^{2+} ions have been shown to be coordinated in this way in NMR and EPR studies [55]. Similar behavior has been described for $[Ru(NH_3)_6]^{3+}$, but the structural details are unknown [56].

Table 4. Cation radii and melting temperatures for HD1-cation complexes

Cation	Ion radius, Å#	Melting temperature, °C†
Sr^{2+}	1.27	63
Ba^{2+}	1.34	58
Pb^{2+}	1.28	55*
K^+	1.33	49
Rb^+	1.47	33
NH_4^+	1.43	29
Na^+	0.97	15
Cs^+	1.57	15
Mg^{2+}	0.66	15
Ca^{2+}	0.99	15
Li^+	0.68	n.m.

Kankia and Marky [51]; Hardin et al. [52, 53]; † Kankia and Marky [51]; * Liu et al. [54]; n.m. – not measurable.

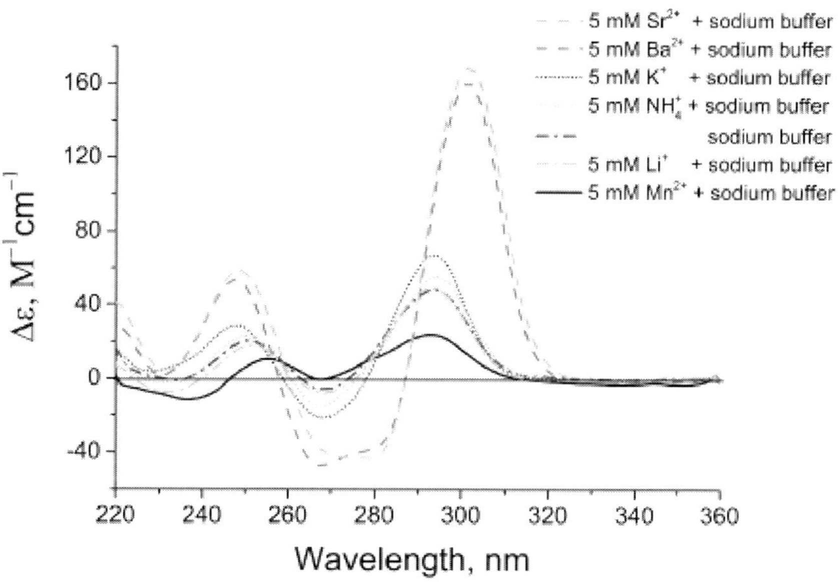

The figure has been made on the basis of the data from Zavyalova et al. [57].

Figure 3. CD spectra for HD1–cation complexes in buffer with 140 mM sodium cations.

Although all studied complexes of HD1 with cations have the same G-quadruplex topology, and all of them have similar CD spectra (Figure 3), their functional activities differ drastically. Thrombin inhibition is possible only with complexes with K^+ and Na^+, whereas Ba^{2+}, Sr^{2+}, NH_4^+ and Mn^{2+} decrease HD1 inhibitory activity in a dose-dependent manner down to its total loss [57]. The aK_i of HD1 complexed with various cations are summarized in Table 5.

Thus, alteration in the nature of the cation coordinated within the G-quadruplex affects aptamer activity more than the majority of single nucleotide mutations. The most likely explanation of the phenomenon is that cation coordination changes G-quadruplex conformation, affecting the conformation of TT-loops that mediate thrombin binding [57]; in other words, it is an excellent example of allosteric interplay within the DNA aptamer structure. Indeed, thorough comparison of the conformations of different HD1–cation complexes has revealed that all functionally active complexes have the same conformation of TT-loops (HD1–K^+, thrombin–HD1–K^+ and thrombin–HD1–Na^+ complexes), whereas HD1–Sr^{2+} and HD1–Mn^{2+} complexes have significantly altered TT-loop conformation [57].

Modulating the Activity of Thrombin Aptamer with Additional Duplex Element

Extension of aptamer HD1 is an obvious and straightforward approach to modify aptamer functional activity. Most common is the addition of a duplex element – two complementary fragments that are connected to the 5'- and 3'-ends of the G-quadruplex with short single-stranded fragments (internal loops/hinge). Extended aptamers have the same or higher inhibitory activity for thrombin than HD1 [40, 58, 59].

Table 5. Apparent inhibition constants (aK_i) for HD1–cation complexes at different cation concentrations

Cation concentration, mM	aK_i, nM				
	K^+	NH_4^+	Ba^{2+}	Sr^{2+}	Mn^{2+}
0	50	50	50	50	50
0.05	–	–	160	150	–
0.5	–	–	>>150	>>150	72
5	14.3	104	>>150	>>150	>>150
10	10.6	150	–	–	–
50	9.1	>>150	>>150	>>150	>>150

Aptamer concentration was 200-400 nM. For all experiments, sodium cation concentration was 140 mM, which is required for thrombin activity [57]; "–" value not determined.

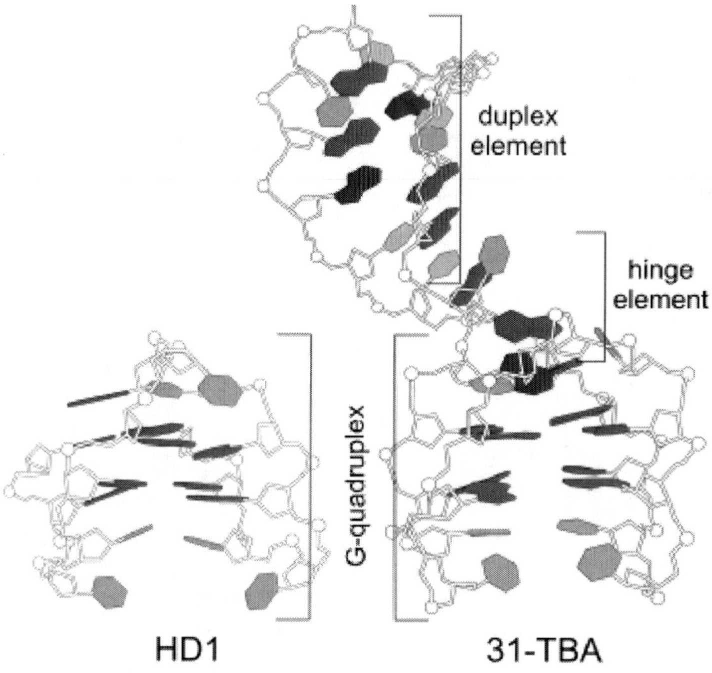

Adopted from the review of Zavyalova et al. [61].

Figure 4. Comparison of the structures of aptamers HD1 and 31-TBA.

Aptamer 31-TBA is a well-studied example of an aptamer with G-quadruplex and duplex elements connected with hinge regions (Figure 4) [59]. The coexistence of G-quadruplex and duplex modules within the molecule was confirmed by Dolinnaya et al. [60]. The G-quadruplex of 31-TBA has the same thermal stability as the HD1 G-quadruplex, but 31-TBA inhibitory activity is 50-fold higher (Table 6).

Duplex elements do not interact with thrombin [66] but, most probably, affect the G-quadruplex conformation, increasing functional activity of the aptamer. The exact sequence of the duplex element has a minor effect on the aptamer inhibitory activity [68]; as seen by comparing aptamers 31-TBA and RE31, partial inversion of the duplex element has a negligible effect on the dissociation and inhibition constants of the aptamers (Table 6).

Another example of a bipartite aptamer is NU172, which has the highest functional activity among the monovalent DNA aptamers to thrombin [67]. This aptamer has a 160-fold higher functional activity than its G-quadruplex element alone (Table 6).

However, not all additional duplex elements increase the functional activity of G-quadruplex aptamers. The extra activity is supported by the existence of modules that include the structural element and hinge regions. A module made up from a duplex/hinge of 31-TBA (aptamers 31-TBA and jja) and a module made up from a hinge/G-quadruplex of NU172 (aptamers NU172 and jaa) have been described [40]. Probably, the hinge regions affect G-quadruplex conformation, but the hinge has to be structured properly either with duplex or G-quadruplex elements creating the module. In the case of mixed aptamers without the modules (aaj and jaj), inhibitory activity is similar to that of HD1, whereas thermal stability is unusually decreased (Table 6).

One more thrombin aptamer with the duplex element is HD22, which binds thrombin exosite II, contrary to all other known G-quadruplex aptamers [71]. The HD22 sequence is AGTCCGT-ggtAggGCAggttgg-GGTGACT; the main reason for the loss of inhibitory activity is T4A replacement in the aptamer G-quadruplex (see Table 3). The aptamer adopts the structure of an antiparallel G-quadruplex as in HD1 and others [69], but its actual structure is more complicated than that of 31-TBA [70]. The high affinity (0.5-11.9 nM) [71, 72]) is achieved by multiple interactions between thrombin exosite II and nucleotides from both the duplex and G-quadruplex elements of the aptamer [70]. This is the sole example of a G-quadruplex-based aptamer that interacts with thrombin via the duplex element.

Modulating the Activity of Thrombin Aptamer with Additional G-Quadruplex Element

One more way to manipulate HD1 functional activity is to connect two G-quadruplexes with a linker.

Aptamer RA-36 is made up of two HD1 aptamers connected with a thymine linker [39, 40, 74]. The aptamer has an aK_i of 7 nM, which is two-fold lower than that of HD1. Surprisingly, the decrease of the constant is not a result of doubling of the thrombin-binding G-quadruplexes, as the extension of the linker or its deletion increases the constant up to 12-14 nM which is equal to the constant of HD1 (Table 7). Analysis of RA-36 deletion mutants has revealed that two symmetrical parts of the aptamer are non-identical. Aptamers without one, two or three nucleotides at the 5'-end have inhibition constants about 9 nM, whereas aptamers without one, two or three nucleotides at the 3'-end have inhibition constants about 18 nM, which corresponds to the alteration of the active conformation of the aptamer. The sequence without the first and the last nucleotides has a constant of 54 nM that corresponds to the rupture of the active structure of the aptamer.

Table 6. Comparative analysis of thrombin aptamers with additional duplex elements

Aptamer	Sequence	Structural data (pdb ID and refs)	Apparent dissociation and inhibition constants	Melting temperature of G-quadruplex in 10 mM K^+
HD1	ggttggtgtggttgg	NMR of HD1–K^+: 1C32, 1C34, 1C35, 1C38 [62], 148D [28]; X-ray of thrombin–HD1 complex: 1HAP, 1HAO [30], 4DIH, 4DII [31]	K_d = 7.1-88 nM [34, 63-65] K_i = 12-16 nM [39, 40]*	38°C [60]
31-TBA	CACTGGTA-ggttggtgtggttgg-GGCCAGTG	CD and UV melting: coexistence of duplex and G-quadruplex [60]	K_i = 0.34-0.65 nM [39, 40]	39°C [60] 43°C*
RE31	GTGACGTA-ggttggtgtggttgg-GGCGTCAC	X-ray of thrombin–RE31 complex: 5CMX [33]	K_d = 0.56 nM [66]	–
jja	CACTGGTA-*ggttggGTAggGtgg*-GGCCAGTG	CD and UV melting: coexistence of duplex and G-quadruplex*	K_i = 1.2 nM [40]	49°C*
jaa	CACTGG*TA*-*ggttggGTAggGtgg*-*T*CCAGTG		K_i = 1.3 nM [40]	41°C*
aaj	*CGCCTA*-ggttggtgtggttgg-*TGGCG*		K_i = 13.4 nM [40]	32°C*
jaj	CACTGG*TA*-ggttggtgtggttgg-*T*CCAGTG		K_i = 14.2 nM [40]	34°C*
NU172	*CGCCTA*-*ggttggGTAggGtgg*-*TGGCG*		K_d = 0.14 nM [67] K_i = 0.29 nM [40]	37°C*
NU	*ggttggGTAggGtgg*	–	K_i = 46 nM*	–

Complementary sequences are underlined; NU172-related sequences are shown in italics; * unpublished data; "–" value not determined.

Although the exact structure of RA-36 and its thrombin complex are not yet known, it is obvious that the high functional activity of RA-36 is provided by the first G-quadruplex, whereas the second part plays the role of a modulating element. Mixed aptamers can demonstrate the inequality of the two parts of RA-36. Combinations of HD1 and the G-quadruplex part of NU172 have different aK_i depending on the position of the HD1 part. The aptamer HD1-NU with HD1 at the 5'-end has the same constant as HD1, whereas the aptamer NU-HD1 has a two-fold lower inhibitory activity (Table 7). The symmetrical aptamer made up from two G-quadruplex parts of NU172 has the same aK_i as the G-quadruplex part of NU172 alone, so the unusual extra effect is a result of some specific peculiarities of the RA-36 structure. The effect cannot be achieved with a simple combinatory approach; the first G-quadruplex has a two-fold higher activity than the second one.

Table 7. Comparative analysis of thrombin aptamers with an additional G-quadruplex element

Aptamer	Sequence	Apparent inhibition constants	Melting temperature of G-quadruplex in 10 mM K^+
HD1	ggttggtgtggttgg	12-16 nM [39, 40]*	38°C [60]
RA-36	ggttggtgtggttgg-t-ggttggtgtggttgg	7.3-7.5 nM [39, 40]	30°C [73]
15-ttt-15	ggttggtgtggttgg-TTT-ggttggtgtggttgg	11.8 nM*	–
RA-36_del	ggttggtgtggttg__ggttggtgtggttgg	14.3 nM*	–
2-31	_gttggtgtggttgg-t-ggttggtgtggttgg	18 nM*	24°C [73]
1-30	ggttggtgtggttgg-t-ggttggtgtggttg_	9.0 nM*	18°C [73]
2-30	_gttggtgtggttgg-t-ggttggtgtggttg_	54 nM*	n.m. [73]
HD1-NU	ggttggtgtggttgg-t-*ggttggGTAggGtgg*	14.8 nM*	–
NU-HD1	*ggttggGTAggGtgg*-t-ggttggtgtggttgg	31 nM*	–
NU-NU	*ggttggGTAggGtgg*-t-*ggttggGTAggGtgg*	52 nM*	–
NU	*ggttggGTAggGtgg*	46 nM*	–
tttt-HD1	TTTT-ggttggtgtggttgg	26 nM*	–
aaaa-HD1	AAAA-ggttggtgtggttgg	53 nM*	–
HD1-tttt	ggttggtgtggttgg-TTTT	12.8 nM*	–
HD1-aaaa	ggttggtgtggttgg-AAAA	14.7 nM*	–

NU172-related sequences are shown in italics; * unpublished data; "–" value not determined; n.m. – not measurable.

The reasons for the inequality of the 5'- and 3'-parts of the aptamer have already been unraveled. Addition of any fragment to the 5'-end of HD1 leads to a considerable decrease of its functional activity, whereas modification of the 3'-end has little effect on the aptamer activity (Table 7). Similar observations were reported by De Rache et al. [75], and the destabilization effect of 5'-extension was demonstrated by Buff et al. [65].

As in the case of aptamers with additional duplex/hinge elements, comprehension of RA-36 functional activity is to be developed. Obviously, the structure of the G-quadruplex could be easily tuned, which greatly affects the aptamer functional activity.

Selectivity of Thrombin Aptamers toward Thrombin-Related Proteins

Aptamer HD1 was selected to human thrombin; it also inhibits thrombins of other mammals with similar inhibition constants (Table 8). Only rabbit thrombin is inhibited with a 10-fold higher aK_i, which correlates with the highest divergence of this protein toward human thrombin [39]. Additional elements can manipulate the selectivity of the aptamer. For example, the addition of a duplex element yields increased selectivity (Table 8). aK_is for 31-TBA are in good agreement with the divergence of the proteins toward human thrombin.

Table 8. Species specificity of aptamers HD1 and 31-TBA

Protein	aK_i, nM		UniprotKB ID	Sequence identity to human thrombin, %
	HD1	31-TBA		
Human thrombin	16	0.65	P00734	100
Bovine thrombin	14.3	13.9	P00735	86.1
Porcine thrombin	17.2	7.3	Q19AZ8	90.8
Rabbit thrombin	182	3200	G1TB36	84.7
Rat thrombin	16.7	2.4	P18292	88.1
Mouse thrombin	26	1.1	P19221	89.2

Apparent inhibition constants (aK_i) were derived from Zavyalova et al. [39].

Table 9. Selectivity of aptamer HD1 toward thrombin-related proteins

Protein	aK_d, nM	References
Prothrombin	78-86	[63, 77]
Prethrombin 1	86	[77]
Prethrombin 2	66	[77]
Meizothrombin des F1	45	[77]
Meizothrombin	35	[77]
α-Thrombin	7-34	[63, 77]
γ-Thrombin	>2000	[78]

Aptamer HD1 also binds several thrombin-related proteins with similar affinity (Table 9). These proteins include prothrombin, a thrombin proform that is hydrolyzed to thrombin during bleeding; prethrombins and meizothrombins are intermediates of prothrombin conversion to thrombin. Anderson and Bock [76] reported that the peptide hirugen, an exosite I inhibitor of thrombin, binds these proteins with varied affinity; the significant differences were between thrombin and prothrombin, having apparent dissociation constants (aK_d) of the complexes of 22 nM and 3200 nM, respectively. The thrombin/prothrombin selectivity of aptamer HD1 is considerably lower than that of hirugen, which is in the range 2.5-11 [63, 77]. As for the partially hydrolyzed thrombin form, γ-thrombin, it has no affinity to the aptamer, probably due to exosite I disturbance [78]. γ-Thrombin is capable of hydrolyzing thrombin low-molecular substrates but not fibrinogen, as its binding is disrupted [79].

Thrombin/prothrombin selectivity has a considerable impact on thrombin aptamer application. In diagnostics, for thrombin “aptasensors” the selectivity should be as high as possible, as the content of prothrombin in the blood normally predominates over thrombin content. For coagulation cascade inhibition the selectivity can be low, providing both thrombin inhibition and inhibition of prothrombin conversion to thrombin [80].

Aptamer HD22 has absolute selectivity to thrombin as it targets exosite II that is sterically hindered in prothrombin [63, 70]. Aptamer NU172 has the highest known affinity to prothrombin; the aK_d of prothrombin–NU172 complex is only 6 nM. The high affinity to prothrombin gives the aptamer extremely high anticoagulant activity and exponential dose–effect dependence [40].

Recently, we have performed an extensive search for determinants of thrombin/prothrombin selectivity studying known and novel exosite I-targeted G-quadruplex aptamers. A competitive enzymatic assay was used to estimate the aK_d of prothrombin–aptamer complexes. Briefly, a turbidimetric assay [36] was modified by adding a varied amount of prothrombin to the sample with an invariable amount of thrombin, fibrinogen and the aptamer. If prothrombin binds a portion of the aptamer, then it leads to the decrease of aptamer inhibition effect. The aK_d of prothrombin–aptamer complex was calculated by curve fitting with the following equation:

$$IC = 1 + \frac{IC_0 K_d^{prothr}}{P_0 + K_d^{prothr}}$$

where IC is the inhibition coefficient, a numerical result of the turbidimetric assay; IC_0 is the inhibition coefficient for the sample without prothrombin; K_d^{prothr} is the aK_d for the prothrombin–aptamer complex; and P_0 is total prothrombin concentration. This equation was deduced for the case when prothrombin concentration greatly exceeds aptamer concentration, and therefore there is a systematic error for aptamers with high affinity to both thrombin and prothrombin. The aK_i for thrombin and aK_d for prothrombin–aptamer complex were determined in 2 μM fibrinogen. Selectivity was estimated as the ratio of prothrombin aK_d to thrombin aK_i. The resulting data are summarized in Table 10.

The selectivity of HD1 aptamer is 9.0, which correlates with previously determined values [63, 77]. Alterations in the HD1 TGT-loop affect the selectivity slightly; T7C and G8A mutants are the cases with the highest and the lowest selectivity, respectively. On the contrary, alterations of T3 and T12 have a great impact on selectivity. It is unsurprising, as these two nucleotides are in close contact with thrombin and prothrombin exosite I. Changing T3 or T12 for A or C considerably increases the selectivity, whereas changing for G has no effect on selectivity (Table 10). Finally, the G-quadruplex part of aptamer NU172 has the lowest selectivity of 3.0, which explains the high affinity of NU172 to prothrombin.

Additional elements affect the selectivity of HD1. Interestingly, adding single-stranded fragments at the 5'- or 3'-end of HD1 decreases the selectivity. Although thrombin aK_i varies depending on the modified end, the selectivity is determined with the sequence of the fragment. Adenine repeats give a selectivity of 3.0, whereas thymine repeats give a selectivity of about 5.3. Adding a duplex element to HD1 increases the selectivity up to 13-15 in the case of aptamers 31-TBA (Table 10) and RE31 [66]. As for aptamer NU172, the selectivity also increases compared to its G-quadruplex – from 3.0 to 5.3; at the same time, both thrombin aK_i and prothrombin aK_d greatly decrease after adding a duplex element.

An additional G-quadruplex element affects the selectivity of HD1 in different ways; in the case of aptamer RA-36 the selectivity increases up to 28, whereas a similar aptamer 15-ttt-15 has a decreased selectivity of 5.5, as does aptamer HD1-tttt (Table 10). So, the high selectivity of RA-36 is a property of its particular structure, where two G-quadruplexes are united into the integral structure. Aptamers HD1-NU, NU-HD1 and NU-NU were constructed to manipulate the selectivity of RA-36. The inhibition constants increased in the row, whereas the prothrombin constants decreased, leading to a significant decrease of the selectivity down to 1.2.

Table 10. Selectivity of thrombin aptamers estimated as the ratio of prothrombin aK_d to thrombin aK_i

Aptamer	Thrombin aK_i or * aK_d, nM	Prothrombin aK_d, nM	Thrombin/prothrombin selectivity
HD1 mutants			
HD1	12.2	110	9.0
T7C	14.1	220	16
G8A	17.2	120	7.0
T3A	8.5	240	28
T3C	46	>3000	>65
T12A	10.8	225	21
T12C	104	>3000	>29
T12G	9.2	76	8.3
NU	46	136	3.0
Extended HD1			
tttt-HD1	26	140	5.4
aaaa-HD1	53	220	3.0
HD1-tttt	12.8	66	5.2
HD1-aaaa	14.7	44	3.0
31-TBA	3.5	52	15.0
RE31	0.56*,#	7.2#	13.0
NU172	1.13	6.0	5.3
RA-36	6.4	180	28
15-ttt-15	11.8	65	5.5
15-NU	14.8	220	15
NU-15	31	190	6.1
NU-NU	52	60	1.2

Thrombin aK_i and prothrombin aK_d are the apparent constants in 2 μM fibrinogen (except RE31). # Spiridonova et al. [66].

In summary, the selectivity of aptamer HD1 can be efficiently modulated with TT-loop modifications and additional elements. A variety of achievable properties is a result of the possibility to manipulate the G-quadruplex conformation.

APTAMERS TO HEMAGGLUTININ WITH I-MOTIFS

i-Motifs are other unusual DNA structures that can be a scaffold for MoRE, including aptamers. An appropriate example has been found recently among DNA aptamers to influenza hemagglutinin.

More than 25 DNA aptamers to hemagglutinin have been described since 2004. The majority were isolated by SELEX; several of them were further optimized, and two aptamers were designed by rational approach. A comprehensive review has been published recently [81]. Two rationally designed aptamers, BV35r and BV42, have similar structures: the complementary ends and the large hairpin loops having C_5-AAGAA-C_{25} and C_5-

AAGAAGAA-C_{22} sequences, respectively (Table 11) [82]. Musafia et al. did not provide a structural study of the aptamers; they supposed that the aptamer-recognizing elements are not sequence-specific, but are placed within large non-structured internal loops [82].

Both aptamers inhibit hemagglutinin function *ex vivo* and *in vivo*. The aptamers have extremely broad specificity, recognizing various strains of influenza A and B. The estimated IC_{50} of hemagglutinin binding is 4 nM for BV42 and 5 nM for BV35r [82].

Based on the sequence of these two aptamers, we have proposed a duplex element and i-motif as a scaffold, whereas AAGAA, AAGAAGAA and short cytosine loops within the i-motif could be recognizing elements for hemagglutinin.

Recent bioinformatics analysis of the putative structures of all DNA aptamers to hemagglutinin has revealed several common features, like requirements for aptamer flexibility for its functional activity. Possibly, the aptamer structure should not be rigid to allow multicenter binding. Several frequent sequences have also been found with the ClustalW2 Multiple Sequence Alignment tool, namely: 1) the consensus TAA and CGAC sequences in hairpin loops and 2) the consensus TAA, GAA and CC sequences in single-stranded regions. In total, TAA and GAA are the most abundant patterns in unpaired regions of the aptamers. These patterns could comprise the binding sites and other patterns could provide aptamer specificity [81]. These results are in total agreement with our suppositions on BV42 and BV35r structures.

The topology of the i-motif can be described as two cytidine stretches that form a parallel-stranded duplex stabilized with C–H^+–C pairs, and two parallel-stranded duplexes that interact in an antiparallel manner through base pair intercalation. The structure is stable at acidic and slightly acidic pH; the typical melting temperature at pH 5.0 is generally common to those for G-quadruplexes, in the range 40-60°C [83]. The typical features of i-motifs are characteristic CD spectra with a maximum at 290 nm; the spectra have apparent pH dependence – the peak is absent at pH > 8.0 [84].

Here we present the first evidence of the existence of i-motif topology in a DNA aptamer. Aptamer BV42 has a characteristic CD spectrum that clearly depends on pH (Figure 5). The transitional mid-point pH from i-motif to coiled coil is estimated to be about 7.1. Thus, the aptamer i-motif exists at physiological conditions and could provide the aptamer functional activity. This i-motif is unimolecular and consists of four stretches of five cytosines; these data are preliminary and were obtained from the pH dependence of melting temperature and concentration dependence of CD spectra.

Table 11. Sequences of aptamers with putative i-motif structures

Aptamer ID	Aptamer sequence
BV35r	AACGCTCACT-*CCCCC*-AAGAA-*CCCCCCCCCCCCCCCCCCCCCCCCC*-AGTGAGCGTT
BV42	AACGCTCACT-*CCCCC*-AAGAAGAA-*CCCCCCCCCCCCCCCCCCCCCC*-AGTGAGCGTT

Complementary sequences are underlined; blocks of cytosines are shown in italics.

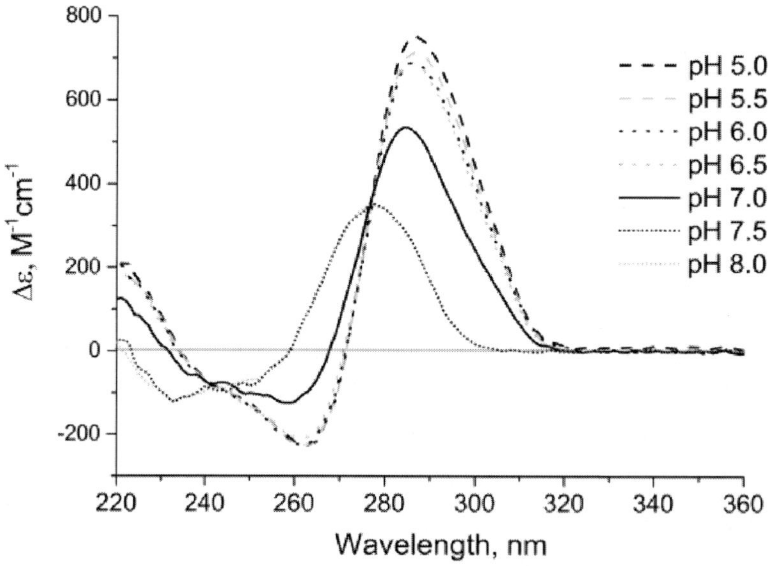

Figure 5. pH dependence of CD spectra of aptamer BV42.

Further studies are needed to obtain data on the precise structure of the aptamer and the determinants of functional activity, but even from these first data there is no doubt that the i-motif is an alternative scaffold to the G-quadruplex for organizing the 3D structure of aptamer MoRE. The novel scaffold could be used for the creation of novel aptamers with novel properties that will depend on pH.

CONCLUSION

DNA aptamers have a certain spatial 3D structure that provides high affinity and high specificity to a target. The structure can be considered as containing recognizing elements and a scaffold that handles the recognizing loops. Two types of scaffold have been described.

Firstly, G-quadruplexes are well-known and characterized scaffolds. A clear structure–activity relationship has been described for thrombin aptamers; the alteration of G-quadruplex conformation affects the conformation of the recognizing loops. Aptamer functional activity can be modified with a variety of factors: coordinated cations, alterations in the sequence and additional structural elements. Besides the functional activity, aptamer selectivity can be also manipulated.

Secondly, the i-motif is a novel example of an aptamer scaffold. An aptamer to influenza hemagglutinin, BV42, has continuous blocks of cytosines that fold into a unimolecular i-motif. The structure is stable at slightly acidic pH, having a pK_a value of 7.1 that provides possibility of application of the aptamer under physiological conditions. The i-motif is a scaffold for hemagglutinin-binding elements that are, most probably, the loops with GAA and CC sequences.

Both G-quadruplexes and i-motifs could be used for the molecular engineering of novel aptamers with desirable properties.

Conflict Of Interest: The authors declare no conflict of interest. The work was supported by the Russian Scientific Foundation (grant no. 15-13-00033) in part of "Aptamers to hemagglutinin with i-motif," and by the Russian Foundation for Basic Research (grant nos. 14-04-01757, 15-33-70003 and 16-03-00136) in part of "G-quadruplex aptamers to thrombin." The funders had no role in study design, data collection and analysis, decision to publish, or preparation of the manuscript.

REFERENCES

[1] Marciniak RA, Garcia-Blanco MA, Sharp PA. Identification and characterization of a HeLa nuclear protein that specifically binds to the trans-activation-response (TAR) element of human immunodeficiency virus. *Proc Natl Acad Sci US*. 1990;87(9):3624–8.

[2] Burgert HG, Ruzsics Z, Obermeier S, Hilgendorf A, Windheim M, Elsing A. Subversion of host defense mechanisms by adenoviruses. *Curr Top Microbiol Immunol*. 2002;269:273–318.

[3] Bouchard PR, Hutabarat RM, Thompson KM. Discovery and development of therapeutic aptamers. *Annu Rev Pharmacol Toxicol*. 2010;50:237–57.

[4] Keefe AD, Pai S, Ellington A. Aptamers as therapeutics. *Nat Rev Drug Discov*. 2010;9 (7):537–50.

[5] Darmostuk M, Rimpelova S, Gbelcova H, Ruml T. Current approaches in SELEX: an update to aptamer selection technology. *Biotechnol Adv*. 2015;33(6 Pt 2):1141–61.

[6] Lao YH, Chiang HY, Yang DK et al. Selection of aptamers targeting the sialic acid receptor of hemagglutinin by epitope-specific SELEX. *Chem Commun (Camb)*. 2014; 50(63):8719–22.

[7] Nitsche A, Kurth A, Dunkhorst A et al. One-step selection of Vaccinia virus-binding DNA aptamers by MonoLEX. *BMC Biotechnol*. 2007;7:48.

[8] Ahmad KM, Oh SS, Kim S et al. Probing the limits of aptamer affinity with a microfluidic SELEX platform. *PLoS One*. 2011;6(11):e27051.

[9] Misono TS, Kumar PK. Selection of RNA aptamers against human influenza virus hemagglutinin using surface plasmon resonance. *Anal Biochem*. 2005;342(2):312–7.

[10] Dupont DM, Andersen LM, Botkjaer KA, Andreasen PA. Nucleic acid aptamers against proteases. *Curr Med Chem*. 2011;18:4139–51.

[11] Shigdar S, Luczo J, Wei MQ et al. Aptamer therapeutics: the 21st century's magic bullet of nanomedicine. *Open Conf Proceedings J*. 2010;1:118–24.

[12] Vallian S, Khazaei MR. Medical applications of aptamers. *Res Pharm Sci*. 2007;2(2): 59–66.

[13] Nimjee SM, Rusconi CP, Sullenger BA. Aptamers: an emerging class of therapeutics. *Annu Rev Med*. 2005;56:555–83.

[14] Kraemer S, Vaught JD, Bock C et al. From SOMAmer-based biomarker discovery to diagnostic and clinical applications: a SOMAmer-based, streamlined multiplex proteomic assay. *PLoS One*. 2011;6(10):e26332.

[15] Gold L, Walker JJ, Wilcox SK, Williams S. Advances in human proteomics at high scale with the SOMAscan proteomics platform. *N Biotechnol*. 2012;29(5):543–9.

[16] Park NJ, Wang X, Diaz A et al. Measurement of cetuximab and panitumumab-unbound serum EGFR extracellular domain using an assay based on slow off-rate modified aptamer (SOMAmer) reagents. *PLoS One*. 2013;8(8):e71703.

[17] Matsunaga K, Kimoto M, Hanson C, Sanford M, Young HA, Hirao I. Architecture of high-affinity unnatural-base DNA aptamers toward pharmaceutical applications. *Sci Rep*. 2015;5:18478.

[18] Kimoto M, Matsunaga K, Hirao I. DNA aptamer generation by genetic alphabet expansion SELEX (ExSELEX) using an unnatural base pair system. *Methods Mol Biol*. 2016;1380:47–60.

[19] Gómez-Outes A, Suárez-Gea ML, Lecumberri R, Terleira-Fernández AI, Vargas-Castrillón E. Specific antidotes in development for reversal of novel anticoagulants: a review. *Recent Pat Cardiovasc Drug Discov*. 2014;9:2–10.

[20] Potti A, Rusconi CP, Sullenger BA, Ortel TL. Regulatable aptamers in medicine: focus on antithrombotic strategies. *Expert Opin Biol Ther*. 2004;4(10):1641–7.

[21] Lee JF, Stovall GM, Ellington AD. Aptamer therapeutics advance. *Curr Opin Chem Biol*. 2006;10:282–9.

[22] Ni X, Castanares M, Mukherjee A, Lupold SE. Nucleic acid aptamers: clinical applications and promising new horizons. *Curr Med Chem*. 2011;18(27):4206–14.

[23] Baird GS. Where are all the aptamers? *Am J Clin Pathol*. 2010;134:529–31.

[24] Zavyalova EG. The aptamers. Whether the paradigm works in therapeutic applications? *Drug Des*. 2015;4:2.

[25] Griffin LC, Toole JJ, Leung LLK. The discovery and characterization of a novel nucleotide-based thrombin inhibitor. *Gene*. 1993;137:25–31.

[26] Bock LC, Griffin LC, Latheam JA, Vermass EH, Toole JJ. Selection of single stranded DNA molecules that bind and inhibit human thrombin. *Nature*. 1992;355:564–6.

[27] Macaya RF, Schultze P, Smith FW, Roet JA, Feigon J. Thrombin-binding DNA aptamer forms a unimolecular quadruplex structure in solution. *Proc Natl Acad Sci US*. 1993;90:3745–9.

[28] Schultze P, Macaya RF, Feigon J. Three-dimensional solution structure of the thrombin-binding DNA aptamer d(GGTTGGTGTGGTTGG). *J Mol Biol*. 1994;235(5): 1532–47.

[29] Wang KY, McCurdy S, Shea RG, Swaminathan S, Bolton PH. A DNA aptamer which binds to and inhibits thrombin exhibits a new structural motif for DNA. *Biochemistry*. 1993;32:1899.

[30] Padmanabhan K, Tulinsky A. An ambiguous structure of a DNA 15-mer thrombin complex. *Acta Cryst D*. 1996;52:272.

[31] Russo Krauss I, Merlino A, Randazzo A, Novellino E, Mazzarella L, Sica F. High-resolution structures of two complexes between thrombin and thrombin-binding aptamer shed light on the role of cations in the aptamer inhibitory activity. *Nucleic Acids Res*. 2012;40:8119–28.

[32] Russo Krauss I, Merlino A, Giancola C, Randazzo A, Mazzarella L, Sica F. Thrombin-aptamer recognition: a revealed ambiguity. *Nucleic Acids Res*. 2011;39(17):7858–67.

[33] Russo Krauss I, Spiridonova V, Pica A, Napolitano V, Sica F. Different duplex/quadruplex junctions determine the properties of anti-thrombin aptamers with mixed folding. *Nucleic Acids Res*. 2016;44(2):983–91.

[34] Nagatoishi S, Isono N, Tsumoto K, Sugimoto N. Loop residues of thrombin-binding DNA aptamer impact G-quadruplex stability and thrombin binding. *Biochimie*. 2011; 93:1231–8.

[35] Smirnov I, Shafer RH. Effect of loop sequence and size on DNA aptamer stability. *Biochemistry*. 2000;39:1462–8.

[36] Zavyalova EG, Protopopova AD, Yaminsky IV, Kopylov AM. Kinetic characterization of inhibition of human thrombin with DNA aptamers by turbidimetric assay. *Anal Biochem*. 2012;421:234–9.

[37] Zavyalova E, Kopylov A. Multiple inhibitory kinetics reveal an allosteric interplay among thrombin functional sites. *Thromb Res*. 2015;135(1):212–6.

[38] Virgilio A, Petraccone L, Scuotto M, Vellecco V, Bucci M, Mayol L, Varra M, Esposito V, Galeone A. 5-Hydroxymethyl-2'-deoxyuridine residues in the thrombin binding aptamer: investigating anticoagulant activity by making a tiny chemical modification. *Chembiochem*. 2014;15(16):2427–34.

[39] Zavyalova E, Golovin A, Timoshenko T, Babiy A, Pavlova G, Kopylov A. DNA aptamers for human thrombin with high anticoagulant activity demonstrate target- and species-specificity. *Curr Med Chem*. 2012;19:5232–7.

[40] Zavyalova E, Golovin A, Pavlova G, Kopylov A. Module-activity relationship of G-quadruplex based DNA aptamers for human thrombin *Curr Med Chem*. 2013;20:4836–43.

[41] Wang KY, Krawczyk SH, Bischofberger N, Swaminathan S, Bolton PH. The tertiary structure of a DNA aptamer which binds to and inhibits thrombin determines activity. *Biochemistry*. 1993;32:11285–92.

[42] Bini A, Mascinic M, Mascinia M, Turner APF. Selection of thrombin-binding aptamers by using computational approach for aptasensor application. *Biosens Bioelectron*. 2011; 26:4411–6.

[43] Borbone N, Bucci M, Oliviero G, Morelli E, Amato J, D'Atri V, D'Errico S, Vellecco V, Cirino G, Piccialli G, Fattorusso C, Varra M, Mayol L, Persico M, Scuotto M. Investigating the role of T7 and T12 residues on the biological properties of thrombin-binding aptamer: enhancement of anticoagulant activity by a single nucleobase modification. *J Med Chem*. 2012;55:10716–28.

[44] Cai B, Yang X, Sun L, Fan X, Li L, Jin H, Wu Y, Guan Z, Zhang L, Zhang L, Yang Z. Stability and bioactivity of thrombin binding aptamers modified with D-/L-isothymidine in the loop regions. *Org Biomol Chem*. 2014;12(44):8866–76.

[45] Scuotto M, Persico M, Bucci M, Vellecco V, Borbone N, Morelli E, Oliviero G, Novellino E, Piccialli G, Cirino G, Varra M, Fattorusso C, Mayol L. Outstanding effects on antithrombin activity of modified TBA diastereomers containing an optically pure acyclic nucleotide analogue. *Org Biomol Chem*. 2014 Jul. 28;12(28):5235–42.

[46] Tsiang M, Gibbs CS, Griffin LC, Dunn KE, Leung LLK. Selection of a suppressor mutation that restores of an oligonucleotide inhibitor for thrombin using *in vitro* genetics. *J Biol Chem*. 1995;270:19370–6.

[47] Mao X, LA Marky, WH Gmeiner. NMR structure of the thrombin-binding DNA aptamer stabilized by Sr^{2+}. *J Biomol Struct Dyn*. 2004;22(1):25–33.

[48] Smirnov I, Shafer RH. Lead is unusually effective in sequence-specific folding of DNA. *J Mol Biol*. 2000;296(1):1–5.

[49] Smirnov IV, Kotch FW, Pickering IJ, Davis JT, Shafer RH. Pb EXAFS studies on DNA quadruplexes: identification of metal ion binding site. *Biochemistry*. 2002;41(40): 12133–9.
[50] Trajkovski M, Sket P, Plavec J. Cation localization and movement within DNA thrombin binding aptamer in solution. *Org Biomol Chem*. 2009;7(22):4677–84.
[51] Kankia BI, Marky LA. Folding of the thrombin aptamer into a G-quadruplex with Sr^{2+}: stability, heat, and hydration. *J Am Chem Soc*. 2001;123(44):10799–804.
[52] Hardin CC, Henderson E, Watson T, Prosser JK. Monovalent cation induced structural transitions in telomeric DNAs: G-DNA folding intermediates. *Biochemistry*. 1991;30
[53] (18):4460–72.
[54] Hardin CC, Watson T, Corregan M, Bailey C. Cation-dependent transition between the quadruplex and Watson-Crick hairpin forms of d($CGCG_3GCG$). *Biochemistry*. 1992;31 (3):833–41.
[55] Liu W, Fu Y, Zheng B, Cheng S, Li W, Lau TC, Liang H. Kinetics and mechanism of conformational changes in a G-quadruplex of thrombin-binding aptamer induced by Pb^{2+}. *J Phys Chem B*. 2011;115(44):13051–6.
[56] Marathias VM, Wang KY, Kumar S, Pham TQ, Swaminathan S, Bolton PH. Determination of the number and location of the manganese binding sites of DNA quadruplexes in solution by EPR and NMR in the presence and absence of thrombin. *J Mol Biol*. 1996;260(3):378–94.
[57] De Rache A, Doneux T, Kejnovská I, Buess-Herman C. On the interaction between $[Ru(NH_3)_6]^{3+}$ and the G-quadruplex forming thrombin binding aptamer sequence. *J Inorg Biochem*. 2013;126:84–90.
[58] Zavyalova E, Tagiltsev G, Reshetnikov R, Arutyunyan A, Kopylov A. Cation coordination alters the conformation of a thrombin-binding G-quadruplex DNA aptamer that affects inhibition of thrombin. *Nucleic Acids Ther. Forthcoming 2016.*
[59] Macaya RF, Waldron JA, Beutel BA et al. Structural and functional characterization of potent antithrombotic oligonucleotides possessing both quadruplex and duplex motifs. *Biochemistry*. 1995;34:4478–92.
[60] Ikebukuro K, Okumura Y, Sumikura K, Karube I. A novel method of screening thrombin-inhibiting DNA aptamers using an evolution-mimicking algorithm. *Nucleic Acids Res*. 2005;33:e108.
[61] Dolinnaya NG, Yuminova AV, Spiridonova VA, Arutyunyan AM, Kopylov AM. Coexistence of G-quadruplex and duplex domains within the secondary structure of 31-mer DNA thrombin-binding aptamer. *J Biomol Struct Dyn*. 2012;30(5):524–31.
[62] Zavyalova E, Golovin A, Pavlova G, Kopylov A. Development of antithrombotic aptamers: from recognizing elements to drugs. Curr Pharm Des. Forthcoming 2016.
[63] Marathias VM, Bolton PH. Structures of the potassium-saturated, 2:1, and intermediate, 1:1, forms of a quadruplex DNA. *Nucleic Acids Res*. 2000;28:1969–77.
[64] Muller J, Freitag D, Mayer G, Potzsch B. Anticoagulant characteristics of HD1-22, a bivalent aptamer that specifically inhibits thrombin and prothrombinase. *J Thromb Haemost*. 2008;6:2105–12.
[65] Dougan H, Hobbs JB, Weitz JI, Lyster DM. Synthesis and radioiodination of a stannyl oligodeoxyribonucleotide. *Nucleic Acids Res*. 1997;25:2897–901.

[66] Buff MCR, Schafer F, Wulffen B, Muller J, Potzsch B, Heckel A, Mayer G. Dependence of aptamer activity on opposed terminal extensions: improvement of light-regulation efficiency. *Nucleic Acids Res*. 2010;38:2111–8.

[67] Spiridonova VA, Barinova KV, Glinkina KA, Melnichuk AV, Gainutdynov AA, Safenkova IV, Dzantiev BB. A family of DNA aptamers with varied duplex region length that forms complexes with thrombin and prothrombin. *FEBS Lett*. 2015;589 (16):2043–9.

[68] Diener JL, Wagner-Whyte J, Fontana D, inventor; Archemix corp. Aptamers that bind thrombin with high affinity. Intern*ational patent PCT WO* 2007/025049 A2. 2006 Aug.

[69] Ikebukuro K, Yoshida W, Noma T, Sode K. Analysis of the evolution of the thrombin-inhibiting DNA aptamers using a genetic algorithm. *Biotechnol Lett*. 2006;28:1933–7.

[70] Lin PH, Chen RH, Lee CH, Chang Y, Chen CS, Chen WY. Studies of the binding mechanism between aptamers and thrombin by circular dichroism, surface plasmon resonance and isothermal titration calorimetry. *Colloids Surf B Biointerfaces*. 2011;88: 552–8.

[71] Russo Krauss I, Pica A, Merlino A, Mazzarella L, Sica F. Duplex–quadruplex motifs in a peculiar structural organization cooperatively contribute to thrombin binding of a DNA aptamer. *Acta Cryst D Biol Crystallogr*. 2013;69:2403–11.

[72] Tasset DM, Kubik MF, Steiner W. Oligonucleotide inhibitors of human thrombin that bind distinct epitopes. *J Mol Biol*. 1997;272:688–98.

[73] Muller J, Wulffen B, Potzsch B, Mayer G. Multidomain targeting generates a high-affinity thrombin-inhibiting bivalent aptamer. *ChemBioChem*. 2007;8:2223–6.

[74] Yuminova AV, Smirnova IG, Arutyunyan AM, Golovin, AV, Pavlova GV, Kopylov AM. The structure of G-quadruplex thrombin binding DNA-aptamer RA36. *Bull Mosc Univ*. 2015;56(1);44–8.

[75] Zavyalova E, Golovin A, Reshetnikov R, Mudrik N, Panteleyev D, Pavlova G, Kopylov A. Novel modular DNA aptamer for human thrombin with high anticoagulant activity. *Curr Med Chem*. 2011;18(22),3343–50.

[76] De Rache A, Kejnovska I, Vorlickova M, Buess-Herman C. Elongated thrombin binding aptamer: a G-quadruplex cation-sensitive conformational switch. *Chem Eur J*. 2012;18:4392–400.

[77] Anderson PJ, Bock PE. Role of prothrombin fragment 1 in the pathway of regulatory exosite I formation during conversion of human prothrombin to thrombin. *J Biol Chem*. 2003;278(45):44489–95.

[78] Kretz CA, Stafford AR, Fredenburgh JC, Weitz JI. HD1, a thrombin-directed aptamer, binds exosite 1 on prothrombin with high affinity and inhibits its activation by prothrombinase. *J Biol Chem*. 2006;281(49):37477–85.

[79] Paborsky LR, McCurdy SN, Griffin LC, Toole JJ, Leung LLK. The single-stranded DNA aptamer-binding site of human thrombin. *J Biol Chem*. 1993;288(28):20808–11.

[80] Pospisil CH, Stafford AR, Fredenburgh JC, Weitz JI. Evidence that both exosites on thrombin participate in its high affinity interaction with fibrin. *J Biol Chem*. 2003;278 (24):21584–91.

[81] Kretz CA, Cuddy KK, Stafford AR, Fredenburgh JC, Roberts R, Weitz JI. HD1, a thrombin- and prothrombin-binding DNA aptamer, inhibits thrombin generation by attenuating prothrombin activation and thrombin feedback reactions. *Thromb Haemost*. 2010;103(1):83–93.

[82] Zavyalova E, Kopylov A. Aptamers to hemagglutinin: a novel tool for influenza virus recognition and neutralization. *Curr Pharm Des*. 2016;22:1.
[83] Musafia B, Oren-Banaroya R, Noiman S. Designing anti-influenza aptamers: novel quantitative structure activity relationship approach gives insights into aptamer – virus interaction. *PLoS One*. 2015;9(5):e97696.
[84] Guéron M, Leroy JL. The i-motif in nucleic acids. *Curr Opin Struct Biol*. 2000;10(3): 326–31.
[85] Fujii T, Sugimoto N. Loop nucleotides impact the stability of intrastrand i-motif structures at neutral pH. *Phys Chem Chem Phys*. 2015;17(26):16719–22.

Biographical Sketch

Alexey M. Kopylov
Chemistry Department,
Lomonosov Moscow State University,
Moscow, Russian Federation

Education: Chemistry Department, Lomonosov Moscow State University

Address: Russian Federation, 119991 Moscow, Leninskiye Gory 1-3, Lomonosov Moscow State University, Chemistry Department

***Research and Professional Experience*:**

Graduated from Chemistry Department, Lomonosov Moscow State University in 1972; PhD in 1976, and Doctor of Chemistry in 1986 from the same Institution. 1979 – 1980 – Post-Doc at University of California, Santa Cruz; 1989-1990 – Visiting Professor at University of Massachusetts, Amherst. Most of research has been dedicated to the study of RNA-protein interactions in the bacterial ribosomes. Then it was gradually moved into studying of artificial examples of nucleic acid – protein recognition, aptamers as an example. Current interests are aptamers to the human thrombin, and to the influenza virus, as well as development of system approaches to study coagulation cascade, and brain cancer signal transduction as examples of complex networks.

Principle Investigator for the International Grants: the Netherlands, Germany, China, etc.

***Professional Appointments*:**

1972-1980: Junior Researcher, Belozersky Laboratory of Molecular Biology and Bioorganic Chemistry, Moscow State University, USSR

1980-1987: Senior Researcher; Division of Chemistry of Natural Compounds, Chemistry Department, Moscow State University, USSR

1981-1987: Vice-Chairman, Division of Chemistry of Natural Compounds, Chemistry Department, Moscow State University, USSR

1987-1988: Leader Researcher, Division of Chemistry of Natural Compounds, Chemistry Department, Moscow State University, USSR

1988-present: Professor, Division of Chemistry of Natural Compounds, Chemistry Department, Group Leader, AN Belozersky Institute Physical-Chemical Biology, Moscow State University, Russian Federation

1979-1980: IREX Scholar, University of California, Santa Cruz, US

1989-1990: Visiting Professor, Department of Biochemistry, University of Massachusetts, Amherst, US

2008-2009: Leader Researcher, Center of Obstetrics, Gynecology, and Perinatology

2008-2012: Professor, Course Instructor of Polytechnic Institute of New York University, New York, US

***Honors*:**

Award of Ministry of High Education of USSR (1976)

Award of International Science Foundation (1993)

The State Stipend for the Distinguished Scientist of Russia (1994-2003)

"Professor of the Year," Moscow (2003, 2005)

IREX Scholar, University of California, Santa Cruz, US (1979-1980)

Visiting Professor, Department of Biochemistry, University of Massachusetts, Amherst, US (1989-1990)

EMBO Scholar of Workshop on RNA, Aarhus University, Denmark (1987)

Member of The RNA Society (1995)

Member of American Society for Biochemistry and Molecular Biology (1998-1999)

Member of the Federation of American Societies for Experimental Biology (1998-1999)

Vice-Chairman of PhD Defense Council for Bioorganic Chemistry, Russian Federation

Member of PhD Defense Council for Molecular Biology and Virology, Russian Federation

Scientific Secretary of "Universities of Russia" Program (1993-2004)

Member of Editorial Board: "Russian Chemical Journal," "Moscow University Chemistry Bulletin," "Haematopoesis Immunology"

Section Chairman for International Conference on Drug Discovery and Therapy

The International Expert for Elsevier

Elena G. Zavyalova

Chemistry Department,
Lomonosov Moscow State University,
Russian Federation

Education: Lomonosov Moscow State University, Chemistry Department

Address: Russian Federation, 119991 Moscow, Leninskiye Gory 1-3, Lomonosov Moscow State University, Chemistry Department

***Research and Professional Experience*:** Graduated from Chemistry Department, Lomonosov Moscow State University in 2010; PhD in 2013. The main scientific interests are in the field of thrombin aptamers and anti-influenza aptamers, as well as studying the coagulation cascade and developing the new approaches for its inhibition.

***Professional Appointments*:**

2012-2013: Junior Researcher, Division of Chemistry of Natural Compounds, Chemistry Department, Moscow State University, Russian Federation

2013-present: Assistant of Division of Chemistry of Natural Compounds, ChemistryDepartment, Moscow State University, Russian Federation

2015-present: Junior Researcher, AN Belozersky Institute Physical-Chemical Biology, Moscow State University, Russian Federation

***Publications Last 3 Years*:**

1. Zavyalova E., Kopylov A. Aptamers to hemagglutinin: a novel tool for influenza virus recognition and neutralization. *Curr Pharm Des*. 2016, in press.
2. Zavyalova E, Golovin A, Pavlova G, Kopylov A. Development of antithrombotic aptamers: from recognizing elements to drugs. *Curr Pharm Des*. 2016, in press.
3. Zavyalova E, Tagiltsev G, Reshetnikov R, Arutyunyan A, Kopylov A. Cation coordination alters the conformation of a thrombin-binding G-quadruplex DNA aptamer that affects inhibition of thrombin. *Nucleic Acids Ther*. 2016, in press.
4. Zavyalova E, Golovin A, Pavlova G, Kopylov A. G-quadruplex aptamers to human thrombin versus other direct thrombin inhibitors: the focus on mechanism of action and drug efficiency as anticoagulants. *Curr Med Chem*. 2016, in press.
5. Zavyalova E., Kopylov A. Exploring potential anticoagulant drug formulations using thrombin generation test. *Biochem Biophys Rep*. 2016, vol. 5, pp. 111-119.
6. Protopopova A.D., Zavyalova E.G., Barinov N.A., Kopylov A.M., Sergienko V.I., Klinov D.V. Visualization of fibrinogen C regions and their arrangement during fibrin network formation by high-resolution AFM. *J Thromb Haemost*. 2015; vol. 13, n. 4, pp. 570-579.
7. Zavyalova E. The Aptamers. Whether the Paradigm Works in Therapeutic Applications? *Drug Des Open Access*. 2015, vol. 4, pp. 2.
8. Zavyalova E., Kopylov A. Multiple inhibitory kinetics reveal an allosteric interplay among thrombin functional sites. *Thromb Res*. 2015, vol. 135, n. 1, pp. 212-216.
9. Zavyalova E., Kopylov A. How does association process affect fibrinogen hydrolysis by thrombin? *Biochimie*. 2014, vol. 107, pp. 216-222.
10. Zavyalova E., Samoylenkova N., Revishchin A., Golovin A., Pavlova G., Kopylov A. Evaluation of antithrombotic activity of thrombin DNA aptamers by a murine thrombosis model. *PLoS ONE*. 2014, vol. 5, n. 9, p. e107113.
11. Zavyalova E., Golovin A., Pavlova G., Kopylov A. Module-Activity Relationship of G-quadruplex Based DNA Aptamers for Human Thrombin. *Curr Med Chem*. 2013, vol. 20, n. 38, pp. 4836-4843.
12. Savchik E.Yu, Kalinina T.B., Drozd N.N., Makarov V.A., Zavyalova E.G., Lapsheva E.N., Mudrik N.N., Babij A.V., Pavlova G.V., Golovin A.V., Kopylov A.M. Aptamer RA36 Inhibits of Human, Rabbit, Rat Plasma Coagulation Activated with Thrombin or Snake Venom Coagulases. *Bull Exper Biol Med*. 2013, vol. 156, n. 1, pp. 44-48.

In: G-Quadruplex Structures, Formation and Role in Biology ISBN: 978-1-63485-512-9
Editor: Hannah Santos

Chapter 3

DOUBLE HELICES FORMING UNUSUAL BINDING MOTIFS: THE CASE OF THE NATURAL ALKALOIDS COPTISINE AND CHELERYTHRINE

***Marta Ferraroni*[1], *Carla Bazzicalupi*[1,*], *Francesco Papi*[1,2] *and Paola Gratteri*[2,*]**
[1]Department of Chemistry "Ugo Schiff", University of Firenze, Sesto Fiorentino, Firenze, Italy
[2]Department of NEUROFARBA – Pharmaceutical and Nutraceutical Section - Laboratory of Molecular Modeling Cheminformatics and QSAR, University of Firenze, Sesto Fiorentino, Firenze, Italy

ABSTRACT

Nucleic acids are highly polymorphic biomolecules: their significant structural variability is not only due to sugar-phosphate backbone flexibility, but even to the ability of purine bases to donate and/or to accept hydrogen-bonding by using different molecular sides. Together with the so-called G-quadruplex structures, abundant in many sequenced genomes, recent evidence of unusual binding motifs containing non-guanine-based quartets and other kinds of mismatches has been reported. Small molecules able to preferentially bind mismatching points, instead of the fully matched DNA stretches, are recognized to have potential as drug candidates or tools for chemical biology *in vivo*.

The two natural isoquinoline alkaloids chelerythrine and coptisine are supposed to be unable to interact through classic intercalative processes with ds-DNAs due to their elongated nature and were selected as probes of unusual binding motifs. The X-ray diffraction analyses of the CGTACG/chelerythrine, CGTACG/coptisine and CGATCG/coptisine adducts actually show chelerythrine and coptisine localized at the junction of adjacent DNA helices, in unusual intercalative binding sites containing G:G homo base mismatches, in which the guanines are hydrogen-bonded through their minor groove faces. In the CGTACG/chelerythrine and CGTACG/coptisine, the intercalation

* To whom correspondence should be addressed. Prof. Carla Bazzicalupi Tel: +390554573286; Fax: +390554574913; Email: carla.bazzicalupi@unifi.it; Prof. Paola Gratteri Tel: +390554573702; Fax: +390554573380; Email: paola.gratteri@unifi.it.

site contains six bases and is defined by two 3'-GpC dinucleotides from different chains and completed by two more guanine residues of the 5'-CpG. Also the CGATCG/coptisine complex evidences the same kind of G:G mismatch, in which two C:G:G:C platforms sandwich the alkaloid. In all the reported crystal structures, the overlapping surface between the ligands and the double stranded non-canonical binding sites, is significantly higher than what is expected for classic intercalative binding.

INTRODUCTION

Nucleic acids are highly polymorphic biomolecules whose 3D arrangement depends not only on the nucleotide sequence, but also on external factors such as the pH or the presence in solution of ions and other organic compounds [1, 2]. Several DNA secondary structures, having or not double helix conformation, have been recently recognized and described in literature [3]. Among them, the so-called quadruplex structures, typically formed in guanine-rich nucleic acids, are extensively studied and deeply structurally characterized as they are potentially given by sequences localized at non-random positions in functionally important genomic regions [4-7].

Recent studies report the existence of non-guanine-based quartets both in newly developed programmable nucleic acid biomaterials and in biologically relevant systems, such as the DNA fragments containing the Fragile X Syndrome triplet repeat, the human chromosome 19 targeted by the adeno-associated parvovirus, the $(CTAGGG)_n$ variant of the human telomeric sequence [8-11], as well as in a promoter sequence in the B-raf gene [8]. As a matter of fact, much experimental evidence for homo-quartets (i.e., quartets formed by the same base: A-, T-, C-, U-), mixed-quartets (formed by two different bases: G:C:G:C, G:T:G:T or A:T:A:T), and even higher order platforms formed by non-guanine residues flanking classic G-quartets, are now present in literature [12-14].

Two different aspects are at the basis of this significant structural variability: the well known sugar-phosphate backbone flexibility of the nucleic acid chains, and the ability of purine bases to donate and/or to accept hydrogen-bonding by using different molecular sides. In fact, adenines have H-bond donor and acceptor atoms on both the so-called Watson-Crick (WC) and Hoogsteen (HS) faces, while guanines can donate H-bonds only on the WC face, but can act as H-bond acceptor on their HS face (Figure 1a). Nowadays it is well established that in forming quadruplex structures, purines hydrogen bond each other through both the WC and HS faces (Figure 1b), as is the case of the G-quartets and non-guanines-quartets [14]. On the other hand, in canonical nucleic acid duplexes, evidence is cumulating for the existence of transient HS base pairs having comparable stability with respect to the WC base pairs (Figure 1c) [15]. Moreover, even the purines' minor groove face can be available for H-bonding (Figure 1d) and a number of interactions of such kind have been reported in literature [16-18].

All these non-classic interaction points can be seen as mismatches featuring a particular nucleic acid structure, even though they do not necessarily induce destabilization. Looking for small molecules able to preferentially bind mismatching points, while having much lower affinity to fully matched DNA stretches, is a well-established research area in view of their potential as drug candidates or tools for chemical biology *in vivo* [19].

Among all the compounds which are known to interact with non-canonical forms of DNA, isoquinoline alkaloids represent an interesting and varied family. They are natural products, with wide biological and pharmacological activities [20-22], including anticancer effects particularly in the case of the protoberberine group [23-25]. Indeed, they are positively charged and almost planar heterocyclic compounds, containing a fused, aromatic ring system, which enables π-stacking interactions with the nucleic acids (Scheme 1) [26-31].

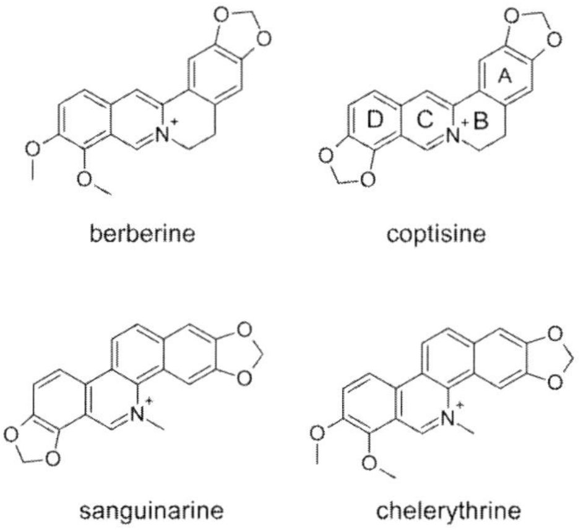

Scheme 1.

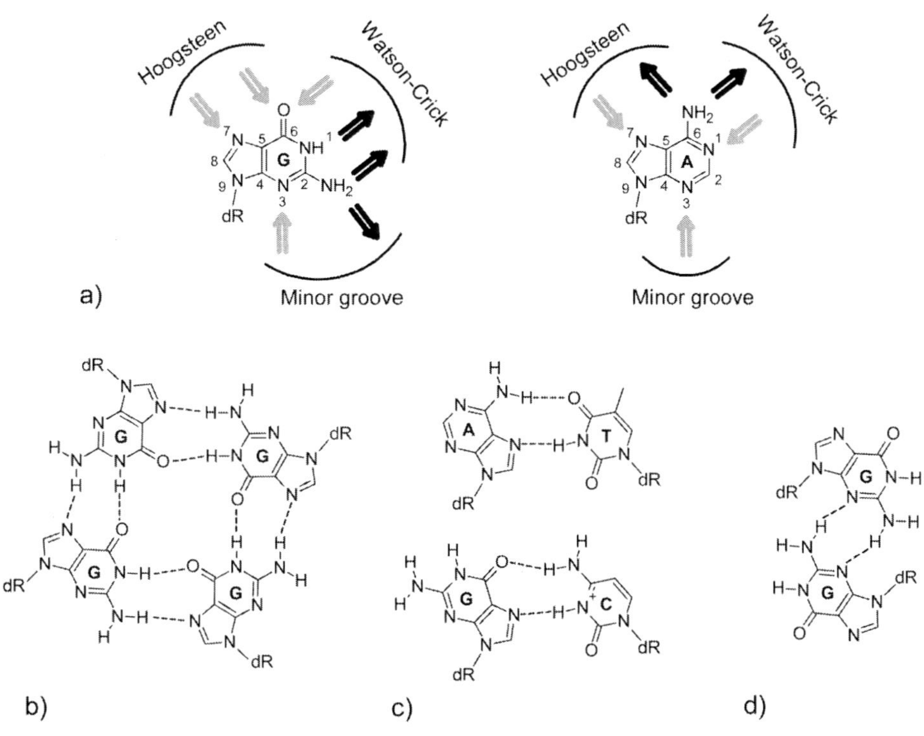

Figure 1. Watson-Crick, Hoogsteen and minor groove faces in purines: (a) non-canonical structural motifs involving purine residues in G-quartet, (b) transient Hoogsteen type base pairs, (c) and G:G homo base pairs involving minor groove faces (d).

During the last few years, we have been studying the interaction with DNA of some protoberberine and benzophenanthridine alkaloids, such as berberine (benzo[g]-1,3-benzodioxolo [5,6a]quinolizinium, 5,6-dihydro-9,10-dimethoxy) and sanguinarine (13-methyl-[1,3]benzodioxolo[5,6-*c*]-1,3-dioxolo[4,5-*i*]phenanthridinium), by using a pool of different techniques [32-34]. It clearly results from those studies that intercalative binding in classic ds-DNA is made difficult owing to the elongated nature of these molecules, and that the binding is achieved only upon a significant deformation of the DNA B-type helix [34]. On the other hand, the overall shape of the compounds is particularly well-suited to bind quadruplex platforms and two molecules were found to stack onto the guanine. The peculiar features of this alkaloid/DNA binding prompted us to expand our study and investigate the behavior of the isoquinoline compounds coptisine and chelerythrine (Scheme1), chemical analogues of berberine and sanguinarine, respectively, with the hope of crystallizing new alkaloid/DNA adducts showing unusual binding modes and/or structures.

Here, we report the X-ray crystal structures, confirmed by *in silico* simulation, of the adducts formed by the natural alkaloids coptisine and chelerythrine with the CGTACG DNA hexamer and by coptisine with the CGATCG sequence. The CGATCG/coptisine adduct is the first structure showing a ligand intercalated between two unusual C:G:G:C platforms where non-WC and non-HS contats are at work.

MATERIAL AND METHODS

Crystallization Experiments

Crystallization screenings were performed for chelerythrine and coptisine with the CGTACG, CGATCG, CGCGCG and GTTGCAAC. Crystals suitable for X-ray diffraction analysis were obtained for the CGTACG/chelerythrine, the CGTACG/coptisine and the CGATCG/coptisine adducts.

The DNA-drug complexes were crystallized at 296 K using the sitting drop vapor diffusion method. CGTACG/chelerythrine crystals were obtained from a solution containing 10% MPD, 40 mM sodium cacodylate pH = 6.5, 12 mM spermine tetra-HCl, 80 mM sodium chloride, 12 mM potassium chloride and 20 mM magnesium chloride. CGTACG/coptisine crystals were obtained from a solution containing 10% MPD, 40 mM sodium cacodylate pH = 6.0, 12 mM spermine tetra-HCl, 80 mM potassium chloride and 20 mM magnesium chloride. CGATCG/coptisine crystals were obtained from a solution containing 10% MPD, 40 mM sodium cacodylate pH = 6.0, 12 mM spermine tetra-HCl, 80 mM potassium chloride and 20 mM barium chloride. In all cases drops were equilibrated against a 30% MPD reservoir.

X-ray Diffraction Analyses

CGTACG/chelerythrine complex: a = 30.22 Å, c = 120.33 Å, trigonal, space group $P3_221$, Resolution = 2.1 Å, Rsym = 0.089, Completeness = 99.0%. CGTACG/coptisine complex: a = 30.28 Å, c = 118.33 Å, trigonal, space group $P3_221$, Resolution = 2.44 Å, Rsym = 0.108, Completeness = 99.5%. CGATCG/coptisine: a = 26.65 Å, c = 77.1 Å, hexagonal, space group *P6₅22*. Resolution = 2.70 Å, Rsym = 0.097, Completeness = 99.1%. A summary of data collections and atomic model refinement statistics is reported in Table 1.

Data collection on crystals of the DNA-drugs complexes were performed at 100 K, using as cryoprotectant the mother liquor solution with an increased concentration of MPD (CGATCG/coptisine complex 20% MPD, CGTACG/chelerythrine and CGTACG/ coptisine 30% MPD). Data for both coptisine adducts were collected on an Oxford Diffraction instrument, equipped with a sealed tube Enhance Ultra (Cu) and a Onyx CCD detector, while for the chelerythrine complex data were collected using synchrotron radiation at Elettra (Trieste, Italy) with a wavelength of 1.00 Å. Data were integrated and scaled using the program XDS [35-36]. Data of the two CGTACG complexes are hemihedrally twinned. The possibility of twinning of the data was suggested by the analysis of the intensity distribution performed using the phenix.xtriage module of the PHENIX program [37].

The structures were solved by the Molecular Replacement technique using the program Molrep [38]. The coordinates of the hexamer duplex d(CGTACG) complexed with berberine (PDB code 3NP6) [34], was used as a search model for both the CGTACG complexes, while the structure with PDB code 3FT6 was used for the CGATCG/coptisine complex [39]. In all cases, atoms belonging to ligands and solvent molecules in the model structures were preliminary deleted. The three models were refined with the program Refmac5 [40] from the CCP4 program suite [41]. The twin option was applied in the refinement for both the CGTACG complexes. Fo-Fc electron density maps revealed a clear density for the chelerythrine and coptisine molecule that were introduced and refined at unitary occupancy.

Table 1. Summary of Data Collection and Atomic Model Refinement Statistics (values in parentheses are for the highest resolution shell).

	CGTACG/chelerythrine	CGTACG/coptisine	CGATCG/coptisine
Data Collection			
Space group	$P3_221$	$P3_221$	$P6_522$
Unit cell (Å)	a=30.22, c=120.33	a=30.28, c=118.33	a = 26.65, c = 77.1
DNA content in asymmetric unit	2 d(CGTACG)$_2$	2 d(CGTACG)$_2$	dCGATCG
Wavelength (Å)	1.000	1.542	1.542
Limiting resolution (Å)	2.10 (2.23-2.10)	2.44 (2.59-2.44)	2.70 (2.87-2.70)
Unique reflections	4064	2642	559
R_{sym}	0.089 (0.226)	0.108 (0.611)	0.097 (3.64)
Multiplicity	5.3 (5.6)	13.7 (13.5)	11.1 (11.6)
Completeness overall (%)	99.0 (97.8)	99.5 (97.3)	99.1 (100.0)
CC(1/2)	-	-	99.9 (84.6)
$<I/\sigma (I)>$	13.39 (5.34)	17.48 (4.01)	16.2 (0.98)
Refinement			
Resolution range (Å)	20.0-2.1	30.0-2.44	14.8- 2.70
Unique reflections	3911/154	2387/254	558
Rfactor (%)	22.0	26.5	26.7
Rfree (%)	31.4	30.7	-
Non-hydrogen atoms	488	520	144
r.m.s.d. bonds(Å)			0.0127
r.m.s.d. angles (°)			2.922

A spheric density was also present in the Fo-Fc electron density maps of the CGTACG/coptisine and chelerythrine complexes that could be interpreted as a metal ion. Magnesium and sodium cations were present in the crystallization conditions, nevertheless after the introduction of these metal ions a residual positive electron density in the Fo-Fc map was present at that position. The electron density in the Fo-Fc map was then interpreted, taking into account also the distances with neighbouring atoms, as a calcium ion, that was present in the MPD used for the crystallization as impurity. For the CGATCG/coptisine structure, no R free was calculated as a 5–10% data set would not constitute a statistically robust cross-validation set [42]. Manual rebuilding of each model was performed using the program Coot [43].

At the end of refinement, agreement factors were R = 0.265, 0.220 and Rfree = 0.3074, 0.314 for CGTACG/coptisine and CGTACG/chelerythrine, respectively, and R = 0.267 for CGATCG/coptisine. Final coordinates and structure factors have been deposited with the Protein Data Bank (PDB accession number 4D9X, 4D9Y, 4L5K).

QM and QM/MM Calculations

Three conformers featuring 0°, 11° and 30°, respectively, for the dihedral angle formed by the phenyl and the quinoline units of the coptisine molecule, were minimized in the gas phase at the DFT/B3LYP level of theory [44-46] by using the 6-311G**+ basis set and the Jaguar software [47].

The QM/MM minimization was carried out on the dCGATC/coptisine adduct in 2:1 ds-DNA:ligand molar ratio by using the Qsite software [48,49]. The starting coordinates were obtained from the crystallographic data, having applied the proper symmetry relationships to the DNA polymer and having deleted the terminal C or G residue of each hexamer of the resulting double helices. The four CpG steps defining the quadruplex and the bound coptisine were used to define the QM region and treated at the DFT/B3LYP level of theory with the LACVP* basis set [44-46, 50]. The remaining DNA fragment was treated at the MM level using Impact [51], OPLS2005 force-field, and the implicit solvent effect ($\varepsilon = 4R$) treatment. The Watson-Crick H-bond contacts in the MM moiety of DNA were restrained to the starting values (default value used for the force constant).

RESULTS

Crystallization screenings were performed for both chelerythrine and coptisine in complex with different double helix DNA sequences in order to investigate the structural features of their adducts. The crystal structures of the CGTACG/chelerythrine, CGTACG/coptisine and CGATCG/coptisine complexes were obtained by X-ray diffraction analysis at 2.1, 2.44 and 2.70 Å resolution, respectively.

Both CGTACG complexes crystallize in the $P3_221$ space group, and show similar crystal packings: the asymmetric unit contains four DNA strands, a drug molecule, few cocrystallized water molecules, and a Ca^{2+} ion. The DNA helices are arranged in continuous columns, growing in the (100), (010) or (1-10) symmetry related directions. These columns are built up by stacked B-like "two-molecules" units, formed by the central four base pairs of each DNA helix (sequence GTAC), which give rise to a continuous array of 8 stacked base pairs (Figure 2a,b).

The terminal 3'-G12 bases at the central junction of the "two-molecule" unit, lie in the minor groove of the adjacent duplex and are well localized in the electron density map, while the 5'-Cs are disordered and were not introduced in the models.

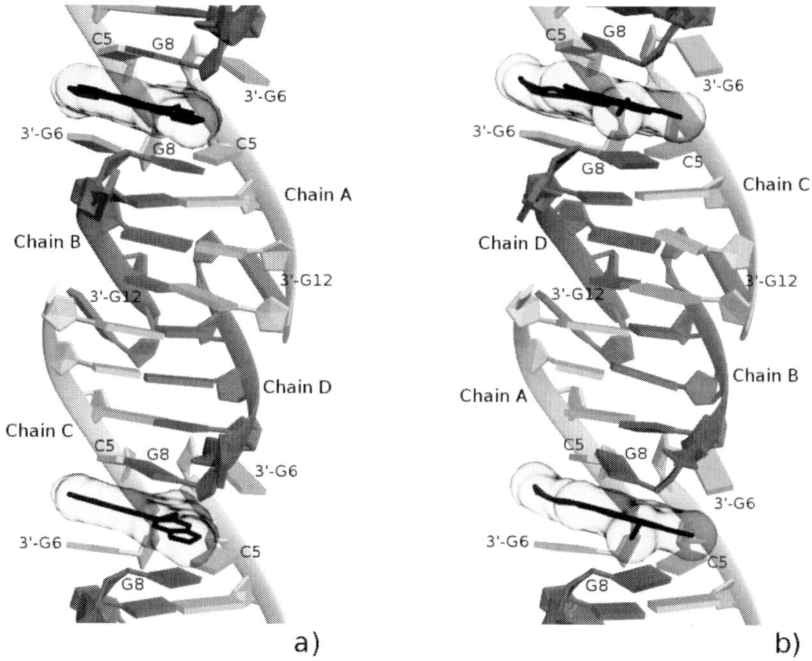

Figure 2. DNA helices forming the "two molecules" units in the CGTACG/alkaloid adducts: (a) coptisine, (b) chelerythrine.

Both chelerythrine and coptisine are found at the interface that joins a couple of "two-molecules" DNA units, located in a six bases non-classic intercalation site, formed by four bases from the two 3'-GpC dinucleotides of chains A and C (A-C5/G6 and C-C5/G6 - Figures 3 and 4), and completed by two more guanine residues of the 5'-CpG end from chains B and D (B-G8 and D-G8). The cytosine groups are flipped out from the duplex and stabilize the crystal packing interacting via H-bonding with symmetry related DNA columns.

Noteworthy, for both compounds the binding site contains the A-G6 and the D-G8 (as well as the C-G6 and B-G8) connected via H-bonds involving the guanines' minor groove faces (H-bond distances ranging from 2.7 to 3.3 Å). The binding of the natural alkaloids is achieved by π-π interactions with the G6:G8 homo base pair given by chains B and C, and with the G8 from chain D (Figures 3 and 4). The π-π stacking distaces between the aromatic rings of coptisine and chelerythrine and the G:G pairs fall in the range 3.3-3.5 Å for both coptisine and chelerythrine. Moreover, chelerythrine is better able than coptisine to interact with a fourth residue, namely cytosine 5 of chain A. In both cases, however, the presence of a positive charged nitrogen on the drug skeleton contributes to the overall stability of the complex. No H-bonds were found between the DNA atoms and the ligands oxygens.

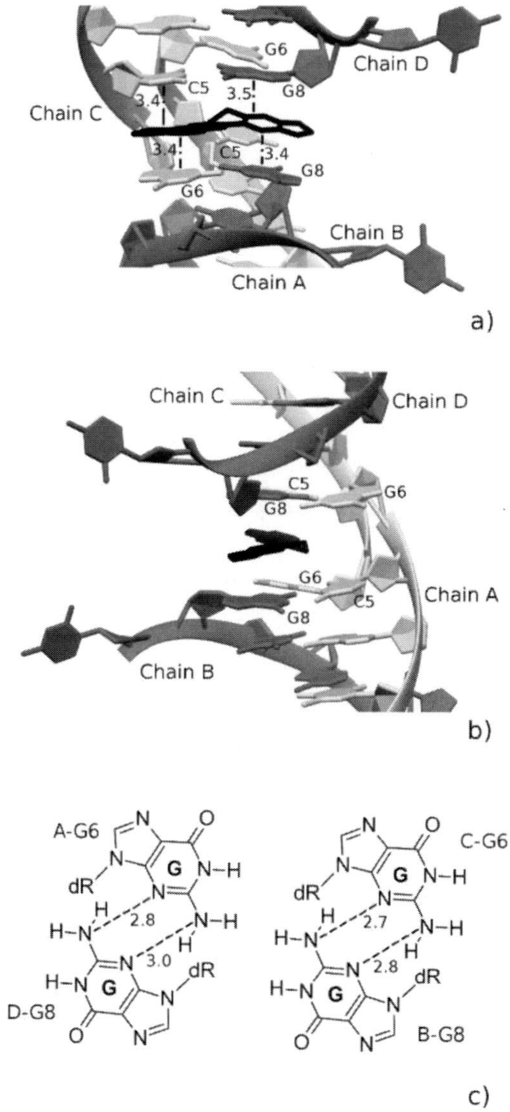

Figure 3. (a) Front and (b) lateral views of the intercalation site at the interface of couples of "two-molecules" DNA units, and (c) details of the G:G homo base pairs for the CGTACG/coptisine adduct.

As far as the CGATCG/coptisine complex is concerned, the experimental data collected have relatively low resolution (2.7 Å), despite several crystal optimization trials and many different crystals tested before data collection. Thus, in order to verify the accuracy of the refined ligand coordinates and the reliability of the measured contacts, a mixed quantum mechanics/molecular mechanics (QM/MM) minimization was carried out and the difference in the experimental (X-ray) vs calculated (QM/MM) RMSD values was measured. Very good agreement was obtained both for the overall complex alignment (0.9Å) and for the binding site (0.07 Å), thus indicating the right positioning of ligand suggested by X-ray diffraction analysis.

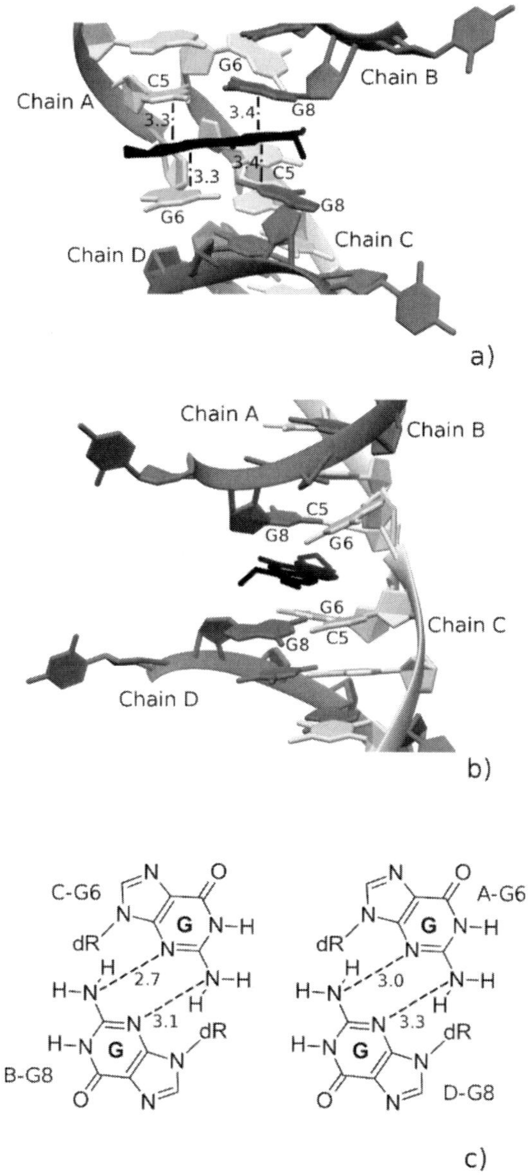

Figure 4. (a) Front and (b) lateral views of the intercalation site at the interface of couples of "two-molecules" DNA units, and (c) details of the G:G homo base pairs for the CGTACG/chelerythrine adduct.

The asymmetric unit contains one single CGATCG strand and one ligand molecule, together with a K^+ ion and few localized water molecules. A crystallographic dyad axis relates two self-complementary hexamers, which run in antiparallel directions, and define two symmetry equivalent binding sites at the CpG steps (see Figure 5a). The ligand is intercalated in each binding site, orthogonal both to the plane formed by the nucleic bases and to the long axis of the DNA helix, which adopts a distorted B-type conformation. Once again the perpendicular intercalative binding mode of the coptisine in this complex is likely due to the overall length of this compound, which prevents the parallel intercalation in the double helix, as already found for the CGTACG/coptisine adduct.

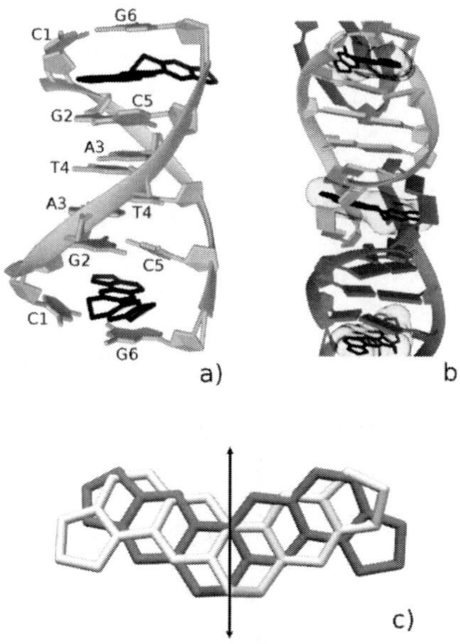

Figure 5. (a) Coptisine molecules intercalated at the CpG steps of the d(CGATCG)$_2$ unit; (b) two symmetry related d(CGATCG)$_2$ units and the coptisine molecule stacked in the binding site (only one of the pseudo-symmetric coptisine per binding sites shown for clarity); (c) C2 crystallographic axis relating the two pseudo-symmetric coptisine molecules in the binding site.

The most interesting feature of the CGATCG/coptisine crystal structure is the presence of a crystallographic 2-fold axis bisecting the ligand molecule along its short axis (Figure 5b,c). This symmetry operator gives rise to two main effects on the local structure: i) the overall binding site is built up by the CpG steps belonging to two adjacent double helices (light and dark gray coloured in Figure 5b and Figure 6a), and the bound ligand is stacked between the two resulting C:G:G:C platforms, where the G2 and G6 guanine residues interact through their minor groove faces (Figure 6b), and ii) two disordered pseudo-symmetric coptisines are virtually sharing the same position (Figure 5c), i.e., in each binding site, coptisine adopts one of the two possible symmetry related orientations which are present at the same probability level in the crystal packing. As a result, the phenyl and quinoline moieties of the alkaloid can interact with either helix, depending on the two possible alternative orientations for the ligand. In both cases, the ligand is placed about 3.5 Å away from the G2 and G6 guanines. Noteworthy, the two C:G base pairs forming a C:G:G:C platform are not coplanar and give a dihedral angle of 46°. The intercalated ligand is bent as well, with a 27° dihedral angle between the phenyl ring A and the quinoline moiety (Scheme 1).

An O···π interaction is established by each benzo[1,3]dioxole and a cytosine per double helix (O···cytosine ring centroid distance 2.9 Å, Figure 6a), strengthening both the ligand/DNA and the duplex/duplex interaction. It is to be underlined that the presence of benzo[1,3]dioxole groups in alkaloid derivatives, was reported to promote the formation of quadruplex structures from human telomeric DNA [52]. However, no similar O···π interaction was recognized in the crystal structure of the berberine/human Tel-23 complex [32]. Similarly to berberine, the binding between coptisine and C:G:G:C platforms is mainly due to van der Waals contacts and to electrostatic interactions involving the positive charged heteroaromatic nitrogen in the alkaloid skeleton. The strength of these interactions depends more on the extension of the overlapped aromatic surfaces than on the orientation assumed by the ligand.

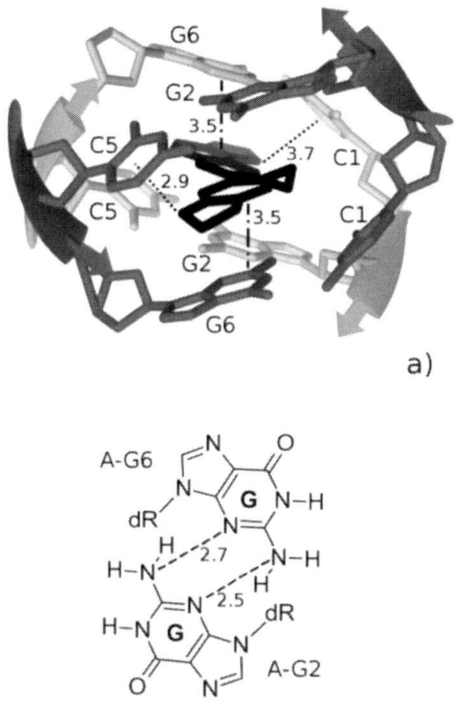

Figure 6. Binding site for the coptisine molecule and C:G:G:C platforms: (a) coptisine localization, π-π and O···π interactions (only one of the two pseudo-symmetric orientations is shown for coptisine), (b) details of the G:G homo base pairs.

Discussion

Despite the difference in the investigated DNA sequences, all the DNA/ligand complexes feature binding sites containing non-classic base pairs, where the interacting guanines are reciprocally H-bonded through their minor groove faces (Figure 3c, 4c and 6b).

Guanines interacting with adenines through the minor groove faces (G:A mismatch), have been found to give very stable structures when present in tandem repeat [17,18], while an isolated single minor groove G:G interaction has been reported only by Takénaka et al. for the crystal structure of the d(GCGAAAGCT) polymer [16]. In this structure a G:G homo base pair, involving the minor groove faces of guanines belonging to CGA stretches, builds up a parallel-stranded duplex with an adjacent A:A homo base pair.

Interestingly, in our CGATCG/coptisine complex the G:G pairs defining the binding sites are present in tandem, and give H-bond contacts ranging from 2.5 to 2.7 Å. A similar quadruplex platform, based on G:G homo base pairs formed through minor groove faces, was recently found in the crystal structure obtained for a DNA sequence containing the CGAA fragment [53]. In that structure two C:G:G:C platforms, built up by adjacent double helices, are stacked on each other, resulting in an overall quadruplex resembling that one found in the CGATCG/coptisine complex. Actually, in the quadruplex core of both structures, it is possible to recognize an antiparallel (up-down-up-down)-stranded arrangement (Figure 6a). Moreover, in the structure reported in ref. 53, the C:G base pairs in each platform are not coplanar and form 30° dihedral angles, similar to that found in our DNA/coptisine. Thus, except for the presence of the intercalated ligand, the highest value of this dihedral angle in the coptisine/DNA complex (46°) represents the most significant difference between the two quadruplex structures, and it is most likely due to the remarkable distortion of the base pairs near the intercalator.

With reference to the CGTACG/ligand complexes, the hexamer assumes conformations very similar to the CGTACG/berberine and the CGTACG/sanguinarine crystal structures [34]. The four complexes feature very similar binding sites, as well as arrangement of the terminal bases. As a matter of fact, the four natural alkaloids in complex with the CGTACG sequence crystallize in the same space group ($P3_221$) and they can be considered almost isostructural systems, with strong analogies in their overall crystal packings.

These structures could demonstrate that the DNA/ligand interactions observed for the four alkaloids with the CGTACG sequence, only depend on the crystal packing forces. However, binding modes resembling those experimentally found for the investigated alkaloids, were previously reported for the CGTACG hexamer in complex with some anthraquinone and acridine derivatives [54]. All these structures are characterized by a different space group and different crystal packings as compared to our alkaloids complexes. This gives strength to the evidence that the positions of the ligands in the structures under discussion are genuinely representative of their actual binding modes.

Moreover, while chelerythrine is found to interact with DNA in its planar and charged iminium form, strictly resembling its structural analogue sanguinarine (Scheme 1 and Figure 7a,b), coptisine is not planar and features an 18° dihedral angle between its phenyl and quinoline moieties (Figure 7c,d). In this aspect, CGTACG/coptisine adduct differs from the analogous CGTACG/berberine complex, where berberine is almost completely planar, clearly indicating peculiar stereochemical features for these two alkaloids, in spite of the similarity of their skeletons (Scheme 1).

Interestingly, an even more bent conformation is assumed by coptisine in its adduct with the CGATCG hexamer (Figure 7e,f). In this adduct, coptisine features a 27° dihedral angle between the phenyl ring A and the quinoline moiety (Scheme 1). This bent conformation is surprising, considering that a CSD (Cambridge Structural Database) analysis evidences that analogous alkaloids, such as berberine or palmatine, usually assume almost planar conformations in the solid state, their dihedral angles ranging from 11° to 17° [55-59]. Nevertheless, extending the CSD analysis to other compounds containing the A-B-C fused ring system (Scheme1), two more conformations are found, featuring dihedral angular values near 0° or 30° [60,61]. Actually, QM calculations carried out for coptisine in 0°, 11° and 30° conformations, highlighted a very low energy gap ($\Delta E = 0.271$ kcal/mol), the minimum value being associated with the 30° conformer. Thus it is possible that, in the protoberberine family, the planar structure usually found by diffraction analysis is stabilized by environment interactions, but also that even the bent conformations can be accessible. It is worth noting that the C:G:G:C platforms are significantly bent also in the absence of the intercalating ligand [53], and this makes the 27° bent conformation of the coptisine molecule particularly suited to interact with the C:G:G:C quartet.

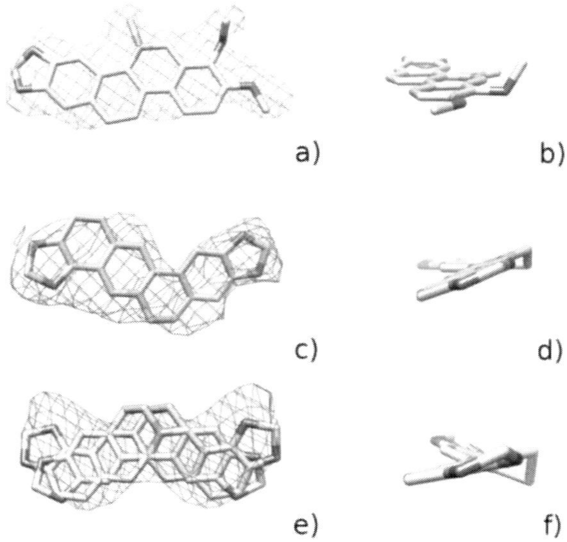

Figure 7. (a) Top and (b) lateral view of the skeleton for chelerythrine (OMIT electron density map contoured at 1.5σ level); (c) top and (d) lateral view of the skeleton for coptisine (CGTACG structure - OMIT electron density map contoured at 1.5σ level); (e) top and (f) lateral view of the skeleton for coptisine (CGATCG structure - OMIT electron density map contoured at 1.5σ level).

CONCLUSION

X-ray diffraction analyses and QM/MM computations were carried out to investigate unusual ligand/ds-DNA binding modes. The two natural isoquinoline alkaloids chelerythrine and coptisine were selected for the investigation, as structural analogs of berberine and sanguinarine, which are not able to interact by classic intercalative processes with ds-DNAs, due to their elongated nature. The overall shape of the compounds seems to be particularly well-suited to bind the extended platforms often formed when the non WC face of nucleic residues are involved in the pairing processes. In the crystal structures solved for the CGTACG/chelerythrine, the CGTACG/coptisine and the CGATCG/coptisine adducts, the ligands are actually positioned at the junction of adjacent DNA helices, resulting in unusual intercalative binding modes.

In the CGTACG/chelerythrine and CGTACG/coptisine, the drug molecules are found in a six bases intercalation site, given by two 3'-GpC dinucleotides from different chains and completed by two more guanine residues of the 5'-CpG, forming G:G homo base pairs, where the two guanines are hydrogen-bonded through their minor groove faces. The same kind of mismatch is present in the CGATCG/coptisine structure. This interaction determines the formation of an unusual quadruplex where the two resulting C:G:G:C platforms sandwich the alkaloid. To our knowledge, the CGATCG/coptisine crystal structure is the first example of ligand intercalated in this kind of quadruplex, joining the two involved symmetry related ds-DNA molecules. In all the structures here reported, the available ligand/DNA contact surface is significantly higher as compared to that one available in classic intercalative binding mode. The main interactions are of electrostatic, van der Waals and $\pi \cdots \pi$-stacking types. Additional $O \cdots \pi$ interactions are established by coptisine with the CGATCG hexamer. Furthermore, in the latter crystal structure, coptisine adopts a bent conformation, which leads to an optimal conformational match with the C:G:G:C quadruplex.

In conclusion, the elongated, positively charged and almost planar fused ring system featuring the natural isoquinoline alkaloids make them particularly prone to giving π-stacking interactions with mismatched nucleic acid systems featuring non WC contacts. Thus the investigated molecules can be considered good lead compounds for the design of new potential drug candidates or interesting tools for chemical biology *in vivo*.

FUNDING

Ente Cassa di Risparmio di Firenze, Italy, is gratefully acknowledged for a grant to F.P. (2014.0309).

REFERENCES

[1] Bansal, M. (2003). DNA structure: Revisiting the Watson-Crick double helix. *Current Science*, **85(11)**, 1556-1563.

[2] Mills, M., Lacroix, L., Arimondo, P. B., Leroy, J. L., François, J. C., Klump, H. and Mergny, J. L. (2002). Unusual DNA conformations: implications for telomeres. *Current Medicinal Chemistry - Anticancer Agents*, **2(5)**, 627-644.

[3] Zhao, J., Bacolla, A., Wang, G. and Vasquez, K. M. (2010). Non-B DNA structure-induced genetic instability and evolution. *Cellular and Molecular Life Science*, **67(1)**, 43-62, and ref. therein.

[4] Rhodes, D. and Lipps, H. J. (2015). G-quadruplexes and their regulatory roles in biology. *Nucleic Acids Research*, **43(18)**, 8627-8637, and ref. therein.

[5] Burge, S., Parkinson, G.N., Hazel, P., Todd, A.K. and Neidle S. (2006). Quadruplex DNA: sequence, topology and structure. *Nucleic Acids Research*, **34(19)**, 5402-5415.

[6] Lech, C.J., Heddi, B. and Phan, A.T. (2013). Guanine base stacking in G-quadruplex nucleic acids. *Nucleic Acids Research*, **41(3)**, 2034-2046.

[7] Collie, G. W., Campbell, N. H. and S. Neidle. (2015). Loop flexibility in human telomeric quadruplex small-molecule complexes. *Nucleic Acids Res.*, **43(10)**, 4785-4799.

[8] Kettani, A., Kumar, R. A. and Patel, D. J. (1995). Solution structure of a DNA quadruplex containing the fragile X syndrome triplet repeat. *Journal of Molecular Biology*, **254(4)**, 638-656.

[9] Kettani, A., Bouaziz, S., Gorin, A., Zhao, H., Jones, R. A. and Patel, D. J. (1998). Solution structure of a Na cation stabilized DNA quadruplex containing G·G·G·G and G·C·G·C tetrads formed by G-G-G-C repeats observed in adeno-associated viral DNA. *Journal of Molecular Biology*, **282(3)**, 619-636.

[10] Lim, K. W., Alberti, P., Guédin, A., Lacroix, L., Riou, J. F., Royle N. J., Mergny, J. L. and Phan, A. T. (2009). Sequence variant (CTAGGG)n in the human telomere favors a G-quadruplex structure containing a G·C·G·C tetrad. *Nucleic Acids Research*, **37(18)**, 6239-6248.

[11] Leontis, N.B., Stombaugh, J. and Westhof E. (2002). The non-Watson-Crick base pairs and their associated isostericity matrices. *Nucleic Acids Research*, **30(16)**, 3497-3531.

[12] Wei, D., Todd, A. K., Zloh, M., Gunaratnam, M., Parkinson, G. N., Neidle, S. (2013). Crystal structure of a promoter sequence in the *B-raf* gene reveals an intertwined dimer quadruplex. *Journal of the American Chemical Society*, **135(51)**, 19319-19329.

[13] Borbone, N., Amato, J., Oliviero, G., D'Atri, V., Gabelica, V., Pauw, D. E., Piccialli, G. and Mayol, L. (2011). d(CGGTGGT) forms an octameric parallel G-quadruplex via stacking of unusual G(:C):G(:C):G(:C):G(:C) octads. *Nucleic Acids Research*, **39(17)**, 7848-7857.

[14] Tran, P. L., De Cian, A., Gros, J., Moriyama, R. and Mergny, J. L. (2013). Tetramolecular quadruplex stability and assembly. *Topics in Current Chemistry*, **330**, 243-273, and references therein.

[15] Nikolova, E. N., Zhou, H., Gottardo, F. L., Alvey, H. S., Kimsey, I. J. and Al-Hashimi, H. M. (2013). A historical account of Hoogsteen base-pairs in duplex DNA. *Biopolymers*, **99(12)**, 955-968.

[16] Sunami, T., Kondo, J., Kobuna, T., Hirao, I., Watanabe, K., Miura, K. and Takénaka, A. (2002). Crystal structure of d(GCGAAAGCT) containing a parallel-stranded duplex with homo base pairs and an anti-parallel duplex with Watson-Crick base pairs. *Nucleic Acids Research*, **30(23)**, 5253-5260.

[17] Chou, S. H., Chin, K. H. and Wang, A. H. J. (2003). Unusual DNA duplex and hairpin motifs. *Nucleic Acids Research*, **31(10)**, 2461-2474.

[18] Chou, S. H. (2005). Stable and unique duplex motifs. *Nucleic Acids Symposium Series*, **49**, 11-12.

[19] Granzhan, A., Kotera, N. and Teulade-Fichou, M. P. (2014). Finding needles in a basestack: recognition of mismatched base pairs in DNA by small molecules. *Chemical Society Reviews*, **43**, 3630-3665.

[20] Bhadra, K. and Kumar, G. S. (2011). Therapeutic potential of nucleic acid-binding isoquinoline alkaloids: binding aspects and implications for drug design. *Medicinal Research Reviews*, **31(6)**, 821-862.

[21] Issat, T., Jakóbisiak, M. and Golab, J. (2006). Berberine, a natural cholesterol reducing product, exerts antitumor cytostatic/cytotoxic effects independently from the mevalonate pathway. *Oncology Reports*, **16(6)**, 1273-1276.

[22] Kuo, H. P., Chuang, T. C., Tsai, S. C., Tseng, H. H., Hsu, S. C., Chen, Y. C., Kuo, C. L., Kuo, Y. H., Liu, J. Y. and Kao, M. C. (2012). Berberine, an isoquinoline alkaloid, inhibits the metastatic potential of breast cancer cells via Akt pathway modulation. *Journal of Agricultural and Food Chemistry*, **60(38)**, 9649-9658.

[23] Basu, A., Jaisankar, P. and Kumar, G. S. (2012). Synthesis of novel 9-O-N-aryl/aryl-alkyl amino carbonyl methyl substituted berberine analogs and evaluation of DNA binding aspects. *Bioorganic and Medicinal Chemistry*, **20(8)**, 2498-2505.

[24] Goto, H., Kariya, R., Shimamoto, M., Kudo, E., Taura, M., Katano, H. and Okada, S. (2012). Antitumor effect of berberine against primary effusion lymphoma via inhibition of NF-κB pathway. *Cancer Science*, **103(4)**, 775-781.

[25] Wang, J., Liu, Q. and Yang, Q. (2012). Radiosensitization effects of berberine on human breast cancer cells. *International Journal of Molecular Medicine*, **30(5)**, 1166-1172.

[26] Arora, A., Balasubramanian, C., Kumar, N., Agrawal, S., Ojha, R. P. and Maiti, S. (2008). Binding of berberine to human telomeric quadruplex – spectroscopic, calorimetric and molecular modeling studies. *FEBS Journal*, **275(15)**, 3971-3983.

[27] Bhadra, K. and Kumar, G. S. (2011). Interaction of berberine, palmatine, coralyne, and sanguinarine to quadruplex DNA: a comparative spectroscopic and calorimetric study. *Biochimica et Biophysica Acta*, **1810(4)**, 485-496.

[28] Li, J., Li, B., Wu, Y., Shuang, S., Dong, C. and Choi, M. M. (2012). Luminescence and binding properties of two isoquinoline alkaloids chelerythrine and sanguinarine with ctDNA. *Spectrochimica Acta Part A - Molecular and Biomolecular Spectroscopy*, **95**, 80-85.

[29] Li, J., Shi, W., Han, X., Dong, C. and Choi, M. M. (2012). Luminescence Studies on Interaction of an Isoquinoline Alkaloid with Calf Thymus DNA in Aqueous Medium and on Solid Substrate. *Chemical Research in Chinese Universities*, **28(4)**, 614-619.

[30] Ma, L., Song, F., Liu, Z. and Liu, S. (2013). Study on noncovalent complexes of alkaloids with DNA duplex using electrospray ionization mass spectrometry. *Rapid Communications in Mass Spectrometry*, **27(1)**, 51-58.

[31] Sinha, R., Saha, I. and Kumar, G. S. (2011). Protoberberine alkaloids berberine, palmatine, and coralyne binding to Poly(dT) - (Poly(dA) - Poly(dT)) triplex: comparative structural aspects and energetics profiles of the interaction. *Chemistry and Biodiversity*, **8(8)**, 1512-1528.

[32] Bazzicalupi, C., Ferraroni, M., Bilia, A. R., Scheggi, F. and Gratteri, P. (2013). The crystal structure of human telomeric DNA complexed with berberine: an interesting case of stacked ligand to G-tetrad ratio higher than 1:1. *Nucleic Acids Research*, **41(1)**, 632-638.

[33] Bessi, I., Bazzicalupi, C., Richter, C., Jonker, H. R., Saxena, K., Sissi, C., Chioccioli, M., Bianco, S., Bilia, A. R., Schwalbe, H. and Gratteri, P. (2012). Spectroscopic, molecular modeling, and NMR-spectroscopic investigation of the binding mode of the natural alkaloids berberine and sanguinarine to human telomeric G-quadruplex DNA. *ACS Chemical Biology*, **7(6)**, 1109-1119.

[34] Ferraroni, M., Bazzicalupi, C., Bilia, A. R. and Gratteri, P. (2011). X-Ray diffraction analyses of the natural isoquinoline alkaloids Berberine and Sanguinarine complexed with double helix DNA d(CGTACG). *Chemical Communications*, **47**, 4917-4919.

[35] Kabsch, W. (1993). Automatic processing of rotation diffraction data from crystals of initially unknown symmetry and cell constants. *Journal of Applied Crystallography*, **26**, 795-800.

[36] Kabsch, W. (2010). Integration, scaling, space-group assignment and post-refinement. *Acta Crystallographica Section D - Biological Crystallography*, **66**, 133-144.

[37] Adams, P. D., Afonine, P. V., Bunkoczi, G., Chen, V. B., Davis, I. W., Echols, N., Headd, J. J., Hung, L. W., Kapral, G. J., Grosse-Kunstleve, R. W., McCoy, A. J., Moriarty, N. W., Oeffner, R., Read, R. J., Richardson, D. C., Richardson, J. S., Terwilliger, T. C. and Zwart, P. H. (2010). PHENIX: a comprehensive Python-based system for macromolecular structure solution. *Acta Crystallographica Section D - Biological Crystallography*, **66**, 213-221.

[38] Vagin, A. and Teplyakov, A. (1997). MOLREP: an automated program for molecular replacement. *Journal of Applied Crystallography*, **30**, 1022-1025.

[39] Maehigashi, T., Persil, O., Hud, N. V. and Williams, L. D. PDB code: 3FT6.

[40] Murshudov, G. N., Vagin, A. A. and Dodson, E. J. (1997). Refinement of macromolecular structures by the maximum-likelihood method. *Acta Crystallographica Section D - Biological Crystallography*, **53**, 240-255.

[41] Winn, M. D., Ballard, C. C., Cowtan, K. D., Dodson, E. J., Emsley, P., Evans, P. R., Keegan, R. M., Krissinel, E. B., Leslie, A. G. W., McCoy, A., McNicholas, S. J., Murshudov, G. N., Pannu, N. S., Potterton, E. A., Powell, H. R., Read, R. J., Vagin, A. and Wilson, K. S. (2011). Overview of the CCP4 suite and current developments. *Acta Crystallographica Section D - Biological Crystallography*, **67**, 235-242.

[42] Kleywegt, G. J. and Brünger, A. T. (1996). Checking your imagination: applications of the free R value. *Structure*, **4(8)**, 897-904.

[43] Emsley, P., Lohkamp, B., Scott, W. G. and Cowtan, K. (2010). Features and development of Coot. *Acta Crystallographica Section D - Biological Crystallography*, **66**, 486-501.

[44] Becke, A. D. (1993). A new mixing of Hartree-Fock and local density-functional theories. *Journal of Chemical Physics*, **98**, 1372-1377.

[45] Becke, A. D. (1993). Density-functional thermochemistry. III. The role of exact exchange. *Journal of Chemical Physics*, **98**, 5648-5652.

[46] Miehlich, B., Savin, A., Stoll, H. and Preuss, H. (1989). Results obtained with the correlation energy density functionals of becke and Lee, Yang and Parr. *Chemical Physics Letters*, **157(3)**, 200-206.

[47] Bochevarov, A. D., Harder, E., Hughes, T. F., Greenwood, J. R., Braden, D. A., Philipp, D. M., Rinaldo, D., Halls, M. D., Zhang, J. and Friesner, R. A. (2013). Jaguar: a high-performance quantum chemistry software program with strengths in life and materials sciences. *International Journal of Quantum Chemistry*, **113**, 2110-2142.

[48] Murphy, R. B., Philipp, D. M. and Friesner, R. A. (2000). A mixed quantum mechanics/molecular mechanics (QM/MM) method for large-scale modeling of chemistry in protein environments. *Journal of Computational Chemistry*, **21(16)**, 1442-1457.

[49] QSite. (2011) Suite 2011: QSite, version 5.8, Schrödinger, LLC, New York, NY, 2011.

[50] Hay, P. J. and Wadt, W. R. (1985). Ab initio effective core potentials for molecular calculations. Potentials for K to Au including the outermost core orbitals. *Journal of Chemical Physics*, **82**, 299-310.

[51] Banks, J. L., Beard, H. S., Cao, Y., Cho, A. E., Damm, W., Farid, R., Felts, A. K., Halgren, T. A., Mainz, D. T., Maple, J. R., Murphy, R., Philipp, D. M., Repasky, M. P., Zhang, L. Y., Berne, B. J., Friesner, R. A., Gallicchio, E. and Levy R. M. (2005). Integrated Modeling Program, Applied Chemical Theory (IMPACT). *Journal of Computational Chemistry*, **26(16)**, 1752-1780.

[52] Yang, S., Xiang, J. F., Yang, Q. F., Zhou, Q. J., Zhang, X. F., Li, Q., Tang, Y. L. and Xu, G. Z. (2011). An important functional group, benzo[1,3]dioxole, of alkaloids induces the formation of the human telomeric DNA G-quadruplex. *Chinese Science Bulletin*, **56(7)**, 613-617.

[53] Muser, S. E. and Paukstelis, P. J. (2012). Three-Dimensional DNA Crystals with pH-Responsive Noncanonical Junctions. *Journal of the American Chemical Society*, **134(30)**, 12557-12564.

[54] see as an exemple: Yang, X. L., Robinson, H., Gao, Y. G. and Wang, A. H., (2000). Binding of a macrocyclic bisacridine and ametantrone to CGTACG involves similar unusual intercalation platforms. *Biochemistry,* **39(36)**, 10950-10957.

[55] Chahine, J., Saffon, N., Cantuel, M. and Fery-Forgues, S. (2011). Spontaneous formation of fluorescent nanofibers and reticulated solid from berberine palmitate: a new example of aggregation-induced emission enhancement in organic ion pairs. *Langmuir*, **27(6)**, 2844-2853.

[56] Kariuki, B. M. and Jones, W. (1995). Five Salts of Berberine. *Acta Crystallographica Section C - Crystal Structure Communications*, **51**, 1234-1240.

[57] Man, S., Potáček, M., Nečas, M., Žák, Z. and Dostál, J. (2001). Molecular and crystal structures of three berberine derivatives. *Molecules*, **6(5)**, 433-441.

[58] Marek, J., Hulová, D., Dostál, J. and Marek, R. (2003). Berberine formate-succinic acid (1/1). *Acta Crystallographica Section C - Crystal Structure Communications*, **59**, o583-o585.

[59] Tong, H. H., Chow, A. S., Chan, H. M., Chow, A. H., Wan, Y. K., Williams, I. D., Shek, F. L. and Chan, C. K. (2010). Process-induced phase transformation of berberine chloride hydrates. *Journal of Pharmaceutical Sciences*, **99(4)**, 1942-1954.

[60] Allen, F. H. (2002). The Cambridge Structural Database: a quarter of a million crystal structures and rising. *Acta Crystallographica Section B - Structural Science*, **58**, 380-388.

[61] Fletcher, D. A., McMeeking, R. F. and Parkin, D. (1996). The United Kingdom Chemical Database Service. *Journal of Chemical Information and Computer Sciences*, **36**, 746-749.

Reviewed by Prof. Stephen Neidle, School of Pharmacy, University College London, UK

BIOGRAPHICAL SKETCH

Carla Bazzicalupi
Chemistry Dept. University of Florence, Italy

Education: 1996 PhD in Chemical Sciences, University of Florence
1992 Degree in Chemistry with first class honours, University of Florence

Address: Chemistry Dept. "U. Schiff", Via della Lastruccia 3
50019 Sesto Fiorentino, Florence, Italy

Professional Appointments:

2015 – currently – Associate Professor in the Scuola di Scienze Matematiche Fisiche e Naturali dell'Università degli Studi di Firenze, at the Department of Chemistry.
2001 – 2015 – Researcher in the Facoltà di Scienze Matematiche Fisiche e Naturali dell'Università degli Studi di Firenze, at the Department of Chemistry.
1998 – 2001 Technical Officer (VIII administrative level) at the Department of Chemistry at the University of Florence.
1997 - 1998 Post-doctorate grant, University of Florence.
1996 - 1997 – Grant from the Consiglio Nazionale delle Ricerche ISSEC – Istituto per lo studio della Sterochimica ed Energetica dei Composti di Coordinazione, CNR via J. Nardi 39, Florence.
Research and Professional experience:
2014 Chair of the Organizing Committee of the 2nd Joint AIC-SILS (Associazione Italiana di Cristallografia, Società Italiana Luce di Sincrotrone) conference, September 15-18 2014, Florence, Italy
2012 – currently - Coordinator of the Bilateral Agreement of cultural and scientific collaboration between the University of Florence and the University of Zagreb, Croatia, since the 16th May 2012.
2010 – currently - Representative of the Department of Chemistry of the University of Florence at the Centro di Cristallografia dell'Università degli Studi di Firenze.
2005 Member of the Organizing Committee for the XX Congress of the International Union of Crystallography, Florence IUCr 2005, Florence 23rd-31st August 2005.
2005 Member of the Organizing Committee for the VII National Conference of Supramolecular Chemistry, Florence 4th-7th September 2005.
1996 Member of the Organizing Committee for the XXI International Symposium on Macrocyclic Chemistry, Montecatini Terme 23rd-28th June 1996.

Publication Last Three Years:

1. M. Ferraroni, C.Bazzicalupi, F.Papi, G.Fiorillo, L.M. Guaman-Ortiz, A. Nocentini, A.I. Scovassi, P.Lombardi, P.Gratteri. *Solution and Solid-State Analysis of Binding of 13-Substituted Berberine Analogues to Human Telomeric G-quadruplexes. (2016) CHEMISTRY - AN ASIAN JOURNAL, 11, 1107-1115.*

2. C. Bazzicalupi, M.Ferraroni, F. Papi, L.Massai, B.Benoit, L.Messori, P.Gratteri, A.Casini. *Determinants for Tight and Selective Binding of a Medicinal Dicarbene Gold(I) Complex to a Telomeric DNA G-Quadruplex: a Joint ESI MS and XRD Investigation. (2016) ANGEWANDTE CHEMIE, INTERNATIONAL EDITION, 55(13), 4256-4259.*
3. A. Casula, C.Bazzicalupi, A.Bettoschi, E.Cadoni, S.J. Coles, P.N. Horton, F.Isaia, V.Lippolis, L.K.Mapp, G.M.Marini, R. Montis, M.A. Scorciapino, C. Caltagirone.*Fluorescent asymmetric bis-ureas for pyrophosphate recognition in pure water. (2016) DALTON TRANSACTIONS, 45(7), 3078-3085.*
4. C.Righeschi, M.C.Bergonzi, B.Isacchi, C.Bazzicalupi, P.Gratteri, A.R. Bilia. *Enhanced curcumin permeability by SLN formulation: The PAMPA approach. (2016) LWT--FOOD SCIENCE AND TECHNOLOGY, 66, 475-483.*
5. M.Citroni, S.Fanetti, C.Bazzicalupi, K.Dziubek, M.Pagliai, M.M. Nobrega, M.Mezouar, R.Bini. *Structural and Electronic Competing Mechanisms in the Formation of Amorphous Carbon Nitride by Compressing s-Triazine (2015) JOURNAL OF PHYSICAL CHEMISTRY C (2015), 119(51), 28560-28569.*
6. M.A.Olivella, C.Bazzicalupi, A.Bianchi, J.C. del Rio, N.Fiol, I. Villaescusa. *Binding interactions between suberin monomer components and pesticides (2015) SCIENCE OF THE TOTAL ENVIRONMENT, 527-528, 159-164.*
7. F.Del Bello, V.Bargelli, C.Cifani, P.Gratteri, C.Bazzicalupi, E.Diamanti, M.Giannella, V.Mammoli, R.Matucci, M.V.Micioni Di Bonaventura, A.Piergentili, W.Quaglia, M.Pigini. *Antagonism/Agonism modulation to build novel antihypertensives selectively triggering i1-imidazoline receptor activation (2015) ACS MEDICINAL CHEMISTRY LETTERS, 6(5), 496-501.*
8. C.Bazzicalupi, A.Bianchi, C.Giorgi, M.Savastano, F.Morales-Lara. *ATP dephosphorylation can be either enhanced or inhibited by pH-controlled interaction with a dendrimer molecule (2015) CHEMICAL COMMUNICATION, 51(18), 3907-3910.*
9. P.Gratteri, L.Massai, E.Michelucci, R.Rigo, L.Messori, M.A.Cinellu, C.Musetti, C.Sissi, C.Bazzicalupi. *Interactions of Selected Gold(III) Complexes with DNA G Quadruplexes (2015) DALTON TRANSACTIONS, 44(8), 3633-3639.*
10. S.Blasco, B.Verdejo, C.Bazzicalupi, A.Bianchi, C.Giorgi, C.Soriano, E.Garcia-España. *A thermodynamic insight into the recognition of hydrophilic and hydrophobic amino acids in pure water by aza-scorpiand type receptors. (2015) ORGANIC and BIOMOLECULAR CHEMISTRY, 13, 843-850.*
11. M.A. Olivella, C.Bazzicalupi, A.Bianchi, N.Fiol, I. Villaescusa. *New insights into the interactions between cork chemical components and pesticides. The contribution of π-π interactions, hydrogen bonding and hydrophobic effect. (2015) CHEMOSPERE, 119, 863-870.*
12. C.Bazzicalupi, A.Bianchi, T.Biver, C.Giorgi, S.Santarelli, M.Savastano. *Formation of Double-Strand Dimetallic Helicates with a Terpyridine-Based Macrocycle. (2014) INORGANIC CHEMISTRY, 53(22), 12215-12224.*
13. M. Savastano, P.Arranz-Mascaros, C.Bazzicalupi, A.Bianchi, C.Giorgi, M.L.Godino-Salido, M.D.Gutierrez-Valero, R.Lopez-Garzón. *Binding and removal of octahedral, tetrahedral, square planar and linear anions in water by means of activated carbon*

functionalized with a pyrimidine-based anion receptor. (2014) RSC ADVANCE, 4(102), 58505-58513.

14. N.Veiga, J.Torres, C.Bazzicalupi, A.Bianchi, C.Kremer. *The copper(II)-phytate-terpyridine ternary system: the first crystal structures showing the interaction of phytate with bivalent metal and ammonium cations. (2014) CHEMICAL COMMUNICATION, 50(95), 14971-14974.*
15. M.N.El-Kaheli, R.M.El-Mehdawi, R.G.Abuhmaiera, M.M.Ben Younes, F.A. Treish, A.Guerri, C.Bazzicalupi. *Poly[[(μ4-benzene-1,3,5-tricarboxylato-κ4O1:O1':O2:O3) bis(2,2-bipyridine-κ2N,N')(μ2-hydroxido)dicopper(II)] trihydrate. (2014) ACTA CRYSTALLOGRAPHICA, SECTION E: STRUCTURE REPORTS ONLINE, 70(7), m270-m271.*
16. C.Bazzicalupi, A.Bianchi, E.Garcia-España, E.Delgado-Pinar. *Metals in supramolecular chemistry. (2014) INORGANICA CHIMICA ACTA, 417, 3-26.*
17. C. Bazzicalupi, A.Bianchi, C.Giorgi, B.Valtancoli. *Zn(II) enhances nucleotide binding and dephosphorylation in the presence of a poly(ethylene imine) dendrimer. (2014) INORGANICA CHIMICA ACTA, 417, 163-170.*
18. C.Bazzicalupi, C.Caltagirone, C.Zenfeng, C.Qibin, C.Di Natale, A.Garau, V.Lippolis, L.Lvova, L.Honglai, L.Ingemar, M. C.Mostallino, M.Nieddu, R.Paolesse, L.Prodi, M.Sgarzi, N. Zaccheroni. *Multimodal Use of New Coumarin-Based Fluorescent Chemosensors: Towards Highly Selective Optical Sensors for Hg^{2+} Probing. (2013) CHEMISTRY – A EUROPEAN JOURNAL, 19(43), 14639-14653.*
19. H.Eshtiagh-Hosseini, M. Mirzaei, M Biabani, V.Lippolis, M.Chahkandi, C.Bazzicalupi. *Insight into the connecting roles of interaction synthons and water clusters within different transition metal coordination compounds of pyridine-2,5-dicarboxylic acid: experimental and theoretical studies. (2013) CRYSTENGCOMM, 15(34), 6752-6768.*
20. C. Bazzicalupi, A. Bianchi, C. Giorgi, P. Gratteri, P. Mariani, B. Valtancoli. *Anion and ion-pair binding by a G-2 poly(ethylene imine) dendrimer. (2013) DALTON TRANSACTIONS, 42(34), 12130-12138.*
21. S.Puccioni, C.Bazzicalupi, A.Bencini, C.Giorgi, B.Valtancoli, G.De Filippo, V.Lippolis, P.R.Salvi, G.Pietraperzia, R.Chelli, C.Gellini. *Tuning the Emission Properties of Fluorescent Ligands by Changing pH: The Unusual Case of an Acridine-Containing Polyamine Macrocycle. (2013) JOURNAL OF PHYSICAL CHEMISTRY A, 117(18), 3798-3808.*
22. C. Caltagirone, C. Bazzicalupi, F. Isaia, M.E.Light, V.Lippolis, R.Montis, S. Murgia, M.Olivari, G.Picci. *A new family of bis-ureidic receptors for pyrophosphate optical sensing. (2013) ORGANIC and BIOMOLECULAR CHEMISTRY, 11(15), 2445-2451.*
23. C. Bazzicalupi, A. Bianchi, C. Giorgi, P. Gratteri, P. Mariani, B. Valtancoli. *Metal Ion Binding by a G-2 Poly(ethylene imine) Dendrimer. Ion-Directed Self-Assembling of Hierarchical Mono- and Two-Dimensional Nanostructured Materials. (2013) INORGANIC CHEMISTRY, 52(4), 2125-2137.*

24. C.Bazzicalupi, M.Ferraroni, A.R.Bilia, F.Scheggi, P.Gratteri. *The crystal structure of human telomeric DNA complexed with berberine: an interesting case of stacked ligand to G-tetrad ratio higher than 1:1. (2013) NUCLEIC ACIDS RESEARCH, 41(1), 632-638.*
25. P. Arranz-Mascaros, C. Bazzicalupi, A.Bianchi, C. Giorgi, M-L. Godino-Salido, M-D. Gutierrez-Valero, R. Lopez-Garzon, M. Savastano. *Thermodynamics of Anion-π Interactions in Aqueous Solution. (2013) JOURNAL OF THE AMERICAN CHEMICAL SOCIETY, 135(1), 102-105.*

In: G-Quadruplex Structures, Formation and Role in Biology ISBN: 978-1-63485-512-9
Editor: Hannah Santos

Chapter 4

G-QUADRUPLEX-INDUCED FLUORESCENCE SIGNAL

Rajen Kundu*
Department of Chemistry and Biochemistry,
University of Colorado, Boulder, CO, US

ABSTRACT

Fluorescently labeled oligonucleotides are the remarkable achievement for the biochemical science and become the most important probes to detect and localize specific nucleic acid targets in cells. Under certain conditions, repetitive oligonucleotide (DNA or RNA) sequences folded into a non-B form higher order secondary structure such as tetraplex. In particular G-quadruplex is a tetraplex of a single or more oligonucleotide sequence(s) formed by Hoogsteen hydrogen bonding of four guanine units. Such G-quadruplex structure present at the termini of the eukaryotic chromosome and inhibits the activity of telomerase. Moreover, they are also a potential target for novel anticancer drugs. The topology of G-quadruplexes can be parallel, antiparallel, as well as other combinations of strand orientation. The structure and folding of tetraplex have been well studied by fluorescent phenomena such as excimer, exciplex, fluorescence resonance energy transfer (FRET), fluorescence enhancement, fluorescence quenching etc. G-quartet plane is an excellent platform for stacking interaction with many small fluorophores and vice versa. Moreover, fluorophores having large π-aromatic surface and/or positively charged substituents could be an effective G-quadruplex binder. Thus, tetraplex formation leads to a readable fluorescent signal upon stacking interaction between G-quartet and the fluorophore, whether it is external or covalently attached fluorophore. This chapter describes the G-quadruplex induced fluorescent signals that have been employed for the study of their folding and unfolding mechanism and thereby visualisation of their formation and structure.

Keywords: G-quadruplex, DNA, fluorescence, FRET, small molecule

* Corresponding Author address Email: Rajen.Kundu@colorado.edu.

INTRODUCTION

Apart from Watson–Crick duplex form, DNA can adopt many other non-B form higher order secondary structures such as hairpin, triplex, cruciform, left-handed Z-form, tetraplex, i-motif, A-motif, etc. Among those non-B form structures, guanine rich G-quadruplex is an emerging G-tetraplex which is pivotal for many major cellular processes. Single-stranded sequences containing guanine repeat units folded to a higher order G-tetraplex structure which is formed by Hoogsteen-type hydrogen bonds between guanine quartets and in presence of monovalent metal cations (K+ or Na+). The monovalent metal cations coordinate to O6 atoms of guanines in the electron rich G-quartet and stabilized the G-tetraplex structure (Figure 1). It is noteworthy that the formation of tetraplex/G-quadruplex structure needs at least two quartet units and one monovalent metal cation.

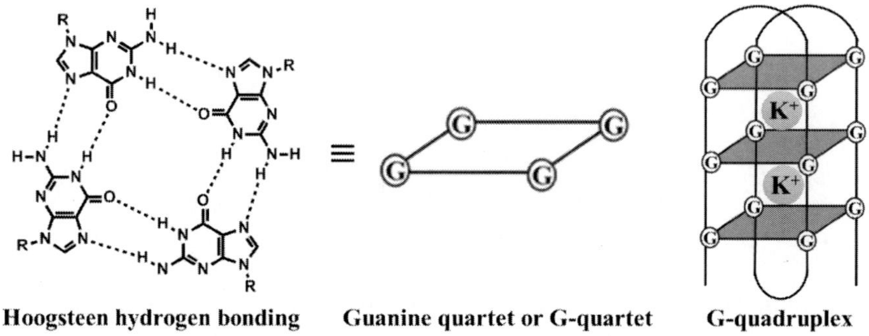

Figure 1. Structure of G-quartet and G-quadruplex.

The topology of G-tetraplex varies with the number of individual guanine repeat strand and the guanine repeat linker bases. Intramolecular G-tetraplex structure is formed by a single guanine repeat strand. Whereas, intermolecular G-tetraplex structure can be formed by two or more guanine repeat strands. These intra/intermolecular G-tetraplex structure can be further denominated into parallel, antiparallel or hybrid-type according to the orientation of the four guanine repeat strand (Figure 2).

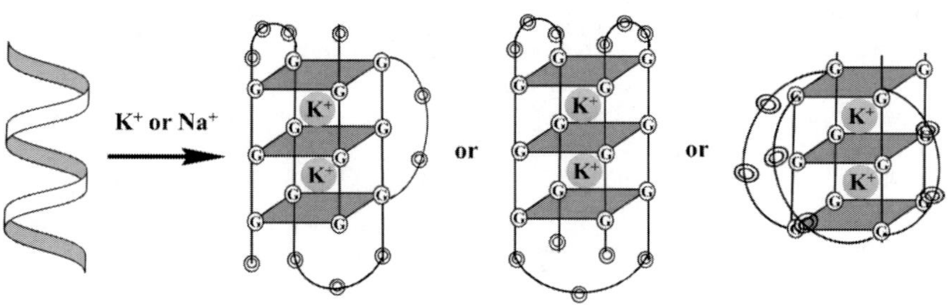

Figure 2. Structure of hybrid (left), antiparallel (centre), and parallel (right) forms of G-tetraplex.

During the last decades, considerable attentions have been paid to the detection of G-quadruplex. In vitro G-quadruplex formation has been nicely studied. However, in vivo evidence of the G-quadruplexes formation still remains a big challenge for the current researchers. Current research results of biochemical and molecular genetic studies have supported the in vivo G-quadruplexes formation [1–5]. However, the direct visualization of G-quadruplex in vivo or in a functional biological system is still a demanding research area.

Fluorescent Probes for G-Quadruplexes

The fluorescent molecule that could signal the presence of a G-quadruplex structure is of great interest. Additionally, fluorescent probes which show enhancement of fluorescent signal upon binding with G-quadruplex structure are of special interest in cellular biology and fluorescent microscopy-based assays [6, 7]. Ideal fluorescent probes for cellular imaging should have high binding affinity and selectivity to its target. Moreover, the probes should have good water solubility and membrane permeability and also should show enhanced fluorescence emission upon target binding. Towards the journey of the quest for a G-quadruplex specific fluorescent sensor, researchers have developed various fluorescent probes which can interact with the G-quartet and display a strong enhancement or a decreased fluorescence signal. The folding and unfolding mechanism of G-quadruplexs and the visualisation of their formation have been studied by fluorescent methods. From the engineering point of view, the fluorescent probe can be either an external or covalently attached molecule (Figure 3). The external probe binds with G-quadruplex via non covalent interaction. Whereas, the covalently attached fluorescent probe interact with the G-quartet upon conformational change from a single strand to G-quadruplex. The engineering of probe design depends upon the modes of accommodation of the probe by G-quadruplex. A tetraplex can accommodate a probe by π-stacking interaction on the external G-quartet or by interaction in the grooves of tetraplex. Thus, probe having planer aromatic rigid structure is the best suitable for π-stacking interaction with the G-quartet [8, 9]. Whereas, nonplaner flexible structured probe prefers to bind in the grooves of tetraplex [10-12].

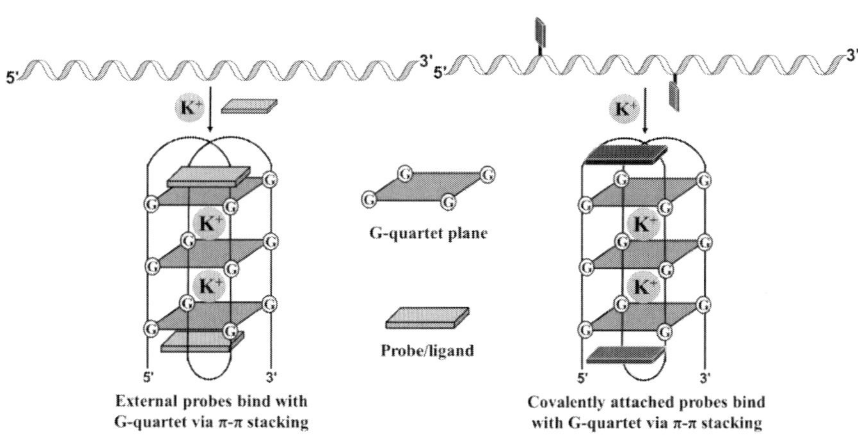

Figure 3. Illustration of external and covalently attached probe bound G-quadruplex.

The major breakthrough work in small molecule recognized G-quadruplex nucleic acid was reported by S. Neidle and L. Hurley et. al [13]. They developed an anthraquinone derivative for targeting G-quadruplex DNA. This work pave the way for the design and synthesis of small molecules which can probe the G-quadruplex formation and visualize the biological functions. Unlike external fluorescent probes, the covalently attached probes show enhancement of fluorescent signal upon conformation change from single strand to a folded tetraplex structure. The embedded probes show strong enhancement of fluorescent signal and/or shifting of wavelength upon stacking with G-quartet plane. In this chapter we will discuss G-quadruplex-induced fluorescence signal both from external as well as covalently attached probe.

NONCOVALENT PROBES FOR G-QUADRUPLEXES

The work reported by S. Neidle and L. Hurley et. al [13] is the starting point towards the development of noncovalent or external probe for visualization of G-quadruplex structure and functions. However, in that communication, they have not reported any fluorescence properties of the G-quadruplex bound probe.

Fluorescent Intercalator Displacement

Thiazole orange (TO) is well-known for G-quadruplex binder fluorescent probe which shows high binding affinity. Moreover, TO is a stronger G-quadruplex binder, compared to ethidium bromide which is known as weak G-quadruplex DNA binder but a strong duplex DNA binder. The fluorescent intensity of TO enhanced by ~500 to ~3000 fold upon binding with G-quadruplex DNA. Consequently, the displacement of TO from TO-quadruplex complex by some external ligand will show the reverse result, that is, decreased of fluorescent intensity. With this hypothesis, D. Monchaud and co-workers [14] developed a fluorescent intercalator displacement (FID) assay based on the displacement of TO from TO-quadruplex complex by quadruplex binding ligand. In this method they have screened and characterized various types of G-quadruplex binders such as quinacridine-based ligands, macrocyclic bisquinacridine BOQ_1 and pyridodicarboxamide ligands etc. [14-16] (Figure 4).

The major advantage of this FID assay is that it requires neither modified fluorescent labeled oligonucleotides nor specific equipment. It is an isothermal experiment and just a mix and read strategy based on the displacement of an on/off fluorescent probe, Thiazole Orange (TO). This assay also works in microplates. They further extended the study of FID assay on to the evaluation of binding affinity of various ligands towards quadruplex-DNA over duplex-DNA [15, 16]. The mechanism of FID assay is described schematically in the following Figure 5.

FRET

Fluorescence resonance energy transfer, so called FRET, is an emerging technique widely used not only in material science but also in chemistry and biology [17-21]. FRET is a dipole-dipole resonance interaction between two different fluorophores (say donor and acceptor pair). When these donor-acceptor pairs are placed within a proper distance (10–50

Å), FRET could be possible from one fluorophore (called donor) to another fluorophore (called acceptor) and thus, it is a distance dependent phenomenon [22]. The donor molecule of a FRET pair absorbs the light (hν) at low wavelength region and emits a long wavelength light. Meanwhile the acceptor molecule absorbs the long wavelength light emitted from the donor molecule and emits in the visible region (λ_{em}). Thus, FRET depends upon the spectral overlap, emission and absorption of the donor and the acceptor respectively.

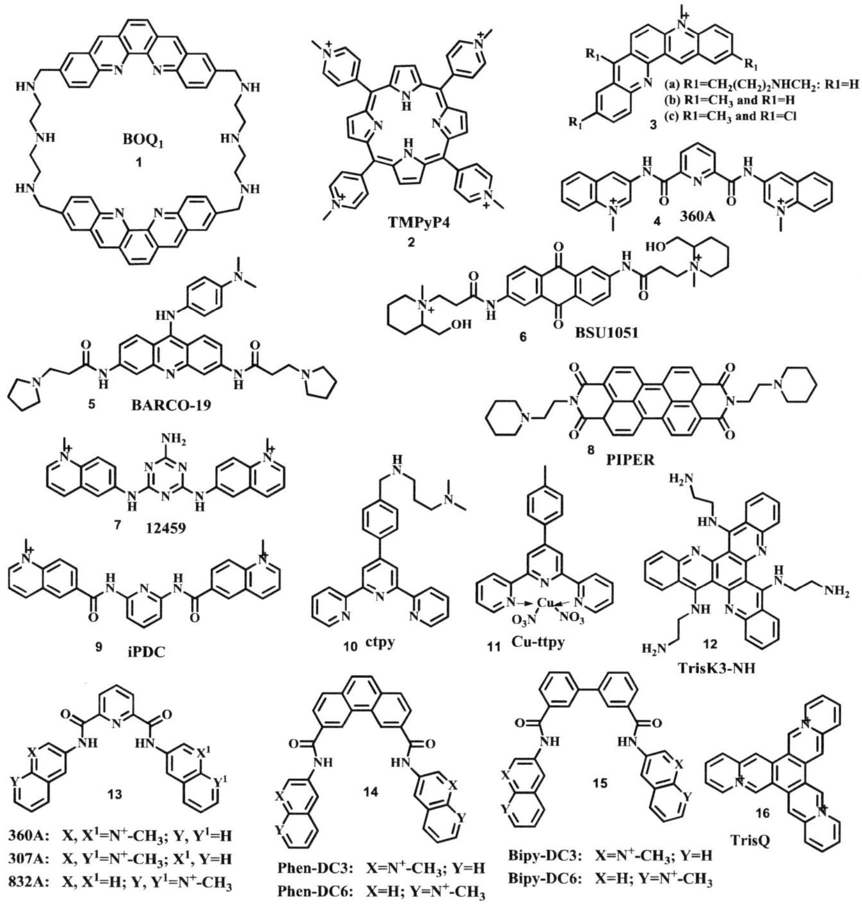

Figure 4. FID assay based G-quadruplex binding ligand.

During the last few years, several groups have focused on FRET phenomena within the context of DNA to establish the FRET engineering, to study the DNA hybridization and/or various biophysical study of DNA [23-28]. However, most of the efforts on DNA-mediated FRET processes have been focused on duplex DNA system or B-DNA with Watson–Crick base pairing. Very few reports have appeared in the literature [26-28] where non B-DNAs including G-quadruplexes platform is used to demonstrate FRET phenomena. It is also noteworthy that, the two external G-quartets of G-quadruplexe DNA offer two well-defined binding sites for positioning a FRET donor/acceptor pair. Thus, G-quadruplex DNA represents more attractive scaffold for FRET study than a duplex DNA. Furthermore, most biological sequences forming G-quadruplexes are composed of three to four stacks of G-quartets which provide a distance close to the lower range of the Forster distance (10-15 Å).

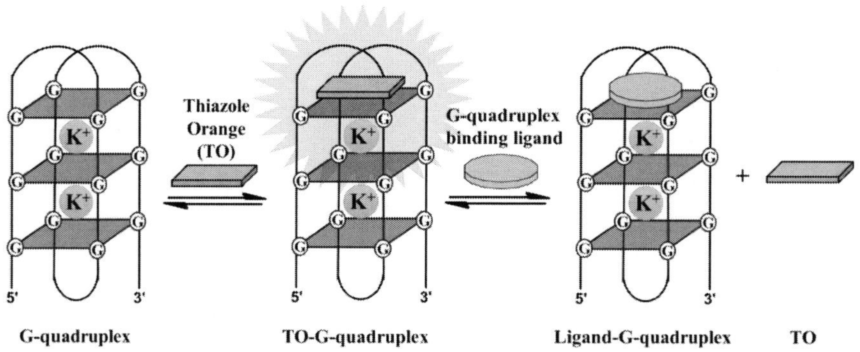

Figure 5. Schematic presentation of FID assay.

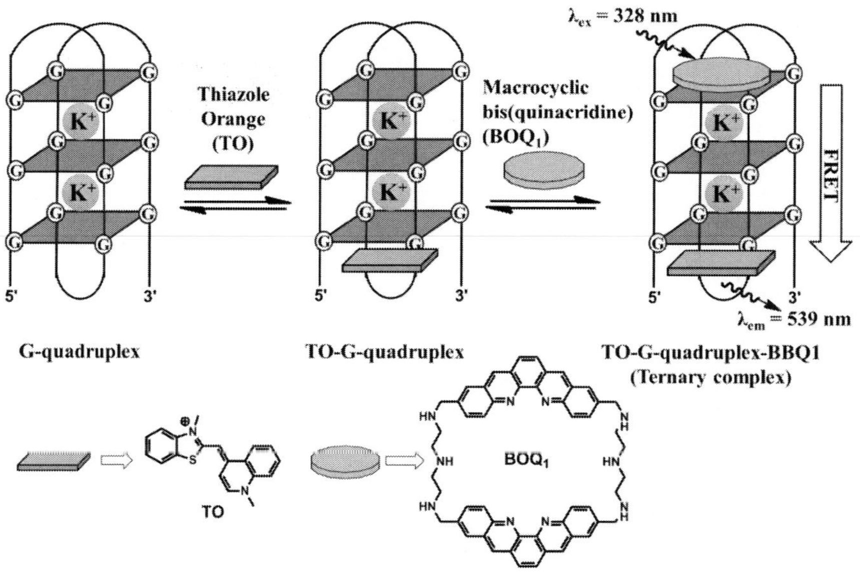

Figure 6. Schematic presentation of G-Quadruplex DNA templated FRET between BOQ_1 and TO.

G-Quadruplex DNA templated FRET by the formation of a ternary complex (FRET donor – G-Quadruplex – FRET acceptor) was reported by Teulade-Fichou and co-workers [28]. They used macrocyclic bis(quinacridine) BOQ_1 or monomeric quinacridine MMQ_1 as an energy donor and thiazole orange (TO) as an energy acceptor. This system is based on the generation of an unprecedented specific ternary complex between the G-quadruplex DNA and the FRET donor/acceptor (quinacridine and thiazole orange, Figure 6). Upon mixing of the two dyes in 1:1 ratio and the addition of G-Quadruplex DNA showed energy transfer from BOQ_1 to TO when excited at 328 nm (absorption maxima of BOQ_1). Addition of G-quadruplex DNA promotes the energy donor (BOQ_1) and acceptor (TO) to bind into DNA in relative positions (outer plane of each terminal G-quartes) suitable for energy transfer. Thus, the ternary complex emits at 539 nm (emission maxima of TO) when excited at 328 nm.

Hui Xu and co-workers reported FRET from a water soluble cationic conjugated polymer to G-quadruplex bounded SYBR Green I (SG) [29]. Fluorescence intensity of SG increased when potassium ion was added into a mixture of SG and single stranded DNA (ssDNA) with G-rich sequence due to the intercalation of SG into G-quadruplex. They further amplified the fluorescence intensity of SG by 6-fold through fluorescence resonance energy transfer from a water soluble cationic conjugated polymer (CCP) to SG. The negatively charged G-quadruplex strongly interacts with the positively charged CCP and thus, the energy donor and acceptor partner come in close proximity for energy transfer (Figure 7). However, this phenomena was not observed for ssDNA in the absence of K^+. Thus, this complex system (ssDNA/SG/CCP) can be used as a label-free selective potassium sensor.

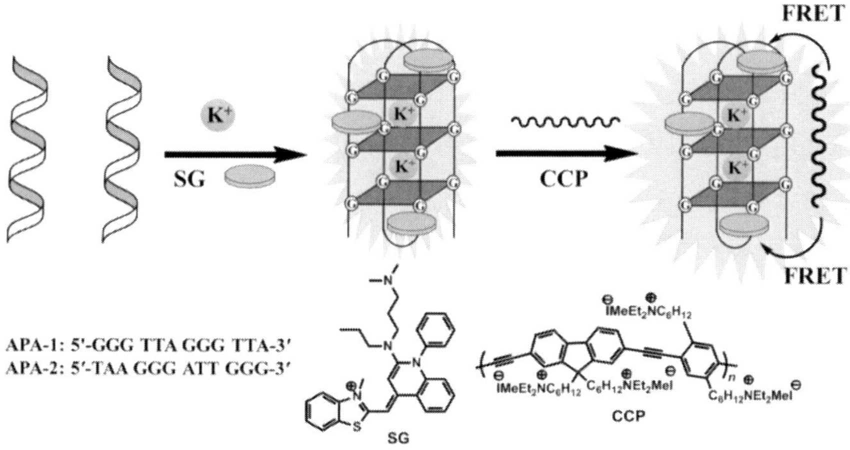

Figure 7. Schematic presentation of FRET from cationic conjugated polymer to G-quadruplex bounded SYBR Green I.

Small Molecules

Ethidium bromide (EB) and Hoechst are well known examples of double stranded DNA binding dye (Figure 8). Both the free dyes are weakly fluorescent or non-fluorescent in aqueous solution. However, their fluorescent intensity dramatically increase upon binding with double stranded DNA. Due to the flexible structure, Hoechst binds into the minor groove region of the DNA. Whereas, the rigid ethidium bromide intercalates between the DNA base pairs. Hoechst binds to the minor groove of double-stranded DNA preferentially in adenine and thymine rich region. Although it can bind to all nucleic acids, however, the considerable fluorescence enhancement is observed from AT-rich double-stranded DNA strands. In the case of ethidium bromide, the fluorescence is almost entirely quenched upon binding to poly d(G)–poly d(C), whereas binding to poly d(GC) and AT-rich duplexes results in a strong fluorescence enhancement. The fluorescence enhancements observed for ethidium are highly dependent on DNA conformation: G-quadruplex < triplex < duplex.

Interestingly, both the dyes bind with G-quadruplex DNA too [30, 31]. In vitro binding of Hoechst 33258 into the human c-myc gene, d(GG GGAGGG TGG GGA GGG TGG GGA AGG TGG GG) in presence of potassium ion was investigated by S. Chowdhury and co-workers with the help of equilibrium absorption, fluorescence, and kinetic surface plasmon resonance methods [30]. Hoechst 33258 binds to the G-quadruplex with a high binding affinity of the order of 10^6 M^{-1}. The G-quadruplex binding was confirmed by hypochromic effect in UV-visible absorption spectra and blue shift in fluorescence emission maxima. Due to the flexible structure of Hoechst, it mainly interacts with the AAGGT loop of the G-quadruplex instead of G-quartet of the G-quadruplex.

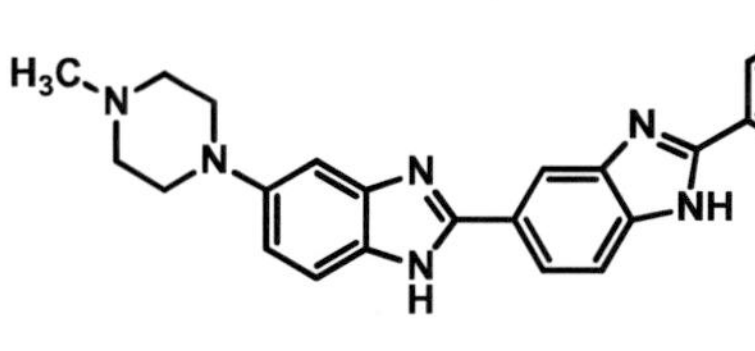

Hoechst 33258
17

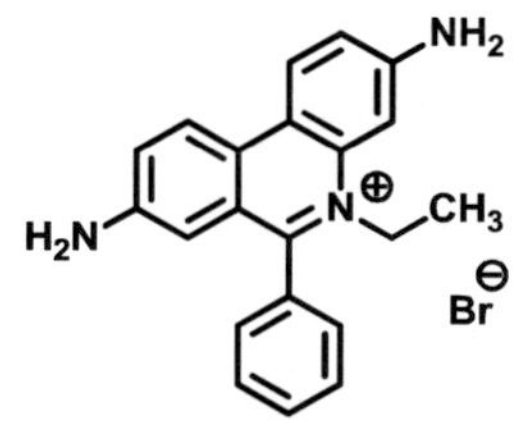

Ethidium bromide
18

Figure 8. Chemical structure of Hoechst 33258 and Ethidium bromide.

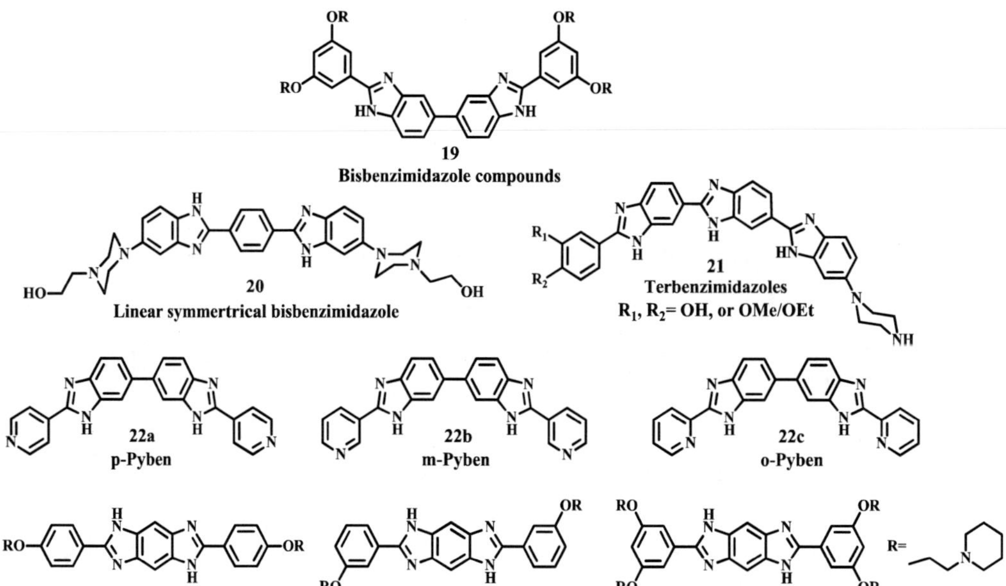

Figure 9. Chemical structure of flexible ligands for G-quadruplex DNA.

Kallenbach and co-workers observed that four dT4G4 single strand DNAs form tetramer in presence of sodium or potassium ions [31]. This tetramer binds ethidium bromide with an intercalative mechanism. However, the 4-fold tandem repeat of this subunit, d(T4G4)4, does not interact with ethidium bromide in the presence of sodium ions. The structural difference might be the responsible for such kind of binding discrepancy. In the tetramer, the guanine bases from each monomer strands form four G-quartets and stabilize in the presence of Na+ or K^+ ion. The hanging thymine bases are free from any backbone strain. Whereas, the thymine bases of 4-fold tandem repeat subunit, d(T4G4)4 imposed backbone strain which prevents the proper folding. But, flexible sequence with increased number of T bases, for example dT4G4(T7G4)3 allows the formation of four G-quartets and also binds with ethidium bromide in the presence of Na^+ and K^+ ions.

Several non-rigid flexible ligands are also available in literature, which strongly interact with the G-Quadruplexe DNA. These flexible ligands neither intercalate into the G-quartet plane nor bind with the external G-quartet via π-π interaction. However, like nonintercalating duplex-DNA minor groove binding ligands, they strongly interact with the loop of the G-quadruplex DNA. Figure 9 shows some selective nonintercalating G-quadruplex DNA binding ligands [32].

Cyanine Dyes

Shafer and co-workers reported that the carbocyanine dye DODC (Figure 10) binds to dimeric hairpin G-quadruplex in vitro [33]. Later on, Tang and co-workers synthesized the cyanine derivative ETC which showed a 70-fold fluorescence enhancement upon binding with parallel G-quadruplexes [34]. It is noteworthy that, based on the guanine rich sequence, different kinds of G-quadruplexes possess different structures and functions. Therefore, selective recognition of specific G-quadruplexes need special attention on the ligand design. Recently, Xuening Fei and co-workers reported a benzindole substituted carbazole cyanine dye (9E PBIC) which can selectively recognize parallel c-myc G-quadruplexes via the enhancement of fluorescent signal [35]. This dye is weakly or non-fluorescent with single stranded or double stranded DNA under aqueous buffer. However, the fluorescence intensity increased over 100 fold in the presence of c-myc G-quadruplex and 10-30 fold in the presence of other G-quadruplexes. This dye showed excellent selectivity to c-myc 2345 G-quadruplexes over single or double-stranded DNA. Due to the flexible structure, the dye interacts with the c-myc 2345 G-quadruplex by using the end-stack binding model.

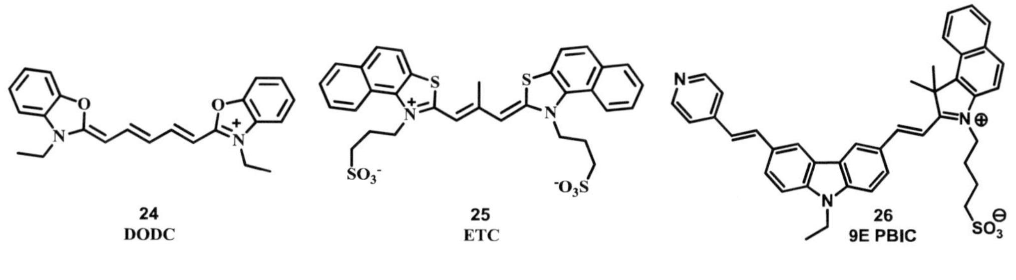

Figure 10. Chemical structure of Cyanine based dyes.

V-Shaped Planer Molecule

G-quadruplex ligands usually consist of a planar aromatic core with an extended delocalized π-electron system and flexible side chains. The planar aromatic core interacts with the external G-quartets of G-quadruplex by π–π stacking and the flexible side chains interact with the DNA bases and phosphate backbone of the loops by hydrogen bonds and/or electrostatic interaction. Therefore, the selectivity of G-quadruplex ligands depends on the size of the planar aromatic core and the structure of side chains. Electron deficient or positively charged aromatic core strongly interact with the external G-quartet due to charge transfer electrostatic interaction with the electron rich G-quartet plane.

Benzimidazole derivatives have been widely used for probing biological activity and extensively investigated as drugs designed for targeting DNA. For example, Hoechst 33258, a well-known minor groove DNA binding ligand contains benzimidazole core. Recently, bisquinolinium [36], bisindole [37], and bisbenzimidazole derivatives [32, 38, 39] also showed good selectivity toward G-quadruplex over duplex DNA (Figure 11). The central part of these bis-molecules are associated with a pyridine or benzene moieties. Substitution at 2,5- and 1,3-positions of pyridine and benzene respectively by the conjugated aromatic core result a V-shaped planar core which shows strong binding affinity to G-quadruplex DNA [40, 41]. Recently, Shangguan and co-worker developed a carbazole containing bisbenzimidazole derivatives (BPBC) where the two benzimidazole derivatives are attached at 3,6-position of carbazole unit [42]. This derivative is non fluorescent or weakly fluorescent in aqueous solution. However, it shows tremendous fluorescence enhancement upon binding with G-quadruplex DNA. Molecular docking model showed that the ligand binds on the surface of the G-quartet via stacking interaction.

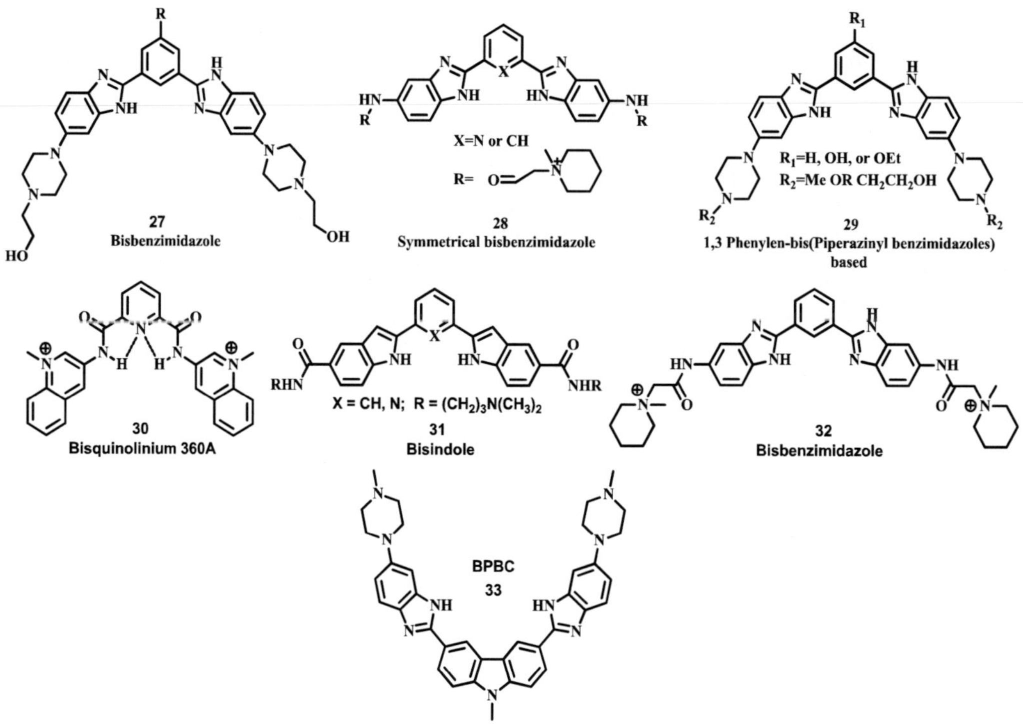

Figure 11. Chemical structure of V-shaped planar ligands for G-quadruplex.

Metal Complexes

Transition metal complexes have attracted as DNA binding ligands [43] due to their π-π stacking interaction with the DNA base pair. For example, bipyridyl (bpy) complexes of ruthenium (Ru) and rhodium (Rh) are well known DNA binding agents. These metal complexes are capable of DNA sequence recognition through intercalation into the π-stack of the DNA duplex. Moreover, these complexes are also capable of recognizing single base mismatches upon changing their shape and functionality. Recently, various research groups have developed varieties of Ru(II) complexes (Figure 12) which show good luminescent selectivity between G-quadruplex and other DNAs [44-48]. Noncoplanar metal complex interacts with tetraplex by end-stacking onto G-quartets. Whereas, square planer complex binds either end-stacking or intercalation onto the G-quartets. Towards this aim, Thomas and co-workers synthesized dinuclear ruthenium(II) complex (34) which functions as a label-free colorimetric sensor for DNA [44]. The interaction of this Ru(II) complex with DNA results in distinctive color changes that are dependent on both DNA sequence and structure. Similarly, Zhou and co-workers also synthesized a dinuclear ruthenium complex, where the two fluorescent groups, Ru(II) complexes (35) were connected by noncyclic crown ethers (Figure 12) [45]. This complex is highly fluorescent in presence of G-quadruplex DNA. Moreover, an iodide-quenching strategy can easily distinguish G-quadruplex structures over duplex structures by the naked eye. Shuo Shi and co-workers synthesized $[Ru(bpy)_2(dppz)]^{2+}$ (36) which has a strong preference for binding to G-quadruplexes over i-motif [46]. Similar Ru complex (37) with two dppz ligand (dppz=dipyrido[3,2-a:2′,3′-c] phenazine) was also available and was found to as a fluorescent light-up probe for G-quadruplex DNA [47].

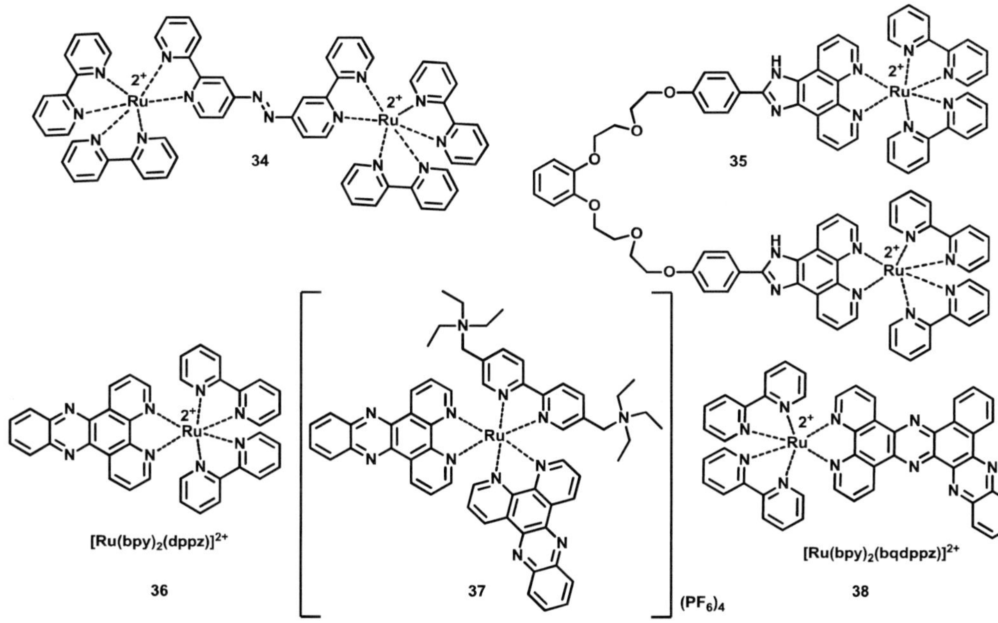

Figure 12. Luminescent ruthenium complexes.

In 2012, Hui Chao and co-workers reported a polypyridyl complex, $[Ru(bpy)_2(bqdppz)]^{2+}$ (38) with specific luminescent selectivity towards hybrid G-quadruplex DNA [48]. They modified the previously reported dppz ligand by replacing one benzene ring of dppz with a phenazine ring and simultaneously extended the lateral area of the ligand bqdppz which cannot effectively intercalate into other forms of DNA and thus increase the selectivity. This complex can easily distinguish hybrid G-quadruplex DNA over other forms of DNA by visual detection.

Square-planar nickel(II) and platinum(II) complexes exhibit interesting spectroscopic properties. For example, Vilar and co-workers reported a nickel–salphen complex (39; Figure 13) with 50-fold selectivity for G-quadruplexes over duplex DNA [49]. However, they have not reported any spectroscopic properties of this complex.

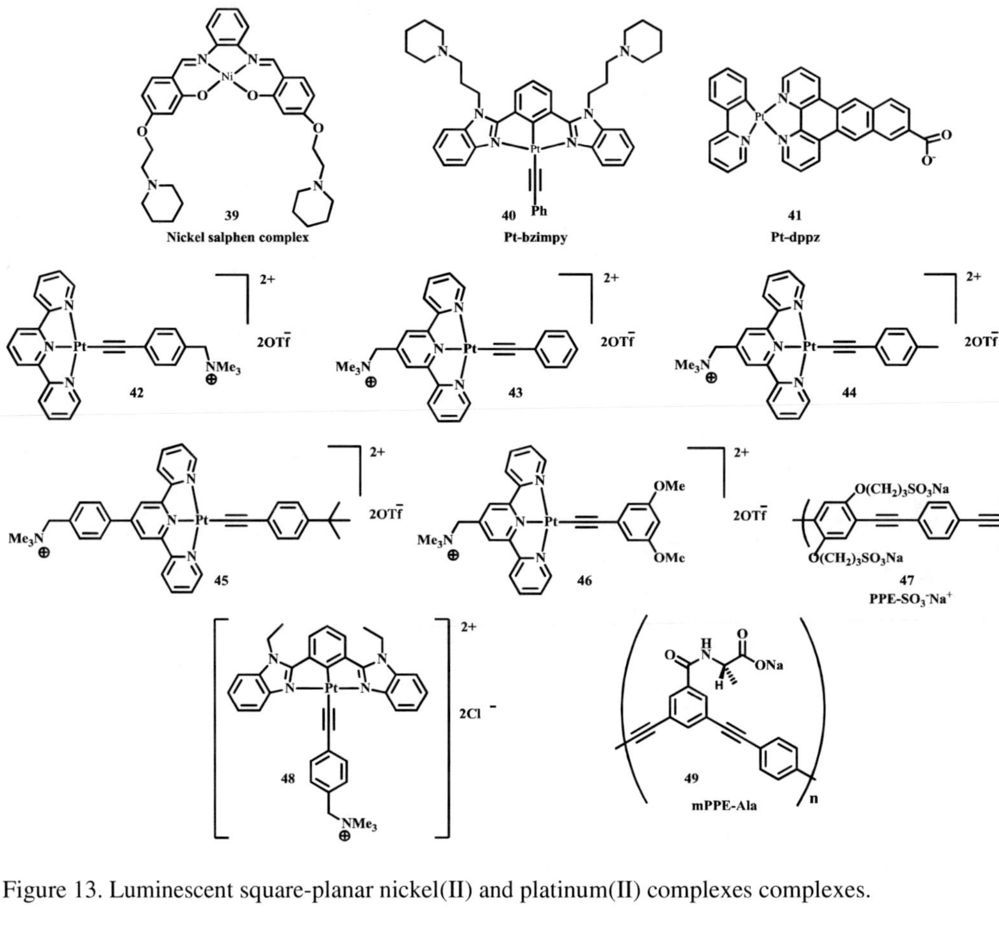

Figure 13. Luminescent square-planar nickel(II) and platinum(II) complexes complexes.

Platinum(II) complexes containing a 2,6-bis-(benzimidazol-2-yl)pyridine (bzimpy) ligand (40, Figure 13) exhibited higher selectivity for binding to G-quadruplex over duplex DNA [50]. Ma and co-workers synthesized platinum–dipyridophenazine complex (41), which exhibited a 293-fold increase in photoluminescence upon G-quadruplex binding [51]. Some particular class of platinum(II) complexes such as alkynylplatinum(II) terpyridine complexes (42-46) have got much attention due to their d^8–d^8 metal–metal interaction and the π–π interaction of the terpyridine ligands which lead to an interesting spectroscopic and

luminescence properties [52-55]. In the presence of biopolymers, the terpyridylplatinum(II) complexes showed remarkable spectroscopic change with emission at near-infrared (NIR) spectral region. Such NIR emissions have great advantages for non-invasive fluorescence microscopy as the emission in the NIR region can minimize photo-damage to cells, increase optical transparency and lessen tissue autofluorescence. Thus, these complexes become an efficient probe for the detection of biomolecules and their related conformational changes, as well as probing the activities of some important biological enzymes. Recently, water-soluble cationic alkynylplatinum(II) complexes are found to be highly desirable class of complex for probing biomolecules. The selectivity of probing has been significantly improved with the use of two-component supramolecular ensembles based on the alkynylplatinum(II) terpyridine complexes and π-conjugated polyelectrolytes (CPE). Yam and co-workers studied the fluorescence resonance energy transfer (FRET) from CPEs to alkynylplatinum(II) terpyridine complexes. G-quadruplex DNA interferes the aggregation of CPE-alkynylplatinum(II) complex and hence spectral change was observed on the FRET from CPE to alkynylplatinum(II) complex. The ratiometric emission of CPE-alkynylplatinum(II) complex distinguishes the G-quadruplex structure formed by human telomeric DNA from those of other G-quadruplex-forming sequences. The alkynylplatinum(II) complexes and the conjugated polyelectrolytes (CPE) are shown in the Figure 13.

Thioflavin T

Thioflavin T is a benzothiazole derivative which is widely used to visualize and quantify the presence of misfolded protein aggregates called amyloid, both in vitro and in vivo. Upon binding with beta sheet-rich structures, such as those in amyloid aggregates, the dye displays enhanced fluorescence and a characteristic red shift of its emission spectrum. Apart from visualization of amyloid, recently, Thioflavin T was found to as selective fluorescent sensor for the G-Quadruplex DNA. For example, Mohanty and co-workers demonstrated for the first time the application of Thioflavin T which induces quadruplex folding in the 22AG human telomeric DNA, both in the presence and absence of Tris buffer/salt [58]. They observed 2100-fold fluorescence enhancement of the Thioflavin T in the visible region upon binding with human telomeric DNA (Figure 14). Later on Gabelica and co-workers conducted the detailed investigations of the ligand binding mode and fluorescence response upon Thioflavin T binding with human telomeric DNAs of various nucleotide sequences. According to their conclusion, multiple Thioflavin T molecules bind to the 5′ G-quartet cooperatively. The first bound Thioflavin T occurs in a binding pocket formed with the A1 and 5′ G-quartet and produces the brightest fluorescence light-up signal [59].

Shao and co-workers nicely explored the fluorescence enhancement of G-quadruplex DNA bounded Thioflavin T [60]. In the excited state, the rotation of benzothiazole (BZT) and dimethylaminobenzene (DMAB) rings are restricted due to binding of Thioflavin T and the G-quatet and thus, fluorescence enhancement was observed (Figure 14). They also confirmed this molecular rotor based fluorescence enhancement using a nonrotor analogue of Thioflavin T. Thioflavin T binding does not disturb the native G-quadruplex structures preformed in Na^+ and K^+ solutions. Thus, this system could be used as a rotor-based fluorescent probe for K^+ sensor.

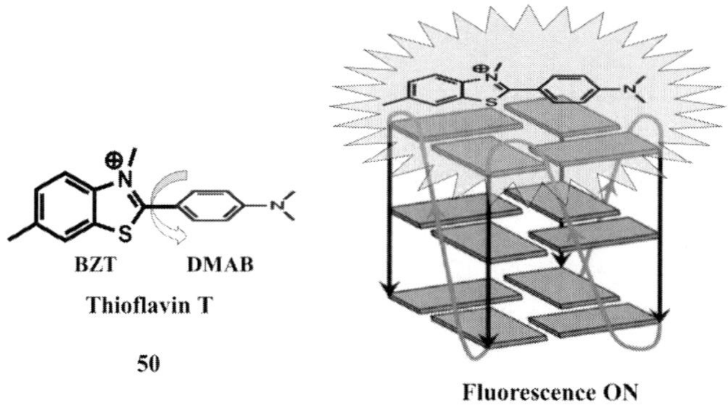

Figure 14. Schematic presentation of fluorescence enhancement of Thioflavin T.

Triphenylmethane Dyes

Malachite green (MG) is a class of triarylmethane dye [61] extensively used in pigment industry. This dye also got much attention for biological applications such as photodynamic therapy [62], site-specific inactivation of RNA transcripts [63], RNA-aptamer-based fluorescence sensors [64, 65] etc. In 2007, Bhasikuttan and co-workers reported that the fluorescence intensity of malachite green enhanced by ~100-fold when interacted with guanine-rich DNA [66]. Furthermore, they observed a bathochromic shift of approximately 18 nm in UV-visible spectra of MG in presence of G-rich DNA. This observation indicates that MG may be used as a sensitive G-quadruplex biosensor. Like MG, crystal violet (CV) is also a family of triphenylmethane dye. The chemical structure of CV and MG is very similar and the only difference between the two dyes is the number of dimethylamino substituents. The dimethylamino substituents provide potential for hydrogen-bonding and cation-dipole interactions with the negatively charged phosphate backbone and loops of G-quadruplexes. In 2009, Kong and co-workers reported the discrimination of G-quadruplexe DNAs from duplex and single-stranded DNAs by fluorescence and energy-transfer fluorescence spectra of malachite green (MG) and crystal violet (CV) dyes [67]. Thus, both the triphenylmethane dyes (MG and CV) can be used as a potential G-quadruplex biosensor.

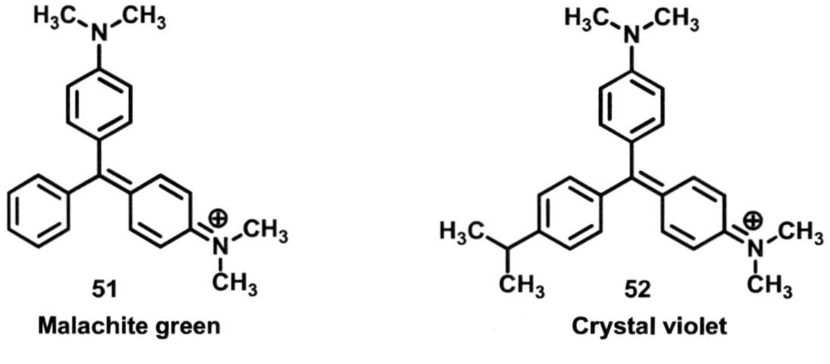

Figure 15. Chemical structure of malachite green and crystal violet.

Carbazole Derivatives

Chang and co-workers synthesized a fluorescent carbazole derivative, 6-bis(1-methyl-4-vinylpyridium)carbazole diiodide (BMVC, Figure 16) for recognizing specific quadruplex structures specially quadruplex of human telomeric sequence d(T_2AG_3)$_4$ [68]. The fluorescence of BMVC can distinguish between duplex and quadruplex DNAs. Teulade-Fichoua and co-workers reported two-photon absorption N-phenyl carbazole-based fluorescent light-up probes for quadruplex selectivity over duplex DNA [69]. The vinylic derivatives (Cbz-2Py, Cbz-3Py) display strong fluorescence enhancement when bound to various duplex- and quadruplex-forming oligonucleotides.

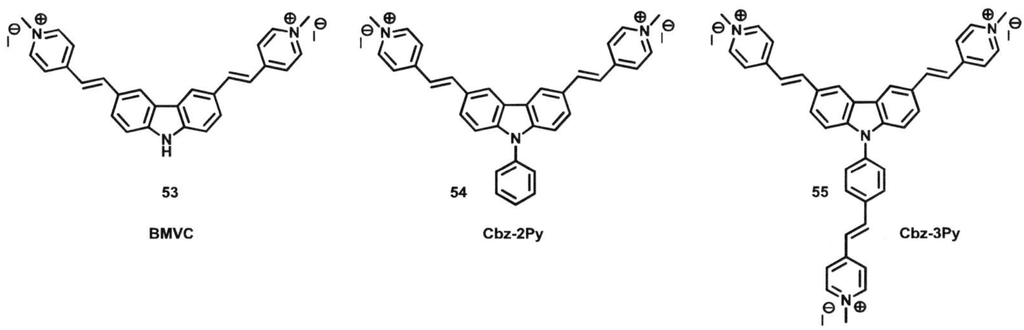

Figure 16. G-quadruplex binding carbazole derivatives.

Aggregation-Induced Emission

Tetraphenylethene derivatives are well known for aggregation-induced emission (AIE). Tang and co-workers observed fluorescence enhancement from tetraphenylethene derivativs, 1,1,2,2-Tetrakis[4-(2-bromoethoxy)phenyl]ethane (TBEPE) and 1,2,2-tetrakis[4-(2-triethylammonioethoxy)phenyl]ethene tetrabromide (TTAPE) when they aggregate and bind with guanine-rich DNA strand [70]. When TTAPE binds with guanine-rich DNA via electrostatic attraction, its intramolecular rotation is restricted and its emission is turned on. This dye can discriminate the folding of G-quadruplex from other DNA structures by a bathochromic shift of the emission spectra.

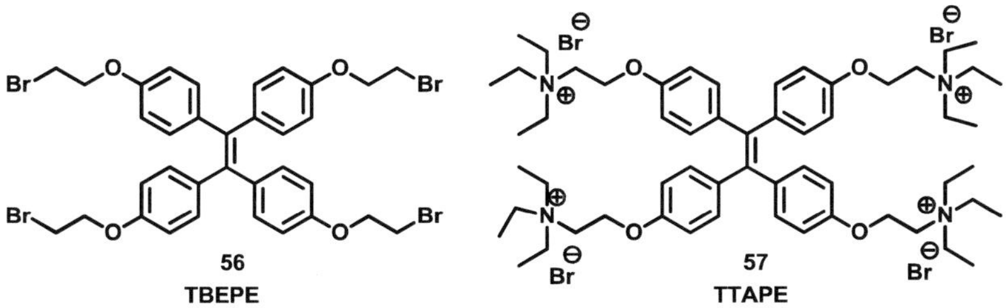

Figure 17. G-quadruplex binding tetraphenylethene derivatives.

Porphyrin Derivatives

Porphyrin dyes are heterocyclic macrocycle composed of four pyrrole subunits interconnected at their α-carbon atoms via methine bridges (=CH−). Porphyrin ring is the core unit of chlorophyll, which converts the light energy to chemical energy. It is found that the porphyrin dyes are highly fluorescent in the presence of G-quadruplex DNA but not duplex DNA. Thus, these dyes can be used as a specific detection of quadruplex DNA under physiological conditions. For example, Bolton and co-workers synthesized a series of porphyrin derivatives [71]. Surprisingly, one of the porphyrin derivative (T4, Figure 18) showed interesting fluorescence properties. They found that edge and crossover type quadruplex DNAs can significantly enhance the fluorescence of T4, while parallel quadruplex DNAs do not. Duplex DNA enhances the fluorescence more than the crossover and parallel quadruplex DNAs. Similarly, Hurley and co-workers synthesized porphyrin derivative (TMPyP2, Figure 18) and studied the interaction with an antiparallel quadruplex DNA [72]. They found that porphyrin stabilizes quadruplex DNA. However, they did not provide any fluorescence result.

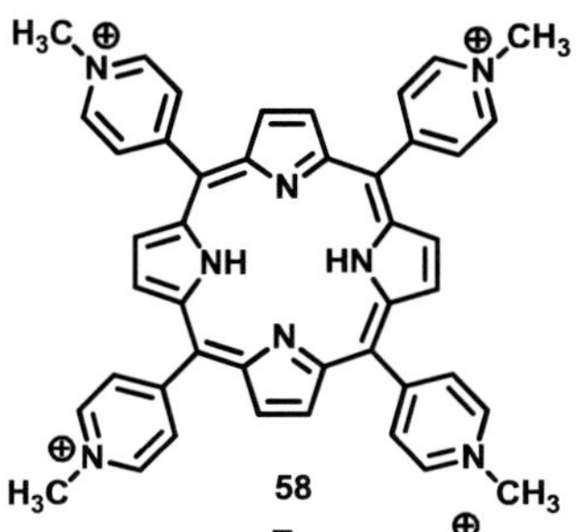
58
T_4
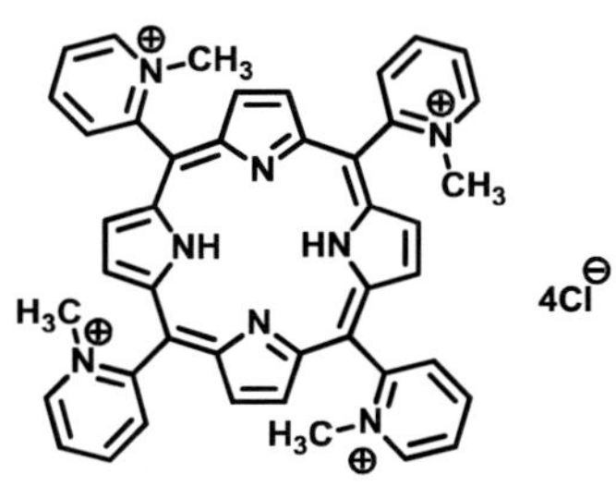
59
$TMPYP_2$

Figure 18. G-quadruplex binding porphyrin derivatives.

Covalent Probes for G-Quadruplexes

Environment sensitive fluorescent nucleoside analogs are attractive probes for investigating nucleic acid structure and folding/unfolding kinetics. Pyrene-modified oligonucleotides exhibits excimer emission (~480 nm) upon pyrene–pyrene self-association or exciplex emission (~480 nm) upon the formation of excited state complex with DNA base. These excimer or exciplex emissions are approximately 100 nm red shifted from the monomer emissions (~380 nm). For example, Kim and co-workers developed quencher-free endstacking ethynyl pyrene–adenosine modified G-rich oligonucleotide which can probe the stable G-Quadruplex transition via exciplex emission [73]. Conformational change from a single strand to folded G-quadruplex upon addition of potassium ion (K^+) bring the pyrene to the close proximity of hydrophobic G-quartet plan and forms an exciplex via π-π stacking (Figure 19). The exciplex emission increased with a red shifting with increasing K^+ ion concentration.

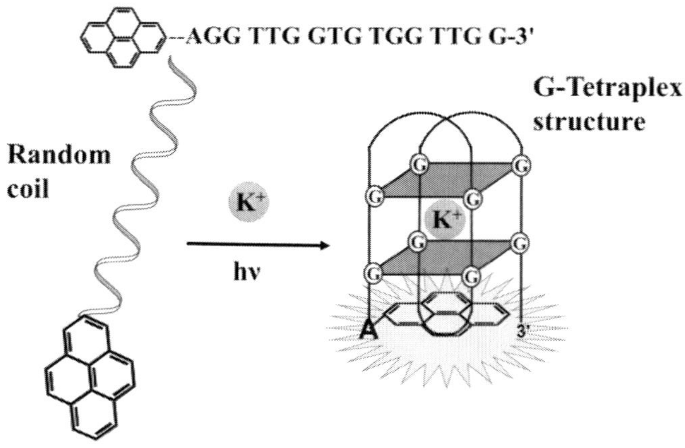

Figure 19. Schematic presentation of quencher-free end-stacking G-quadruplex probing system.

Recently, we have studied G-quadruplex induced FRET within telomeric repeat sequences using nucleoside-tethered FRET pair embedded in the human telomeric G-quadruplex forming sequence 5′-A $(G)_n$ $TT^{Py}A$ $(G)_n$ $TT^{Per}A$ $(G)_n$ TTA $(G)_n$-3′; where $n = 2, 3, 4$; Py = pyrene and Per = perylene [74]. In this case the G-quartet spacer is directly involved to the FRET engineering (Figure 20). Conformational change from a single strand to an anti-parallel G-quadruplex leads to FRET from energy donor (^{Py}A) to acceptor (^{Per}A). The distance between the FRET donor/acceptor partners was controlled by changing the number of G-quartet spacer units. Initially, the conformational change from a single strand ("ss") to an anti-parallel G-quadruplex upon addition of potassium ion (K^+) was confirmed by circular dichroism (CD) spectroscopy which showed characteristic negative and positive CD bands of G-quadruplex DNA at around 265 nm and 293 nm respectively. The hydrophobic staking interaction between the fluorophores (pyrene and perylene) and the external G-quartets [75] was reflected on the UV-visible absorption spectra which showed significant hypochromic shifts (~ 25%) in the chromophoric region.

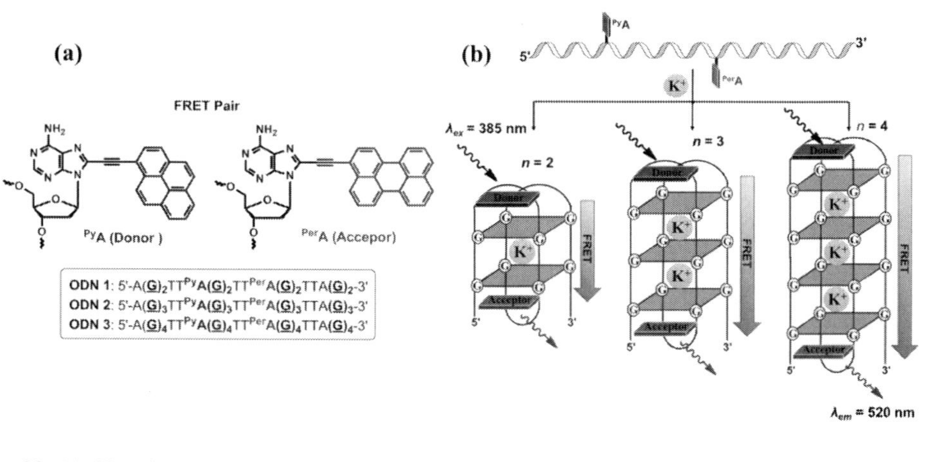

Figure 20. (a) Chemical structure of FRET pair and FRET-pair-embedded G-quadruplex- forming ODN sequences. (b) Schematic presentation of donor (^{Py}A)/acceptor (^{Per}A) tethered G-quadruplex formation and G-quartet spacer-controlled fluorescence resonance energy transfer (FRET).

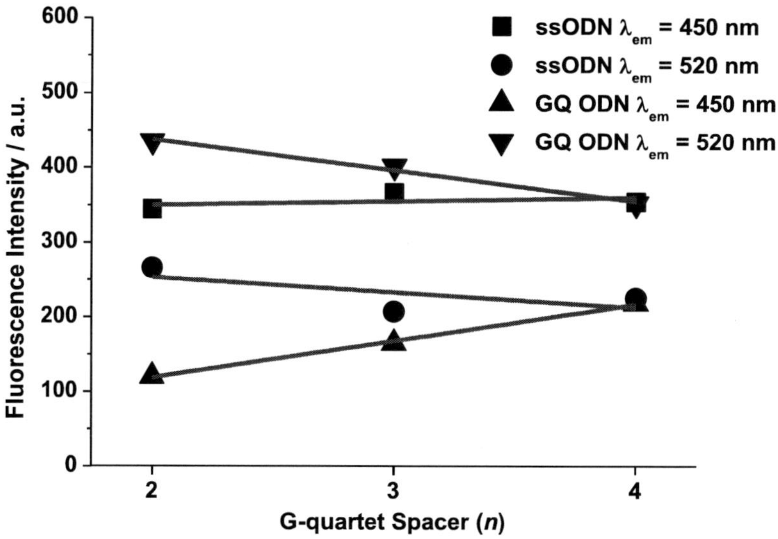

Figure 21. Change of fluorescence intensities of donor at 450 nm and acceptor at 520 nm as a function of number of G-quartet spacer (n).

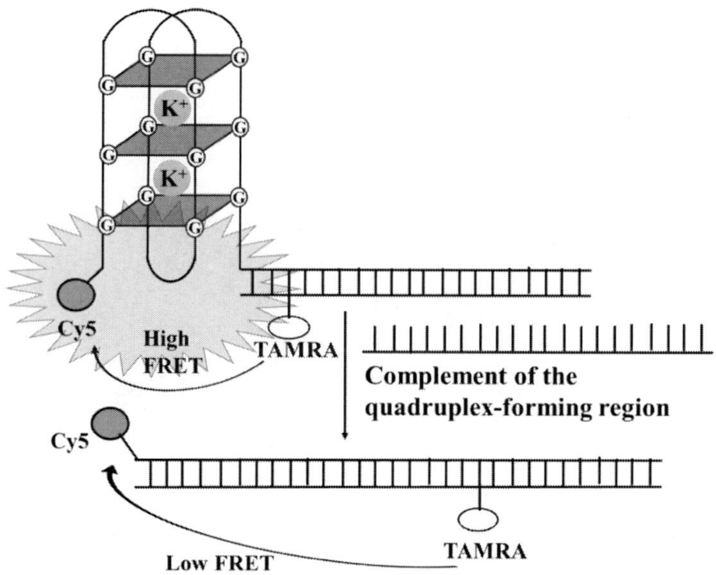

Figure 22. Schematic representation of the unfolding of the G-quadruplex system in presence of complement of the quadruplex-forming region.

Interesting fluorescence emission behavior was observed while adding KCl solution into the ODNs. Decrease of emission intensity of the band at 450 nm (emission maxima of ethynyl pyrene chromophore) and increase of emission intensity of the band at 520 nm (emission maxima of ethynyl perylene chromophore) were observed due to FRET for the ODNs upon conformational change from a single strand to an anti-parallel G-quadruplex (Figure 21). The FRET efficiency decreases with increase in G-quartet units. Our designed system showed that, conformational change from single strand to G-quadruplex structure leads to the sacking interaction between the fluorophore and external G-quartet. However, stacking interaction onto the external G-quartet tends to slip, and thus the orientation of the FRET pairs was not fully controlled during energy transfer in G-quadruplex structure. Thus, the observed FRET of this designed system might be due to both distant-dependent as well as orientation-dependent phenomena.

Balasubramanian and co-workers have investigated the structure and unfolding kinetics of the human telomeric intramolecular G-quadruplex by using singlemolecule fluorescence resonance energy transfer [6]. They used Cy5 and TAMRA (5-Carboxytetramethylrhodamine) FRET pair. Quadruplex formation will bring the two FRET pair in close proximity for energy transfer. Whereas, the unfolding of the quadruplex system will prevent the FRET from TAMRA to Cy5 as the distance between the FRET pair increased and thus low FRET (Figure 22).

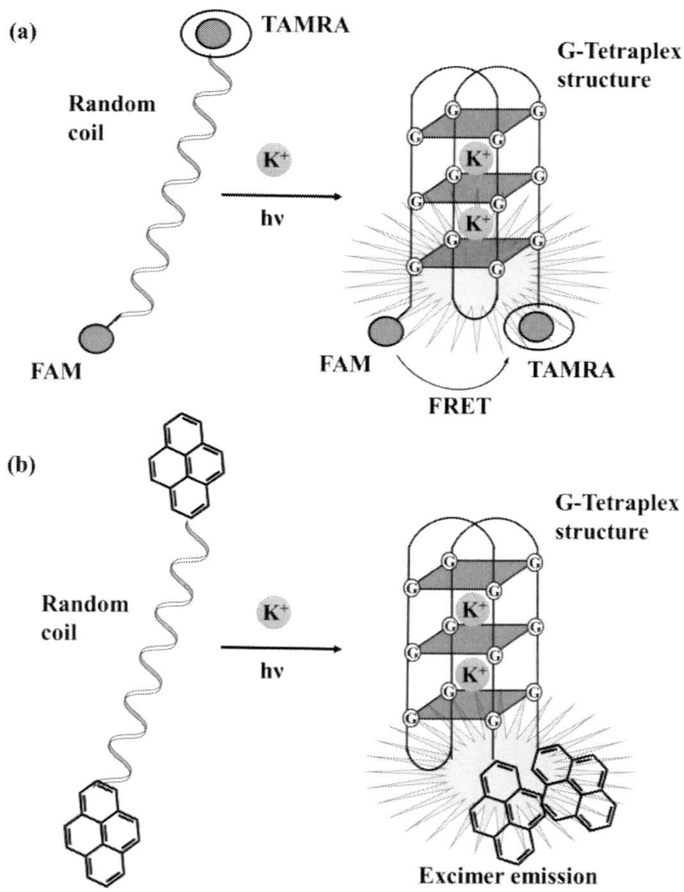

Figure 23. Schematic representation of G-quadruplex templated (a) FRET emission based and (b) excimer emission based potassium ion sensing.

Takenaka and co-workers constructed G-quadruplex templated FRET based potassium ion sensor by attaching fluorophores 6-FAM and 6-TAMRA to the 5′- and 3′-termini of d(GGG TTA GGG TTA GGG TTA GGG), respectively (Figure 23a) [76]. Fluorescence resonance energy transfer from 6-FAM to 6-TAMRA was observed only in the presence of K^+. Moreover, the affinity of K^+ was 43000 times greater than that of Na^+ and thus, this system can selectively detect the K^+ ion in presence of Na^+ ion. Later on they developed G-quadruplex templated pyrene-pyrene excimer based potassium sensor works under physiological conditions [77]. In this approach, they modified the single strand G-rich oligonucleotide with pyrene group at the 5′-end and 3′-end. G-tetraplex formation can bring the two pyrene groups into close proximity and thereby resulting in excimer emissions (Figure 23b).

Conclusion and Perspectives

G-tetraplex is an alternative structure of DNA and is important in the mammalian genome for transcriptional regulation, DNA replication and genome stability. Recent research confirms the existence of G-quadruplex structures and their persistence in human genomic DNA. It is also found that G-quadruplex-stabilizing ligands can modulate transcription. Thus, the design and synthesis of G-quadruplex-stabilizing ligands are still a demanding research area in chemistry and biology. In this chapter we focused on the fluorescent probe whose signal changes upon binding with G-quadruplex DNA and thus results a visual observation. A large number of in vitro G-quadruplex-stabilizing fluorescent probes have been synthesized for academic interest as well as for practical applications. Some recent reports on G-quadruplex selective square planer luminescent metal complexes specially Pt(II) have got particular attention. Moreover, the fluorescent “ON” probes are especially important as they reduce or overcome the background auto fluorescence from the biological sample and thus could be a choice for cell imaging. However, in vivo probes are rarely reported in literature. So, it is a challenging task for us to develop in vivo applicable G-quadruplex selective fluorescent “ON” probes. It may be also helpful if we pay considerable attention on the designing of probes which will exhibit quadruplex specific variable emission wavelength or distinct fluorescence lifetime.

Acknowledgments

The author thanks University of Colorado, Boulder.

References

[1] Cahoon, L. A., Seifert, H. S., (2009). An Alternative DNA Structure Is Necessary for Pilin Antigenic Variation in Neisseria gonorrhoeae. *Science* 325, 764.

[2] Maizels, N., (2006). Dynamic roles for G4 DNA in the biology of eukaryotic cells. *Nat Struct Mol Biol* 13, 1055.

[3] Paeschke, K., Capra, J. A., Zakian, V. A., (2011). DNA replication through G-quadruplex motifs is promoted by the Saccharomyces cerevisiae Pif1 DNA helicase. *Cell* 145, 678.

[4] Beaudoin, J-D., Perreault, J-P., (2010). 5′-UTR G-quadruplex structures acting as translational repressors. *Nucleic Acids Res* 38, 7022.

[5] Lipps, H. J., Rhodes, D., (2009). G-quadruplex structures: in vivo evidence and function. *Trends Cell Biol* 19, 414.

[6] Fernandez-Suarez, M., Ting, A. Y., (2008). Fluorescent probes for super-resolution imaging in living cells. *Nat Rev Mol Cell Biol* 9, 929.

[7] Kobayashi, H., Ogawa, M., Alford, R., Choyke, P. L., Urano, Y., (2009). New Strategies for Fluorescent Probe Design in Medical Diagnostic Imaging. *Chem Rev* 110, 2620.

[8] Neidle, S., (2009). The structures of quadruplex nucleic acids and their drug complexes. *Curr Opin Struct Biol* 19, 239.

[9] Patel, D. J., Phan, A. T., Kuryavyi, V., (2007). Human telomere, oncogenic promoter and 5'-UTR G-quadruplexes: diverse higher order DNA and RNA targets for cancer therapeutics. *Nucleic Acids Res* 35, 7429.

[10] Cosconati, S., Marinelli, L., Trotta, R., Virno, A., Mayol, L., Novellino, E., Olson, A. J., Randazzo, A., (2009). Tandem Application of Virtual Screening and NMR Experiments in the Discovery of Brand New DNA Quadruplex Groove Binders. *J Am Chem Soc* 131, 16336.

[11] White, E. W., Tanious, F., Ismail, M. A., Reszka, A. P., Neidle, S., Boykin, D. W., Wilson, W. D., (2007). Structure-specific recognition of quadruplex DNA by organic cations: influence of shape, substituents and charge. *Biophys Chem* 126, 140.

[12] Hamon, F., Largy, E., Guedin, A., Rouchon-Dagois, M., Sidibe, A., Monchaud, D., Mergny, J-L., Riou, J-F., Nguyen, C-H., Teulade-Fichou, M-P., (2011). An acyclic oligoheteroaryle that discriminates strongly between diverse G-quadruplex topologies. *Angew Chem Int Ed* 50, 8745.

[13] Sun, D., Thompson, B., Cathers, B. E., Salazar, M., Kerwin, S. M., Trent, J. O., Jenkins, T. C., Neidle, S., Hurley, L. H., (1997). Inhibition of Human Telomerase by a G-Quadruplex-Interactive Compound. *J. Med. Chem.* 40, 2113-2116.

[14] Monchaud, D., Allain, C., Teulade-Fichou, M-P., (2006). Development of a fluorescent intercalator displacement assay (G4-FID) for establishing quadruplex-DNA affinity and selectivity of putative ligands. *Bioorg. Med. Chem. Lett.* 16, 4842–4845.

[15] Monchaud, D., Allain, C., Bertrand, H., Smargiasso, N., Rosu, F., Gabelica, V., De Cian, A., Mergny, J.-L., Teulade-Fichou, M.-P., (2008). Ligands playing musical chairs with G-quadruplex DNA: A rapid and simple displacement assay for identifying selective G-quadruplex binders. *Biochimie* 90, 1207-1223.

[16] Tran, P. L. T., Largy, E., Hamond, F., Teulade-Fichou, M-P., Mergny, J-L., (2011). Fluorescence intercalator displacement assay for screening G4 ligands towards a variety of G-quadruplex structures. *Biochimie* 93, 1288-1296.

[17] Cheglakov, Z., Cronin, T. M., He, C., and Weizmann, Y., (2015). Live Cell MicroRNA Imaging Using Cascade Hybridization Reaction. J. Am. Chem. Soc., 137, 6116.

[18] Preus, S., Wilhelmsson, L. M. (2012). Advances in quantitative FRET-based methods for studying nucleic acids. *ChemBioChem* 13, 1990.

[19] Choi, J., Kim, S., Tachikawa, T., Fujitsuka, M., and Majima, T., (2011). pH-Induced Intramolecular Folding Dynamics of i-Motif DNA. *J. Am. Chem. Soc.* 133, 16146.

[20] Sapsford, K. E., Berti, L., Medintz, I. L., (2006). Materials for fluorescence resonance energy transfer analysis: beyond traditional donor-acceptor combinations. *Angew. Chem. Int. Ed.* 45, 4562.
[21] Jares- Erijman, E. A., Jovin, T. M., (2003). FRET imaging. *Nat. Biotechnol.* 21, 1387.
[22] Förster, T., (1946). Energiewanderung und Fluoreszenz. *Naturwissenschaften* 6, 166.
[23] Masuko, M., Ouchi, S., Sode, K., Ohtani, H., Shimadzu, A., (2000). Fluorescence resonance energy transfer from pyrene to perylene labels for nucleic acid hybridization assays under homogeneous solution conditions. *Nucleic Acids Res.* 28, e34.
[24] Grunwell, J. R., Glass, J. L., Lacoste, T. D., Deniz, A. A., Chemla, D. S., Schultz, P. G., (2001). Monitoring the Conformational Fluctuations of DNA Hairpins Using Single-Pair Fluorescence Resonance Energy Transfer. *J. Am. Chem. Soc.* 123, 4295.
[25] Marras, S. A. E., Kramer, F. R., Tyagi, S., (2002). Efficiencies of fluorescence resonance energy transfer and contact-mediated quenching in oligonucleotide probes. *Nucleic Acids Res.* 30, e122.
[26] Ueyama, H., Takagi, M., Takenaka, S., (2002). A Novel Potassium Sensing in Aqueous Media with a Synthetic Oligonucleotide Derivative. Fluorescence Resonance Energy Transfer Associated with Guanine Quartet−Potassium Ion Complex Formation. *J. Am. Chem. Soc.* 124, 14286.
[27] Ying, L., Green, J. J., Li, H., Klenerman, D., Balasubramanian, S., (2003). Studies on the structure and dynamics of the human telomeric G quadruplex by single-molecule fluorescence resonance energy transfer. *Proc. Natl. Acad. Sci. USA* 100, 14629.
[28] Allain, C., Monchaud, D., Teulade-Fichou, M.-P. (2006). FRET Templated by G-Quadruplex DNA: A Specific Ternary Interaction Using an Original Pair of Donor/Acceptor Partners. *J. Am. Chem. Soc.* 128, 11890.
[29] Xu, H., Gao, S., Yang, Q., Pan, D., Wang, L., and Fan, C., (2010). Amplified Fluorescent Recognition of G-Quadruplex Folding with a Cationic Conjugated Polymer and DNA Intercalator. *Applied Materials and Interfaces*, 2, 3211–3216.
[30] Maiti, S., Chaudhury, N. K., and Chowdhurya, S., (2003). Hoechst 33258 binds to G-quadruplex in the promoter region of human c-myc. *Biochem. Biophys. Res. Commun.* 310, 505–512.
[31] Guo, Q., Lu, M., Marky, L. A., and Kallenbach, N. R. (1992). Interaction of the Dye Ethidium Bromide with DNA Containing Guanine Repeats. *Biochemistry* 31, 2451-2455.
[32] Jain, A. K., and Bhattacharya, S., (2011). Interaction of G-Quadruplexes with Nonintercalating Duplex-DNA Minor Groove Binding Ligands. *Bioconjugate Chem.* 22, 2355−2368.
[33] Chen, Q., Kuntz, I. D., Shafer, R. H., (1996). Spectroscopic recognition of guanine dimeric hairpin quadruplexes by a carbocyanine dye. *Proc. Natl. Acad. Sci. USA*, 93, 2635.
[34] Yang, Q., Xiang, J., Yang, S., Zhou, Q., Li, Q., Tang, Y., Xu, G., (2009). Verification of specific G-quadruplex structure by using a novel cyanine dye supramolecular assembly: I. Recognizing mixed G-quadruplex in human telomeres. *Chem. Commun.* 1103.
[35] Lin, D., Fei, X., Gu, Y., Wang, C., Tang, Y., Li, R., and Zhou, J., (2015). A benzindole substituted carbazole cyanine dye: a novel targeting fluorescent probe for parallel c-myc G-quadruplexes. *Analyst*, 140, 5772.

[36] Monchaud, D.; Yang, P.; Lacroix, L.; Teulade-Fichou, M. P.; Mergny, J. L. (2008). A metal-mediated conformational switch controls G-quadruplex binding affinity. *Angew. Chem., Int. Ed.* 47, 4858–4861.

[37] Dash, J.; Shirude, P. S.; Balasubramanian, S. (2008). G-quadruplex recognition by bis-indole carboxamides. *Chem. Commun.* 3055–3057.

[38] Jain, A. K.; Reddy, V. V.; Paul, A.; Muniyappa, K.; Bhattacharya, S. (2009). Synthesis and Evaluation of a Novel Class of G-Quadruplex-Stabilizing Small Molecules Based on the 1,3-Phenylene-Bis(piperazinyl benzimidazole) System. *Biochemistry*, 48, 10693–10704.

[39] Li, G. R.; Huang, J.; Zhang, M.; Zhou, Y. Y.; Zhang, D.; Wu, Z. G.; Wang, S. R.; Weng, X. C.; Zhou, X.; Yang, G. F. (2008). Bis(benzimidazole)pyridine derivative as a new class of G-quadruplex inducing and stabilizing ligand. *Chem. Commun.* 4564–4566.

[40] De Cian, A.; Mergny, J. L. (2007). Quadruplex ligands may act as molecular chaperones for tetramolecular quadruplex formation. *Nucleic Acids Res.* 35, 2483–2493.

[41] Bhattacharya, S.; Chaudhuri, P.; Jain, A. K.; Paul, A. (2010). Symmetrical Bisbenzimidazoles with Benzenediyl Spacer: The Role of the Shape of the Ligand on the Stabilization and Structural Alterations in Telomeric G-Quadruplex DNA and Telomerase Inhibition. *Bioconjugate Chem.* 21, 1148– 1159.

[42] Jin, B., Zhang, X., Zheng, W., Liu, X., Qi, C., Wang, F., and Shangguan, D., (2014). Fluorescence Light-Up Probe for Parallel G-Quadruplexes. *Anal. Chem.* 86, 943−952.

[43] Georgiades, S. N., AbdKarim, N. H., Suntharalingam, K., Vilar, R., (2010). Interaction of Metal Complexes with G-Quadruplex DNA. *Angew. Chem. Int. Ed.* 49, 4020.

[44] Gonzalez, V., Wilson, T., Kurihara, I., Imai, A., Thomas, J. A., and Otsuki, J., (2008). A dinuclear ruthenium(II) complex that functions as a label-free colorimetric sensor for DNA. *Chem. Commun.*, 1868–1870.

[45] Xu, L., Zhang, D., Huang, J., Deng, M., Zhang, M., and Zhou, X., (2010). High fluorescence selectivity and visual detection of G-quadruplex structures by a novel dinuclear ruthenium complex. *Chem. Commun.*, 46, 743–745.

[46] Shi, S., Geng, X., Zhao, J., Yao, T., Wang, C., Yang, D., Zheng, L., Ji, L., (2010). Interaction of $[Ru(bpy)_2(dppz)]^{2+}$ with human telomeric DNA: Preferential binding to G-quadruplexes over i-motif. *Biochimie* 92, 370-377.

[47] Sun, J., An, Y., Zhang, Li, Chen, H-Y., Han, Y., Wang, Y-J., Mao, Z-W, Ji, L-N., (2011). Studies on synthesis, characterization, and G-quadruplex binding of Ru(II) complexes containing two dppz ligands. *J. Inorg. Biochem.* 105, 149–154.

[48] Liao, G-L., Chen, X., Ji, L-N., and Chao, H., (2012). Visual specific luminescent probing of hybrid G-quadruplex DNA by a ruthenium polypyridyl complex. *Chem. Commun.*, 48, 10781–10783.

[49] Reed, J. E., Arnal, A. A., Neidle, S., Vilar, R., (2006). Stabilization of G-Quadruplex DNA and Inhibition of Telomerase Activity by Square-Planar Nickel(II) Complexes. *J. Am. Chem. Soc.* 128, 5992.

[50] Wang, P., Leung, C.-H., Ma, D.-L., Yan, S.-C., Che, C.-M., (2010). Structure-Based Design of Platinum(II) Complexes as c-myc Oncogene Down-Regulators and Luminescent Probes for G-Quadruplex DNA. *Chem. Eur. J.* 16, 6900.

[51] Ma, D.-L., Che, C.-M., Yan, S.-C., (2009). Platinum(II) Complexes with Dipyridophenazine Ligands as Human Telomerase Inhibitors and Luminescent Probes for G-Quadruplex DNA. *J. Am. Chem. Soc.* 131, 1835.

[52] Yam, V. W. W., Chan, K. H. Y., Wong, K. M. C., and Chu, B. W. K., (2006). You have full text access to this content Luminescent Dinuclear Platinum(II) Terpyridine Complexes with a Flexible Bridge and "Sticky Ends." Angew. Chem., Int. Ed., 45, 6169.

[53] Yu, C., Chan, K. H. Y., Wong, K. M. C., and Yam, V. W. W., (2006). Single-stranded nucleic acid-induced helical self-assembly of alkynylplatinum(II) terpyridyl complexes. Proc. Natl. Acad. Sci. U. S. A., 103, 19652.

[54] Chung, C. Y. S., Chan, K. H. Y., and Yam, V. W. W., (2011). "Proof-of-principle" concept for label-free detection of glucose and a -glucosidase activity through the electrostatic assembly of alkynylplatinum(II) terpyridyl complexes. Chem. Commun., 47, 2000.

[55] Chung, C. Y. S., and Yam, V. W. W., (2011). Induced Self-Assembly and Förster Resonance Energy Transfer Studies of Alkynylplatinum(II) Terpyridine Complex Through Interaction With Water-Soluble Poly(phenylene ethynylene sulfonate) and the Proof-of-Principle Demonstration of this Two-Component Ensemble for Selective Label-Free Detection of Human Serum Albumin. J. Am. Chem. Soc., 133, 18775.

[56] Chung, C. Y. S., and Yam, V. W. W., (2013). Selective label-free detection of G-quadruplex structure of human telomere by emission spectral changes in visible-and-NIR region under physiological condition through the FRET of a two-component PPE-$SO3^{-}$–Pt(II) complex ensemble with Pt…Pt, electrostatic and π-π interactions. *Chem. Sci.*, 4, 377.

[57] Chan, K., Chung, C. Y. S., and Yam, V. W. W., (2016). Parallel folding topology-selective label-free detection and monitoring of conformational and topological changes of different G-quadruplex DNAs by emission spectral changes via FRET of mPPE-Ala–Pt(II) complex ensemble. *Chem. Sci.*, 7, 2842.

[58] Mohanty, J., Barooah, N., Dhamodharan, V., Harikrishna, S., Pradeepkumar, P. I., and Bhasikuttan, A. C., (2013). Thioflavin T as an Efficient Inducer and Selective Fluorescent Sensor for the Human Telomeric G-Quadruplex DNA. *J. Am. Chem. Soc.* 135, 367−376.

[59] Gabelica, V., Maeda, R., Fujimoto, T., Yaku, H., Murashima, T., Sugimoto, N., and Miyoshi, D., (2013). Multiple and Cooperative Binding of Fluorescence Light-up Probe Thioflavin T with Human Telomere DNA G-Quadruplex. *Biochemistry* 52, 5620−5628.

[60] Liu, L., Shao, Y., Peng, J., Huang, C., Liu, H., and Zhang, L., (2014). Molecular Rotor-Based Fluorescent Probe for Selective Recognition of Hybrid G-Quadruplex and as a K^+ Sensor. *Anal. Chem.* 86, 1622−1631.

[61] Duxbury, D. F., (1993). The photochemistry and photophysics of triphenylmethane dyes in solid and liquid media. *Chem. Rev.* 93, 381–433.

[62] Kandela, I. K., Bartlett, J. A., Indig, G. L., (2002) *Photochem. Photobiol. Sci.* 1, 309–314.

[63] Liao, J., Roider, J., Jay, D. G., (1994). Chromophore-assisted laser inactivation of proteins is mediated by the photogeneration of free radicals. *Proc. Natl. Acad. Sci. USA* 91, 2659–2663.

[64] Babendure, J. R., Adams, S. R., Tsien, R. Y., (2003). Aptamers Switch on Fluorescence of Triphenylmethane Dyes. *J. Am. Chem. Soc.* 125, 14716–14717.

[65] Kolpashchikov, D. M. (2005). Binary Malachite Green Aptamer for Fluorescent Detection of Nucleic Acids. *J. Am. Chem. Soc.* 127, 12442–12443.

[66] Bhasikuttan, A. C., Mohanty, J., and Pal, H., (2007). Interaction of Malachite Green with Guanine-Rich Single-Stranded DNA: Preferential Binding to a G-Quadruplex. *Angew. Chem. Int. Ed.*, 46, 9305–9307.

[67] Kong, D-M., Ma, Y-E., Wu, J., and Shen, H-X., (2009). Discrimination of G-Quadruplexes from Duplex and Single-Stranded DNAs with Fluorescence and Energy-Transfer Fluorescence Spectra of Crystal Violet. *Chem. Eur. J.* 15, 901– 909.

[68] Chang, C-C., Wu, J-Y., Chien, C-W., Wu, W-S., Liu, H., Kang, C-C., Yu, L-J., and Chang, T-C., (2003). A Fluorescent Carbazole Derivative: High Sensitivity for Quadruplex DNA. *Anal. Chem.* 75, 6177-6183.

[69] Dumat, B., Bordeau, G., Faurel-Paul, E., Mahuteau-Betzer, F., Saettel, N., Bombled, M., Metgé, G., Charra, F., Fiorini-Debuisschert, C., Teulade-Fichou, M.-P., (2011). N-phenyl-carbazole-based two-photon fluorescent probes: Strong sequence dependence of the duplex vs quadruplex selectivity. *Biochimie* 93, 1209-1218.

[70] Hong, Y., Haubler, M., Lam, J. W. Y., Li, Z., Sin, K. K., Dong, Y., Tong, H., Liu, J., Qin, A., Renneberg, R., and Tang, B. Z., (2008). Label-Free Fluorescent Probing of G-Quadruplex Formation and Real-Time Monitoring of DNA Folding by a Quaternized Tetraphenylethene Salt with Aggregation-Induced Emission Characteristics. *Chem. Eur. J.* 14, 6428–6437.

[71] Arthanari, H., Basu, S., Kawano, T. L., and Bolton, P. H., (1998). Fluorescent dyes specific for quadruplex DNA. *Nucleic Acids Res.* 26, 3724–3728.

[72] Han, F. X., Wheelhouse, R. T., and Hurley, L. H., (1999). Interactions of TMPyP4 and TMPyP2 with Quadruplex DNA. Structural Basis for the Differential Effects on Telomerase Inhibition. *J. Am. Chem. Soc.* 121, 3561-3570.

[73] Seo, Y. J., Lee, I. J., Yi, J. W., and Kim, B. H., (2007). Probing the stable G-quadruplex transition using quencher-free endstacking ethynyl pyrene–adenosine. *Chem. Commun.*, 2817–2819.

[74] Kundu, R., (2016). G-Tetraplex-Induced FRET within Telomeric Repeat Sequences Using PyA-PerA as Energy Donor–Acceptor Pair. *Chem. Asian J.* 11, 198-201.

[75] Kawai, T., Ikegami, M., and Arai, T., (2004). Exciplex formation between pyrene and guanine in highly polar solvents. *Chem. Commun.* 824.

[76] Ueyama, H., Takagi, M., Takenaka, S., (2002). A Novel Potassium Sensing in Aqueous Media with a Synthetic Oligonucleotide Derivative. Fluorescence Resonance Energy Transfer Associated with Guanine Quartet−Potassium Ion Complex Formation. *J. Am. Chem. Soc.* 124, 14286.

[77] Nagatoishi, S., Nojima, T., Juskowiak, B., Takenaka, S., (2005). A Pyrene-Labeled G-Quadruplex Oligonucleotide as a Fluorescent Probe for Potassium Ion Detection in Biological Applications. *Angew. Chem. Int. Ed.* 44, 5067.

In: G-Quadruplex Structures, Formation and Role in Biology ISBN: 978-1-63485-512-9
Editor: Hannah Santos

Chapter 5

G-Quadruplex Interacting Small Molecules and Their Biological Implications

Sebastian Müller*
Institut Curie Research Center,
Organic Synthesis and Cell Biology Group, Paris, France

Abstract

Much of what we have learned about G-quadruplexes stems from the research conducted by chemists, who designed specific molecules to target these structures *in vitro* and *in vivo*. Over the past decades there has been an increasing success in selectively targeting G-quadruplex DNA and RNA structures with small molecules, leading to elucidating some of their implications in biological processes *in vivo*. Biological processes, where G-quadruplexes have implications, include telomere maintenance, the DNA damage response, gene transcription and translation and genetic and epigenetic control mechanisms. This has led to the development of potential therapeutic agents, particularly as a novel anti-cancer strategy, to specifically target aspects involving these processes. This chapter deals first with the historical aspect of how major biological roles of G-quadruplexes have been uncovered by the use of specifically designed small molecule probes. This is used to highlight how fundamental research has contributed to identifying biological functions of G-quadruplexes and their potential as therapeutic targets. This will be accompanied with an overview of the major classes of small molecules in this research field. Finally, this chapter discusses the future applications of G-quadruplex interacting small molecules, including the development as potential drug candidates.

Keywords: G-quadruplex, cancer, small molecule, genetics, DNA, RNA, chemical biology

---

* Corresponding Author address Email: Sebastian.muller@curie.fr.

Introduction

Chromatin is a large nucleoprotein complex organized into distinct structures, which define specific gene expression and cell lineage commitment. While complex protein networks are implicated in maintaining chromatin structure and function (Chen and Dent, 2014; Filipescu et al., 2014; Müller and Almouzni, 2014; Müller and Rodriguez, 2014; Papamichos-Chronakis and Peterson, 2013; Swygert and Peterson, 2014; van Steensel, 2011), DNA itself can also adopt distinct secondary structures besides the standard B-helix (Ghosh and Bansal, 2003; Zhao et al., 2010), which adds to the structural repertoire that can influence genome accessibility and function. Certain G-rich sequences can adopt supramolecular structures called G-quadruplexes, composed of tetrads of Hoogsten hydrogen-bonded guanine tetrads that stack on top of each other via π-π interactions (Gellert et al., 1962; Gilbert and Feigon, 1999). Small molecules can target G-quadruplexes in a *structure-specific* manner, a powerful means of site-specific chemical intervention in the genome (Figure 1). G-quadruplexes have been studied by nuclear magnetic resonance (Phan et al., 2004b) and X-ray crystallography (Parkinson et al., 2002) and they can adopt various topologies, depending on primary sequence, numbers of strands involved and strand orientation (Davis, 2004). G-quadruplex-forming sequences (PQS) are prevalent in the human genome and are enriched in promoter regions as revealed by bioinformatic analyses (Huppert and Balasubramanian, 2005, 2007; Todd et al., 2005). There is evidence supporting their existence in cells, such as pull-down of G-quadruplexes from cells using tagged small molecules (Müller et al., 2010), the visualization of them in cells using fluorescently labelled small molecules (Rodriguez et al., 2012) and life-time imaging (Shivalingam et al., 2015), and G-quadruplex specific antibodies *in cellulo* (Biffi et al., 2014; Biffi et al., 2013; Henderson et al., 2014; Lam et al., 2013; Schaffitzel et al., 2010). Using a G-quadruplex specific antibody in conjunction with deep sequencing, G-quadruplex sites were mapped throughout the genome in adenocarcinoma cells (Lam et al., 2013). There is functional evidence of G-quadruplex nucleic acids in *in cellulo* and promoter G-quadruplexes may play a role in gene regulation (Balasubramanian et al., 2011; Rangan et al., 2001; Seenisamy et al., 2004). ATR-X, which is mutated in alpha thalassemia/mental retardation syndrome X-linked binds to G-quadruplex DNA to regulate the expression of key genes and genome-wide sequencing revealed that ATR-X binds to DNA with long G-repeats that can adopt G-quadruplexes (Law et al., 2010). In chicken DT40 cells, G-quadruplex DNA is poorly replicated in REV1-deficient cells leading to altered histone turnover at G-quadruplex forming *loci* (Sarkies et al., 2010). The human pathogen *Neiseria gonorhoeae* uses a 16-nucleotide G-rich sequence that can adopt a G-quadruplex conformation for the initiation of recombination to promote antigenic variation (Cahoon and Seifert, 2009). Thus, G-quadruplexes have possible implications in a range of biological functions, including transcription, translation, telomere maintenance, replication, the DNA damage response, genome rearrangements, and they are involved in epigenetic control mechanisms (Baral et al., 2013; Bochman et al., 2012; Lipps and Rhodes, 2009; Maizels and Gray, 2013; Murat and Balasubramanian, 2013). Targeting G-quadruplexes with small molecules has emerged as a strategy to design (a) chemical tools to interfere with biological processes and (b) potential drugs that could be used to combat disease, in particular cancer (Balasubramanian et al., 2011; Balasubramanian and Neidle, 2009; Miller and Rodriguez, 2011; Neidle, 2010; Shalaby et al., 2013; Wu and Brosh, 2010). A variety of G-quadruplex interacting small molecules have been developed and many interesting chemical tools have emerged, which have contributed greatly to our understanding of the underlying biological mechanisms involving G-quadruplexes in cells. Some of these small molecules have been developed further as potential therapeutic agents. This chapter discusses recent advances concerning biological tools that have been devised to study G-quadruplexes to delineate associated biological processes, and it discusses the development of next generation G-quadruplex interacting small molecules aimed at therapeutic intervention. This chapter does not constitute a discussion of the extensive list of scaffold previously reported, but instead it provides an insight of the mechanistic aspects associated with some well-established G-quadruplex interacting compounds, focussing on recent findings.

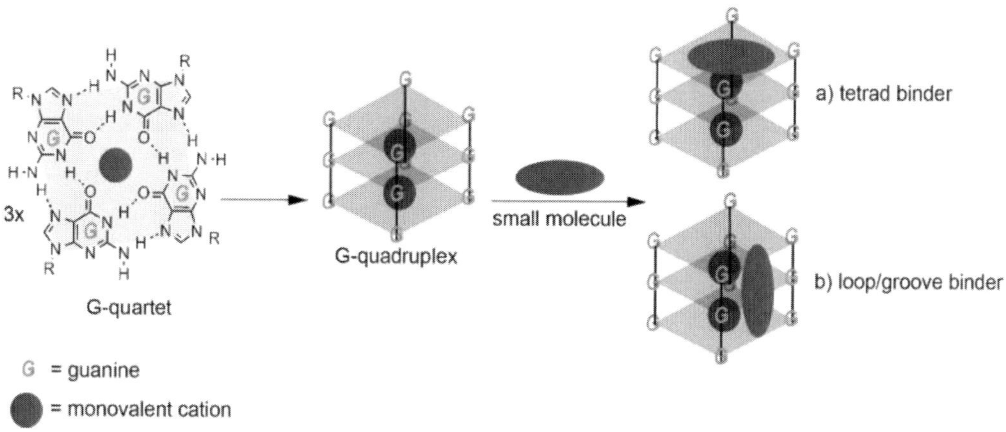

Figure 1. G-quadruplex formation and interaction modes with small molecules. G-tetrads are formed by four guanines bonded via Hoogsteen base pairing. G-quadruplexes comprise three stacked G-tetrads connected by loops. They are stabilized by monovalent cations coordinated in the central cavity. The interactions of small molecules with G-quadruplex nucleic acids can be (a) end-to-end stacking (on the top and/or bottom tetrad), (b) binding to the loops and grooves.

STUDYING G-QUADRUPLEX BIOLOGY WITH SMALL MOLECULES

G-quadruplexes have reported implications in various biological processes including: (1) telomere biology and DNA damage, (2) gene transcription and translation and (3) other mechanisms controlling genome integrity. Small molecules represent remarkable tools to study biological processes and this has been exploited widely. This section describes the knowledge gained about the biology involving G-quadruplexes from small molecule studies. An overview of the structures of selected G-quadruplex interactors discussed in detail can be found in Figure 2. An illustrated summary of the major cellular responses observed upon treatment with different G-quadruplex interacting small molecules is depicted in Figure 3.

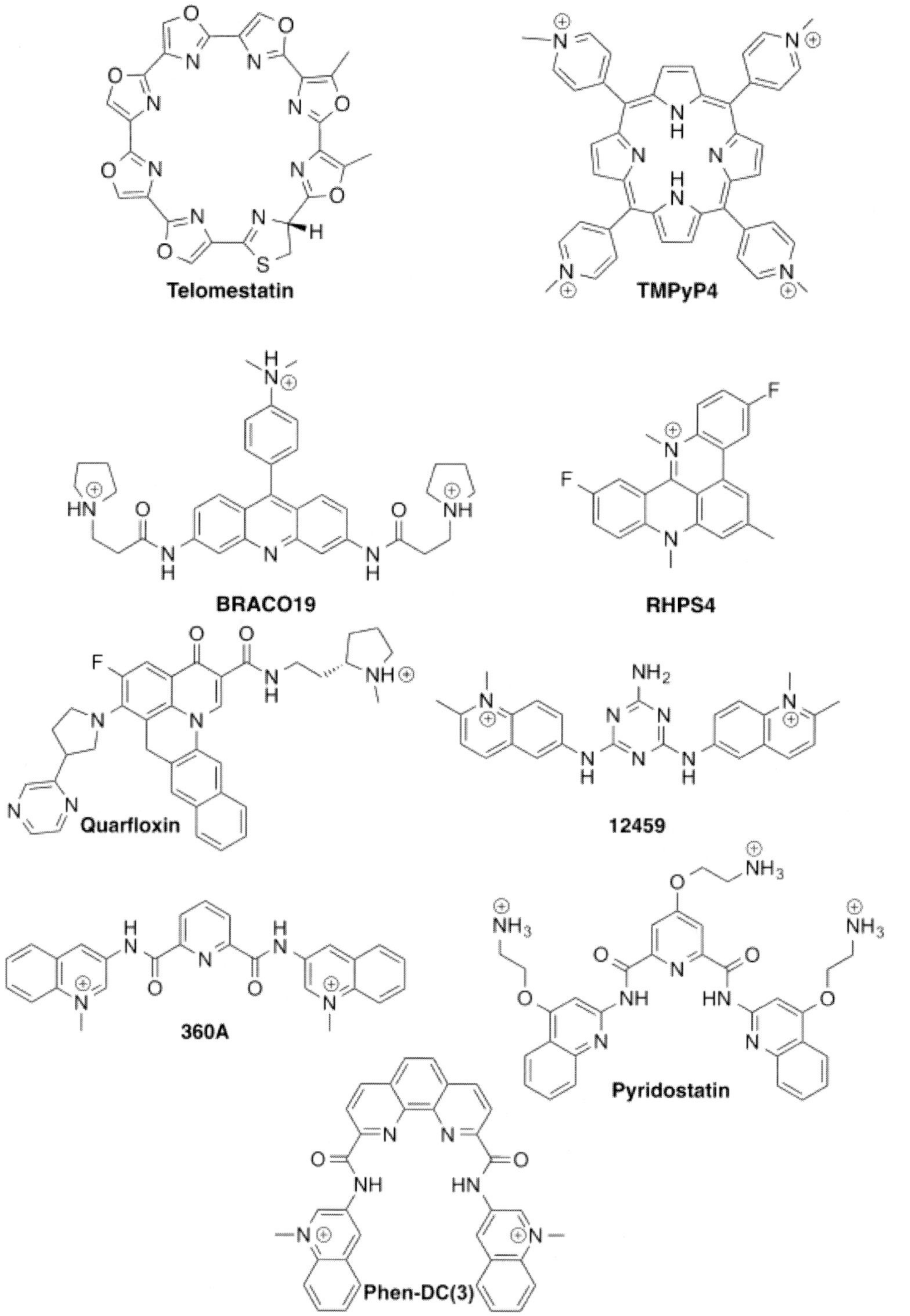

Figure 2. Selection of G-quadruplex interacting small molecules. Telomestatin is a natural product. The other molecules are synthetic molecules that have been developed to study the biological mechanisms of G-quadruplexes.

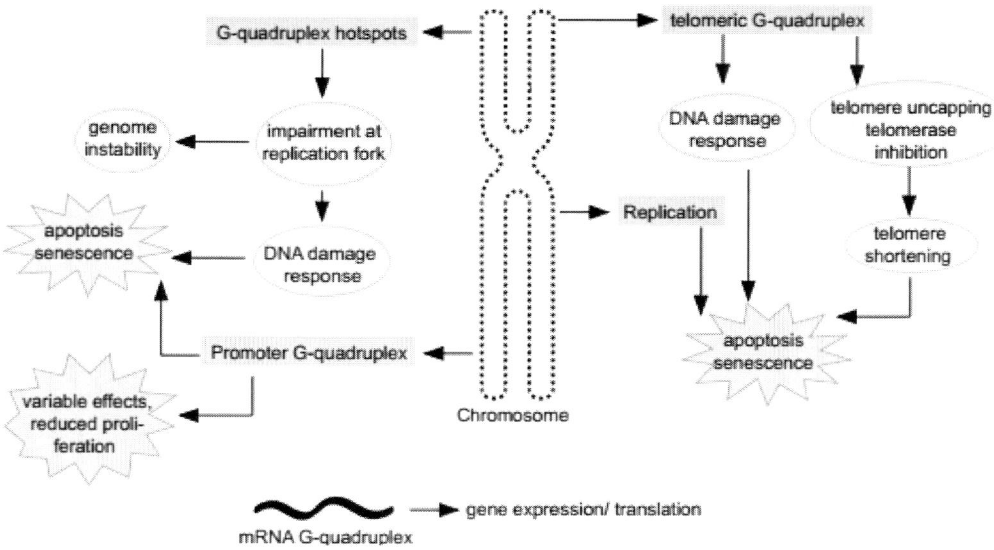

Figure 3. Cellular implications of G-quadruplexes. Overview of mechanistic implications of G-quadruplexes in the genome.

TELOMERE MAINTENANCE AND THE DNA DAMAGE RESPONSE

Human telomeres are chromosomes ends comprising a long stretch of double-stranded tandem repeats and a short single-stranded G-overhang with the sequence $(TTAGGG)_n$ (Hemann and Greider, 1999; Zakian, 1995). Using X-ray crystallography and nuclear magnetic resonance (NMR) it was demonstrated that the telomeric sequence could form G-quadruplexes (Dai et al., 2007; Parkinson et al., 2002; Phan et al., 2007; Wang and Patel, 1993; Williamson et al., 1989; Zhang et al., 2010). Shelterin protects human telomeres (de Lange, 2005), and the three protein subunits POT1, TRF1 and TRF2 directly bind the DNA repeat sequence. Telomerase has reverse transcriptase activity and synthesizes TTAGGG repeats to the single stranded telomeric overhang to maintain telomere length. Given that telomerase is overexpressed in about 85% of cancers (Kim et al., 1994; Shay and Wright, 2006), telomere shortening due to reduced telomerase expression could be a tumour suppressor mechanism (Hanahan and Weinberg, 2011; Stewart and Weinberg, 2006). Inhibition of the catalytic subunit of telomerase (hTERT) in humans, by antisense siRNA or small molecules, can inhibit cancer cell growth, which suggests that induction of telomere shortening could be used as a potential therapeutic strategy (Erlitzki and Fry, 1997; Neidle, 2010). The formation of G-quadruplexes inhibits the function of telomerase (Zahler et al., 1991), by rendering the 3'-end of the single stranded DNA overhang inaccessible for hybridization with the telomerase RNA template. G-quadruplex formation from telomeric overhangs may contribute to suppression of DNA damage signals, by protecting them from replication protein A (RPA) binding (Ray et al., 2014). Interestingly, RPA also prevents G-quadruplex formation at lagging-strand telomeres to allow maintenance of chromosome ends (Audry et al., 2015). 2,6-diamidoanthraquinone was identified as a stabilizer of the telomeric G-quadruplex leading to telomere inhibition (Sun et al., 1997). Later, the natural product telomestatin was reported as a potent stabilizer for telomeric G-quadruplexes (Shin-ya et al., 2001). Telomestatin can stabilize the single stranded DNA G-quadruplexes, which competes with POT1 binding, leading to the activation of the DNA damage response, caused by the presence of the naked single stranded overhang (Gomez et al., 2006; Temime-Smaali et al., 2009). Interestingly, a panel of synthetic G-quadruplex interacting small molecules has the same effect. The triazine 12459 (Gomez et al., 2003; Riou et al., 2002) induces telomere shortening and POT1 uncapping, leading to cellular senescence. The acridine BRACO19, one of the very first promising synthetic G-quadruplex interacting small molecules, is a telomerase inhibitor (Gunaratnam et al., 2007). RHPS4 induces telomere-uncapping leading to a telomere-induced DNA damage response (Leonetti et al., 2004; Phatak et al., 2007; Salvati et al., 2007). Treatment of cultured cells with 360A did not induce telomere shortening but lead to telomere end-fusion and anaphase bridges in telomerase negative cells, which maintain telomere length by the alternative lengthening of telomeres (ALT) mechanism (Granotier et al., 2005; Pennarun et al., 2005). 360A also induces a DNA damage response at telomeres through an ATM-dependent mechanism (Pennarun et al., 2008). However, another study on NSC746364 (Lee et al., 2013) highlighted also that an ATR-dependent DNA damage pathway can be initiated in cultured cells upon treatment with a G-quadruplex interactor (Chung et al., 2014b). The ligand N-methyl mesoporphyrin IX (NMM) causes an increase in telomeric G-quadruplexes, and telomerase colocalizes with a subset of telomeric G-quadruplexes *in vivo* (Moye et al., 2015). Pyridostatin, a very good stabilizer of G-quadruplexes with high selectivity over duplex DNA *in vitro*, also uncaps POT1 from human telomeres and triggers a DNA damage response (Rodriguez et al., 2008). A plethora of synthetic small molecules synthetically more accessible than telomestatin have high G-quadruplex affinity including the telomeric G-quadruplex (Cosconati et al., 2012; De Cian et al., 2007; Jantos et al., 2006; Mergny et al., 2001; Monchaud and Teulade-Fichou, 2008; Müller et al., 2009; Perry et al., 1999). A study using a pyridostatin derivative carrying with a biotin affinity tag isolated telomeric DNA from cells (Müller et al., 2010), and treating cultured cells with pyridostatin or some of its analogues leads to telomere shortening and senescence (Müller et al., 2010; Müller et al., 2012). Taken together, small molecules designed to stabilize these structures may cause (a) an inhibition of telomerase and (b) DNA damage responses analogous to those of other mediators of telomere damage (d'Adda di Fagagna et al., 2003), leading to replicative senescence or apoptosis. Whole-genome sequencing of γH2AX, which is present at double strand DNA breaks, upon treatment with pyridostatin, revealed hotspots all over genomic DNA that contain clusters of PQS (Rodriguez et al., 2012), providing evidence that this small molecule also targets G-quadruplexes at other genomic sites. Fluorescent labelling of the clickable analogue pyridostatin-α revealed a visual means to map interacting sites of this small molecule in cells and provide evidence for G-quadruplex formation in genomic DNA. Other probes to directly visualize G-quadruplexes are also in development (Monchaud et al., 2007; Yang et al., 2009). Indeed, there was considerable overlap of labelled pyridostatin-α with Pif1a, a helicase that binds to and unfolds G-quadruplex DNA (Bochman et al., 2010; Paeschke et al., 2013; Zhang et al., 2006). Enrichment of γH2AX across the gene body of *SRC* correlates with genomic regions containing clusters of PQS. This study revealed the sites of interaction of this small molecule throughout the genome and highlighted that the mechanistic effect is not limited to telomeres, but rather suggests a pleiotropic cellular response. Interestingly, a study in *C.*

elegans revealed a specific polymerase theta-dependent repair pathway suppressing extensive genomic instability at G-quadruplex DNA sites (Koole et al., 2014).

GENE TRANSCRIPTION AND TRANSLATION

The binding of ATR-X, which is mutated in alpha thalassemia/mental retardation syndrome X-linked, to G-quadruplex DNA to regulate key genes (Law et al., 2010) may pose an interesting therapeutic target for G-quadruplex interacting small molecules. Indeed, genome-wide deep sequencing of ATR-X binding sites revealed that ATR-X binding correlates with long PQS-enriched DNA. Gene promoters have drawn their attention as potential targets for G-quadruplex interacting molecules. This was initiated by a study that suggested that the small molecule TMPyP4 could target a PQS in the promoter of the *MYC* oncogene leading to gene repression (Siddiqui-Jain et al., 2002). A specific G to T mutation in the *MYC* promoter destabilizing the promoter G-quadruplex resulted in an increase in transcription, which was partially alleviated by TMPyP4. Bioinformatic analyses revealed an enrichment of G-quadruplexes in gene promoters (Huppert and Balasubramanian, 2007) and in 5'-UTRs of mRNAs (Huppert et al., 2008). Therefore, G-quadruplexes might be (a) involved in influencing transcription and (b) exert control over translation (Kumari et al., 2007). G-quadruplex interacting molecules can stabilize RNA G-quadruplexes, which can lead to a down-regulation of translational activity (Bugaut et al., 2010; Gomez et al., 2010). A study suggested that G-quadruplexes are involved in the control of translation of Epstein-Barr virus-encoded nuclear antigen 1 (EBNA1) mRNA (Murat et al., 2014). G-quadruplexes forming in mRNA are also required for mRNA binding to the protein TDP-43, which is crucial for the function of neurites (Ishiguro et al., 2016). PQS are found in promoters of a wide range of genes important for cell signaling and some promoter PQS could potentially be interesting targets to study the intervention with small molecules. This includes in particular the *KIT* gene (Fernando et al., 2006; Rankin et al., 2005), which encodes for a tyrosine protein kinase and is a clinically validated drug target for treating gastrointestinal stromal tumors (Heinrich et al., 2003), and in the *KRAS* gene (Cogoi and Xodo, 2006), which encodes for the GTPase KRAS. PQS in promoters can fold into intramolecular G-quadruplexes (Neidle, 2009), which require at least four G-tracts. Promoter G-quadruplexes can exhibit different conformations (Ambrus et al., 2005; Fernando et al., 2006; Phan et al., 2004a; Shklover et al., 2010; Tong et al., 2011; Wei et al., 2012; Wei et al., 2013). In addition, many promoter PQS contain more than four G-tracts giving rise to structural polymorphism. The structural differences between promoter G-quadruplexes may be exploited for the design of small molecules to discriminate between them, which is challenging, given (a) the abundance of promoter PQS throughout the genome, and (b) the high degree of G-quadruplex polymorphism, questioning the premise that G-quadruplexes in certain promoters can be targeted selectively by small molecules. Promoter G-quadruplexes are constrained by the complementary DNA strand of duplex DNA, unlike their telomeric or mRNA counterparts. Therefore, they need to compete with duplex DNA. Maybe negative superhelicity during transcription or G-quadruplex interacting proteins that stabilize them once they are formed may play a role in these processes, which should be addressed in future studies. G-quadruplex interacting proteins in human promoters have been characterized, such as NM23-H2

(Dexheimer et al., 2009) and nucleolin (Gonzalez et al., 2009) in the *MYC* promoter. NM23-H2 is suggested to unwind G-quadruplexes and thus facilitates transcription (Thakur et al., 2009). Nucleolin binds G-quadruplexes of the *MYC* promoter sequence with higher affinity than the nucleolin recognition element, a previously described RNA substrate, and nucleolin suppresses *MYC* transcription by preventing Sp1-induced transcription (Thakur et al., 2009). However, cellular nucleic-acid-binding protein (CNBP) also binds to the parallel conformation of the *MYC* promoter G-quadruplex, but enhances transcription. Interestingly, nucleolin stabilizes G-quadruplex structures folded by the LTR promoter and silences HIV-1 viral transcription (Tosoni et al., 2015). Recently, it was suggested that a complex network of genes with G-quadruplexes in promoters could influence embryonic development in zebrafish (David et al., 2016). The regulatory mechanisms of gene promoters involving G-quadruplexes are complex (Balasubramanian et al., 2011). Further research is required to dissect how this circumstantial evidence actually relates to the regulation of genes by G-quadruplexes, or whether it is a result of a pleiotropic effect of these molecules on the cells. Other compounds that have been studied for their interaction with the promoter G-quadruplex of *MYC* include quindolines (Boddupally et al., 2012; Ou et al., 2011; Ou et al., 2007), actinomycin D (Kang and Park, 2009) and Phen-DC(3) (Chung et al., 2014a). Platinum complexes (Wang et al., 2010) and the synthetic telomestatin derivative S2T1-6OTD (Shalaby et al., 2010) also show significant selectivity for G-quadruplex DNA and destabilize the *MYC* promoter G-quadruplex, leading to down-regulation of expression. However, platinum complexes show an even higher affinity for telomeric G-quadruplexes (Xu et al., 2013), suggesting that these molecules target various genomic sites. Another approach to target *MYC* expression employed a small oligonucleotide sequence Pu27 (Islam et al., 2014; Sedoris et al., 2012), which contains the PQS of the gene promoter, and showed it induces cell and mitochondrial swelling, DNA damage and membrane permeability characteristic of oncotic necrosis. Targeting the *KIT* promoter G-quadruplex has also been under investigation. These studies include for instance benzyl phenoxazines and indenoisoquindolines, trisubstituted isoalloxazines, certain foldamers and the quindoline derivative SYUIQ-5 (Bejugam et al., 2010; Bejugam et al., 2007; Gunaratnam et al., 2009; McLuckie et al., 2011; Müller et al., 2014; Shen et al., 2013). Since the transcription factor SP1 can bind to the promoter PQS *in vitro*, the involvement of promoter G-quadruplexes in the regulation of the *KIT* gene might be more complex (Raiber et al., 2012). Therefore, G-quadruplex interacting small molecules could potentially influence the equilibrium of SP1 binding at the *KIT* promoter and other genes. Besides the *KIT* promoter, SYUIQ-5 also targets the *MYC* promoter G-quadruplex, inhibits telomerase activity and causes telomere damage (Liu et al., 2007). This again highlights the challenge for selectivity between different G-quadruplexes and demonstrates that G-quadruplex interacting small molecules can cause a pleiotropic cellular response by hitting various G-quadruplex targets. Taken together, these studies demonstrate that small molecules designed to target promoter PQS can certainly effect gene expression levels. However, given the fact that a large subset of the small molecules presented here stabilizes an array of promoter G-quadruplexes as well as the telomeric counterpart and PQS in gene bodies (Rodriguez et al., 2012), this one-gene-one-target approach might ultimately fail. Furthermore, the idea of targeting only one gene, usually represented by only two alleles in somatic cells, with relatively high concentrations of small molecules used in these studies, seems very difficult to attain. Changes of expression in single genes, such as *MYC* or *KIT* may be causative of other cellular responses unrelated to G-quadruplexes. To address these issues, future investigations could dissect further the global response to cellular treatments with small molecules, integrating gene expression profiling and ChiP-Seq of DNA damage markers or helicases and polymerases involved in unwinding G-quadruplexes in response to treatment with G-quadruplex interacting small molecules (Rodriguez et al., 2012).

GENOMIC CONTROL MECHANISMS

Interest in G-quadruplexes has expanded into other biological fields, including the study of genome integrity and epigenetic control. Since duplex DNA is opened up and becomes single-stranded during replication when they do not compete with double-stranded DNA, the hypothesis of G-quadruplexes forming during replication has drawn much attention (Eddy et al., 2014; Schwab et al., 2013; Sun et al., 1998). A number of DNA helicases, some of which are genetically linked to human disease, catalytically unwind G-quadruplex DNA (see Table 1 for summary). Among the earliest helicases identified in this process were the Bloom's syndrome helicase (BLM) and the Werner syndrome DNA helicase (WRN) (Brosh and Bohr, 2007; Budhathoki et al., 2015; Damerla et al., 2012; Fry and Loeb, 1999; Mohaghegh et al., 2001; Sun et al., 1998). The G-quadruplex interactors belonging to the family of trisubstituted acridines can stall these helicaes (Li et al., 2001). Recently, a structure of BLM with a G-quadruplex was described (Chatterjee et al., 2014). BLM facilitates telomere replication during leading strand synthesis of telomeres and deficiency of BLM or WRN led to greater increases in G-quadruplex DNA detected in the telomere (Drosopoulos et al., 2015). In addition, regulation of gene expression by BLM correlates with the presence of G-quadruplex DNA motifs in human cells (Nguyen et al., 2014). The FANCJ helicase can also unwind G-quadruplexes *in vitro* (London et al., 2008; Wu et al., 2008; Wu et al., 2009), which can be inhibited by G-quadruplex interacting molecules such as Phen-DC(3) (Bharti et al., 2013). Werner syndrome, Bloom's syndrome and FA complementation group J are disorders that show increased chromosomal instability. This is concomitant with mutations in genes encoding DNA helicases that can unwind G-quadruplex DNA. The G-quadruplex unwinding capacity of helicase Dna2 is important for genome integrity at telomeres (Chai et al., 2013; Lin et al., 2013). In addition, the resolvase G4R1 and RNA helicase RHAU can unwind G-quadruplexes in telomeric DNA and RNA, supporting a view where G-quadruplex formation can cause increased genome instability at telomeres (Creacy et al., 2008; Giri et al., 2011; Lattmann et al., 2011; Meier et al., 2013; Tran et al., 2004; Vaughn et al., 2005). A solution structure of a G-quadruplex with RHAU has been described (Heddi et al., 2015) and the mechanism investigated biochemically (Booy et al., 2015). The ablation of RHAU in mouse germ cells causes the increase of G-quadruplex structures, which results in the decrease of spermatogonial differentiation (Gao et al., 2015). The transcriptional helicases XPB/XPD are highly enriched in gene promoters with PQS. Pif1 is another interesting helicase to consider (Bochman et al., 2010; Budd et al., 2006; Paeschke et al., 2013; Paeschke et al., 2011; Ribeyre et al., 2009; Zhang et al., 2006). It unwinds G-quadruplexes and helps guard genome integrity upon DNA damage, and it was shown to suppress DNA damage at G-quadruplex motifs in yeast. This study suggested that the G-quadruplex-induced damage generated in the absence of Pif1 helicases led to new genetic and epigenetic changes (Paeschke et al., 2013). DT40 cells deficient in the Y-family polymerase REV1 are defective in replicating G-quadruplex DNA. This leads to uncoupling of DNA synthesis from redisposition of histones displaced ahead of the replication fork *in vivo*, which leads to loss of transcriptional repression, likely due to failure to recycle pre-existing repressive histone post-translational modifications (Sarkies et al., 2010). A similar process can also affect transcriptionally active genes through an involvement of the helicase FANCJ (Sarkies et al., 2012), leading to their deactivation. The Warsaw breakage syndrome DNA helicase DDX11 is also very efficient at unwinding G-quadruplexes upon binding to another protein called Tim, which is important for maintaining proper fork progression during S-phase in perturbed conditions (Cali et al., 2016). Epigenetic instability of the BU-1 reporter locus in DT40 cells is potentiated by an intragenic G-quadruplex, which also promotes local H2AX phosphorylation and transient heterochromatinization (Papadopoulou et al., 2015). These studies put histone dynamics in connection with G-quadruplexes, which has previously been surprisingly underappreciated. However, the claims of effects on epigenetic stability in these studies have to be taken with caution. The initial definition of epigenetics by Conrad Waddington in 1957 was a symbiosis of the words epigenesis and genetics (Waddington and Kacser, 1957). Now it is used to describe heritable changes of gene expression or cellular phenotypes that are not caused by changes in the underlying DNA sequence. In fact, many histone modifications are not stably transmitted during mitosis or meiosis and thus do not constitute an epigenetic memory (Lennartsson and Ekwall, 2009). Nevertheless, the aspect concerning histone turnover at PQS is fascinating and should be dissected further. It would be interesting to investigate if G-quadruplex interacting small molecules could alter the balance of histones or histone modifications in this context. Much has been learned about G-quadruplex interacting molecules and their involvement in telomere biology, the DNA damage response and gene modulation. However, studies with such molecules are underrepresented in aspects related to genomic control mechanisms, which should be addressed in future work. Another mechanism where G-quadruplexes could be involved in a biological control mechanism is replication. The chromatin environment is believed to shape DNA replication origin organization and define origin classes. Most origins have G-rich repeat elements potentially forming G-quadruplexes. These coincide with nucleosome-depleted regions located upstream of the initiation sites, which are associated with a labile nucleosome (Cayrou et al., 2015). A study in yeast identified Rif1 as a G-quadruplex-binding protein, controlling replication timing by suppressing replication over long distances (Kanoh et al., 2015). In *D. melanogaster* G-quadruplexes may transiently stall replication forks *in vivo* (Comoglio et al., 2015). G-quadruplex structures form a barrier for DNA replication. After transient stalling, G-quadruplexes are efficiently unwound and replicated. Depletion of the FANCJ helicase causes persistent replication stalling at G-quadruplexes, demonstrating a vital role for this helicase in resolving these structures (Castillo Bosch et al., 2014). It will be interesting to see in future studies, if and to what extend G-quadruplexes participate in replication and the formation of origins of replication. Interestingly Fragile X Mental Retardation Protein (FMRP) binds to an mRNA G-quadruplex, suggesting that G-quadruplexes could be involved in further mechanisms exerting control on genome integrity (Vasilyev et al., 2015), and unravelling the details will be interesting to consider in future studies.

Table 1. G-quadruplex interacting polymerases and helicases. Helicases and polymerases that were identified to unwind G-quadruplexes

Helicase/ polymerase	Involvement in G-quadruplex biology	References
BLM	Unwinds G-quadruplex DNA and can be inhibited by G-quadruplex interacting small molecules. BLM facilitates telomere replication during leading strand synthesis of telomeres. Regulation of gene expression by BLM correlates with the presence of G-quadruplex DNA motifs.	(Mohaghegh et al., 2001; Sun et al., 1998) (Budhathoki et al., 2015; Drosopoulos et al., 2015; Nguyen et al., 2014)
WRN	Unwinds G-quadruplex DNA and can be inhibited by G-quadruplex interacting small molecules	(Brosh and Bohr, 2007; Damerla et al., 2012; Fry and Loeb, 1999; Mohaghegh et al., 2001)
FANCJ	Required to unwind G-quadruplexes in order to maintain genome integrity; important during replication.	(Bharti et al., 2013; Castillo Bosch et al., 2014; Henderson et al., 2013; London et al., 2008; Sarkies et al., 2012; Schwab et al., 2013; Wu et al., 2008; Wu et al., 2009)
Pif1	Suppresses genome instability by unwinding G-quadruplexes, unwinds G-quadruplexes during replication	(Bochman et al., 2010; Paeschke et al., 2013; Zhang et al., 2006) (Budd et al., 2006; Paeschke et al., 2011; Ribeyre et al., 2009)
G4R1/RHAU	Unwinds G-quadruplexes in human telomerase RNA	(Booy et al., 2015; Creacy et al., 2008; Gao et al., 2015; Giri et al., 2011; Heddi et al., 2015; Lattmann et al., 2011; Meier et al., 2013; Tran et al., 2004; Vaughn et al., 2005)
XPB/XPD	Transcriptional helicases highly enriched in genes promoters with PQS	(Gray et al., 2014)
Dna2	G-quadruplex unwinding required for telomere integrity, required to unwind DNA in replication	(Chai et al., 2013; Lin et al., 2013)
RTEL1	Dismantles T-loops and counteracts telomeric G-quadruplex DNA, required for telomere integrity	(Vannier et al., 2012; Vannier et al., 2013)
REV1	Polymerase that disrupts G-quadruplex DNA	(Eddy et al., 2014; Sarkies et al., 2010)
Tim/DDX11	Unwinds G-quadruplexes genetically linked to the chromosomal instability disorder Warsaw breakage syndrome	(Cali et al., 2016)

Table 2. G-quadruplex interacting small molecules and biological responses. Selection of G-quadruplex interacting small molecules and their biological responses in a selection of cell lines and tumor models

G4 interacting small molecule	Cancer cell lines and tumor model	Mechanistic biological responses	References
Telomestatin and S2T1-6OTD	neuroblastoma, multiple myeloma, acute leukaemia, glioma	telomerase inhibition, reduction of telomere length, anti-proliferative, induction of apoptosis, increased chemosensitivity	(Kim et al., 2002; Miyazaki et al., 2012; Nakajima et al., 2003; Shalaby et al., 2010; Shammas et al., 2004; Shin-ya et al., 2001; Sumi et al., 2004; Tauchi et al., 2006)
BRACO19	uterine carcinoma, epithelial carcinoma, breast cancer, vulval carcinoma, prostate cancer	telomere uncapping and dysfunction, senescent and apoptotic phenotypes	(Burger et al., 2005; Gowan et al., 2002; Harrison et al., 2003; Incles et al., 2004; Kelland, 2005; Read et al., 2001; Taetz et al., 2006; Taetz et al., 2008)
TMPyP4	leukaemia, myeloma, cervical cancer, pancreatic cancer, breast cancer, colon cancer, prostate cancer, osteosarcoma, neuroblastoma and retinoblastoma	MYC and hTERT inhibition, telomerase inhibition, antiproliferative activity, RecQ helicase inhibition	(Dixon et al., 2007; Fujimori et al., 2011; Grand et al., 2002; Guliaev and Leontis, 1999; Lee et al., 2003; Liu et al., 2013; Mikami-Terao et al., 2009; Mikami-Terao et al., 2008; Rapozzi et al., 2014; Rha et al., 2000; Shammas et al., 2003; Wu and Maizels, 2001)
Quarfloxin	breast cancer, pancreatic cancer, neuroendocrine/ carcinoid tumors	displacement of nucleolin from nucleoli into the nucleoplasm, MYC down-regulation in some cases	(Brooks and Hurley, 2010; Drygin et al., 2009; Duan et al., 2001)
RHPS4	uterine carcinoma, melanoma, breast carcinoma, medulloblastoma, glioblastoma	antiproliferative effects, telomere injury, apoptosis	(Cookson et al., 2005; Lagah et al., 2014; Leonetti et al., 2004; Phatak et al., 2007; Salvati et al., 2007; Salvati et al., 2010)
307A and 360A	osteosarcoma, glioblastoma, cervical cancer	telomere dysfunction, telomere end-fusions, anaphase bridges, DNA damage response through ATM signalling, induction of apoptosis, MYC down-regulation	(Granotier et al., 2005; Halder et al., 2012; Lemarteleur et al., 2004; Pennarun et al., 2005; Pennarun et al., 2008)
Pyridostatin	osteosarcoma, fibrosarcoma, cervical carcinoma, glioblastoma	Alters shelterin component binding to telomeres (POT1, TRF1), induction of DNA damage, activation of the DNA damage response with genome-wide foci correlating with PQS, induction of senescence, down-regulation of specific genes including the proto-oncogene SRC.	(McLuckie et al., 2013; Müller et al., 2010; Müller et al., 2012; Rodriguez et al., 2012; Rodriguez et al., 2008)
12459	adenocarcinoma	DNA damage response regulated by PPM1D/WIP1 phosphatase, short term apoptosis and long-term senescence, telomere shortening	(Douarre et al., 2013; Gomez et al., 2003; Gomez et al., 2004; Riou et al., 2002)

G-Quadruplex Interacting Small Molecules and Clinical Applications

The cellular responses upon treatment with characterized G-quadruplex interacting small molecules, in particular at telomeres, PQS of promoters, in relation to the DNA damage response or interference with helicases, could potentially be exploited to develop drugs. Since G-quadruplexes have potential roles in such a plethora of mechanisms and small G-quadruplex interacting small molecules cause complex cellular pleiotropic responses, a major challenge will be how to translate this knowledge from the bench to the bedside. This part discusses the current state of development of some G-quadruplex interacting molecules for potential clinical applications and relates this information to the mechanistic studies of small molecules and G-quadruplexes presented in the previous sections.

Telomestatin is a natural product isolated from the bacteria *Streptomyces analatus* and is a potent inhibitor of telomerase (Shin-ya et al., 2001). It is composed of a planar ring structure with a very good overlap of the tetrads of G-quadruplexes. Telomestatin has a high selectivity over duplex DNA and preferentially binds intramolecular over intermolecular G-quadruplexes (Kim et al., 2002). The molecule displays promising anticancer activities in some cancer cell lines and cancer models. Treatment of multiple myeloma cell lines with telomestatin for several weeks causes telomere shortening due to inhibition of telomerase and ultimately apoptotic cell death (Shammas et al., 2004). Telomestatin impairs glioma stem cell survival and growth through the disruption of telomeric G-quadruplexes and an inhibition of the proto-oncogene *c-Myb* (Miyazaki et al., 2012). Telomere shortening and apoptotic cell death was also observed in primary and established acute leukaemia cell lines (Nakajima et al., 2003; Sumi et al., 2004). In xenograft models on acute leukaemia treated with telomestatin, tumor tissue from treated animals exhibited marked apoptosis without noticeable toxicity (Tauchi et al., 2006). Telomestatin also kills self-renewable glioma stem cells, potentially providing a method to selectively targeting cells of a glioblastoma tumor, causing relapse after treatment (Hasegawa et al., 2016). However, the total synthesis of telomestatin is very lengthy, lowering the synthetic feasibility as a potential drug candidate (Doi et al., 2006; Linder et al., 2011). Thus, synthetically more accessible telomestatin-like compounds have been developed, for instance macrocyclic polyoxazoles (Iida et al., 2009a; Iida et al., 2009b; Petenzi et al., 2012). S2T1-6OTD is one of these derivatives and down-regulates *MYC* expression in childhood medulloblastoma and atypical teratoid/rhabdoid tumor cell lines and causes growth arrest by senescence and telomere shortening (Shalaby et al., 2010).

TMPyP4 was one of the first synthetic small molecules studied for its ability to interact with G-quadruplexes (Siddiqui-Jain et al., 2002). It has a pleiotropic mode of action in cells, which includes *MYC* down-regulation to telomerase inhibition (Grand et al., 2002) and RecQ helicase inhibition (Wu and Maizels, 2001). TMPyP4 also exhibits considerable ability to bind to duplex (Guliaev and Leontis, 1999) and triplex (Lee et al., 2003) DNA, and thus successful attempts were made to modify the molecule to increase its G-quadruplex selectivity (Dixon et al., 2007). Still, the compound shows remarkable growth inhibitory properties in an array of cell lines and cancer models. The molecule significantly impairs the growth of retinoblastoma (Mikami-Terao et al., 2009) and K562 leukaemia cells (Mikami-Terao et al., 2008), causing telomerase inhibition and a telomere-induced DNA damage response. It also induces telomere shortening in osteosarcoma cells, which leads to apoptosis (Fujimori et al., 2011). TMPyP4 enhances susceptibility to radiation treatment in a cervical cancer xenograft (Liu et al., 2013) and in a melanoma xenograft (Rapozzi et al., 2014).

BRACO19 belongs to a series of trisubstituted acridine derivatives designed to interact with G-quadruplexes. It was identified from this family of molecules after extensive computer modeling for G-quadruplex interactions, comparing different substitutions on the scaffold (Harrison et al., 2003; Read et al., 2001). BRACO19 showed relatively low cytotoxicity and promising telomerase inhibition, and was found to trigger senescence and chromosomal end fusions in human prostate cancer cells, possibly through uncapping of Shelterin components (Incles et al., 2004). Treatment of 21NT human breast cancer cells with BRACO19 resulted in a marked reduction in cell growth after 15 days. BRACO19 also induced significant tumor regression within about a week of treatment in a prostate cancer xenograft (Kelland, 2005). Intraperitoneal administration of nontoxic doses of BRACO19 to mice bearing advanced stage A431 human vulval carcinoma subcutaneous xenografts, which had previously been treated with taxol induced a significant increase in anti-tumoral effects (Gowan et al., 2002). Furthermore, the compound caused a stark inhibition of growth inhibition in a xenograft model derived from uterus carcinoma cells, which was paralleled by a decrease of hTERT expression and telomere dysfunction (Burger et al., 2005). However, BRACO19 has a small therapeutic window and low membrane permeability. It is not very stable in physiological media resulting in decomposition products with decreased binding properties towards G-quadruplexes (Taetz et al., 2008). These findings have been limiting for the further development of BRACO19 as a clinical agent. Nevertheless, these studies have been very valuable to show that a G-quadruplex interacting small molecule can have promising therapeutic potential (Taetz et al., 2006). Related compounds are still under investigation and mono- and bis-pyrimidinoacridines have been shown to be potent G-quadruplex interacting small molecules. Whether these can provide the basis for the development of therapeutic agents will have to be elucidated (Debray et al., 2009). Interestingly, BRACO-19 has an anti-HIV activity and acts both at the reverse transcription and the post-integration level during the HIV life cycle, by targeting a G-quadruplex present at the 3'-end of the virus RNA genome (Perrone et al., 2014).

RHPS4 very selectively binds G-quadruplex over duplex DNA and in addition it is a very potent inhibitor of telomerase. Long-term exposure of 21NT breast cancer or A431 vulval cancer cells to nonacute cytotoxic concentrations of RHPS4 resulted in a significant decrease of cell growth after 15 days and in G2/M phase arrest. Polynucleated cells and telophase bridges were also observed (Leonetti et al., 2004). The small molecule induces POT1 uncapping from telomeres in melanoma cells and human transformed fibroblasts leading to a DNA damage response. Treatment of UXF1138L cells with RHPS4 causes hTERT displacement from the nucleus and chromosome end-fusions. In mice, RHPS4 exerted an antitumor effect in CG5 breast xenografts and M14 melanoma xenograft by telomere injury and tumor cell apoptosis. Interestingly, tumors overexpressing POT1 or TRF2 were resistant to RHPS4 treatment (Phatak et al., 2007; Salvati et al., 2007). The small molecule was also tested in combination therapy with doxorubicin or taxol, which increased sensitivity of MCF-7 cells (Cookson et al., 2005). RHPS4 induces telomeric dysfunction and enhances radiosensitivity in glioblastoma cells (Berardinelli et al., 2015). In addition, various brain tumor cell lines are particularly susceptible to RHPS4 (Lagah et al., 2014). Poly-adenosine diphosphate (ADP) ribose polymerase 1 (PARP1) is recruited to telomeres in response to treatment with RHPS4. Knocking-down PARP1 prevents repairing of RHPS4-induced telomere DNA breaks, leading to increases in chromosome abnormalities and eventually inhibition of tumor cell growth both *in cellulo* and in xenografts. Thus, a multi-component strategy with this G-quadruplex interacting small molecule and PARP inhibitors could be promising (Salvati et al., 2010). This could be applied to exploit synthetic lethality in certain cases.

12459 belongs to a family of disubstituted triazines with a very potent inhibitory activity against telomerase, leading to telomere shortening and senescence in a panel of cultured cancer cell lines (Riou et al., 2002), through a mechanism that does not rely predominantly on telomere inhibition (Gomez et al., 2003). 12459 shows a strong stabilization potential for RNA G-quadruplexes in hTERT and the decrease of telomere length in A549 adenocarcinoma cells is mediated by hTERT RNA alternative splicing upon treatment with 12459 (Gomez et al., 2004). A cross-talk between the apoptotic pathway and BCL-2 was observed, suggesting that pathways involving BCL-2 are at least partially responsible for the apoptotic short-term response. However, the long-term proliferative effect upon treatment with 12459 was independent of BCL-2, suggesting that the small molecule has a broad mechanism of action. Indeed, 12459 also induces a DNA damage signalling modulated by the PPM1D/WIP1 phosphatase (Douarre et al., 2013).

Quarfloxin is a molecule that was derived from a family of fluoroquinolones, which exhibited G-quadruplex as well as topoisomerase II interactions (Duan et al., 2001). In an initial study it was found that it binds to G-quadruplex DNA over duplex DNA with great specificity, and that it induces apoptosis in specific cancer cell lines. It was subsequently optimized to eliminate topoisomerase II inhibition and maximize selectivity for G-quadruplexes. It moved on to Phase II clinical trials for the treatment of neuroendocrine/carcinoid tumors and exhibits a wide therapeutic window and an *in vivo* efficacy in a broad range of tumors (Drygin et al., 2009). Quarfloxin was found to concentrate in nucleoli and inhibit Polymerase I dependent transcription, which could be due to the disruption of the interaction between nucleolin and G-quadruplexes, leading to a redistribution of nucleolin to the nucleoplasm. Nucleolin has a high affinity for the *MYC* promoter G-quadruplex (Gonzalez et al., 2009), and thus a displacement of nucleolin may influence *MYC* expression. Indeed, *MYC* mRNA expression in HCT-116 tumors isolated from mice after treatment with quarfloxin was considerably reduced (Brooks and Hurley, 2010). However, quarfloxin may also interact with other G-quadruplexes and the changes in *MYC* expression might be linked to other mechanistic aspects. 307A and 360A are closely related 2,6-pyridine-dicarboxamide derivatives, which can stabilize telomeric and *MYC* promoter G-quadruplexes (Lemarteleur et al., 2004). Both compounds induce apoptosis in cultured glioblastoma cells, induced by a telomere-specific mechanism (Granotier et al., 2005; Pennarun et al., 2005). SAOS-2 osteosarcoma cells are also affected, although their telomere maintenance does not depend on telomerase but on the ALT mechanism. Interestingly, the compounds do not induce telomere shortening, but telomere end-fusion and anaphase bridges (Granotier et al., 2005). 360A induces a DNA damage response at telomeres through an ATM-dependent mechanism (Pennarun et al., 2008) and it causes wide-spread transcriptome changes in HeLa S3 cells, with considerable correlative overlap with genes carrying PQS (Halder et al., 2012). Again, this highlights the complexity of the cellular responses caused by a G-quadruplex interacting small molecule.

Pyridostatin is a synthetic small molecule that very selectively stabilizes G-quadruplexes over duplex DNA (Müller et al., 2012; Rodriguez et al., 2008). Although it was initially characterized for its effect on POT1 uncapping from telomeres, the whole-genome sequencing of pyridostatin-induced γH2AX revealed a widespread mode of interaction of this compound, correlated with PQS throughout the genome (Rodriguez et al., 2012). Pyridostatin induces growth inhibition in osteosarcoma, fibrosarcoma and cervical carcinoma cell lines, which can lead to senescence (Müller et al., 2010; Müller et al., 2012). Derivatives have been created in an attempt to improve selectivity between G-quadruplexes and between different cell lines (Müller et al., 2012). One derivative has been used to successfully inhibit translation of the mRNA of the *NRAS* proto-oncogene *in vitro* (Bugaut et al., 2010). Treatment with pyridostatin increases telomere fragility in BRCA2-deficient cells, suggesting that G-quadruplex formation drives telomere instability (Zimmer et al., 2016). Pyridostatin has been made commercially available, which has prompted follow-up studies investigating its growth-inhibitory and pharmacokinetic properties. ATRX-mediated ALT suppression is dependent on the histone chaperone DAXX and re-expression of ATRX is associated with a reduction in replication fork stalling, a known trigger for homologous recombination, which can be partially reversed by pyridostatin (Clynes et al., 2015).

Phen-DC(3) and Phen-DC(6) are very closely related bisquinolinium compounds and belong to a class of very potent G-quadruplex interacting small molecules (De Cian et al., 2007). They cause genetic instability in *S. cerevisiae* by inhibiting helicase activity. Pif1-deficient cells are more sensitive to the small molecules, suggesting that these compounds directly impair the function of this helicase (Piazza et al., 2010). Phen-DC(6) also causes telomere uncapping in budding yeast, demonstrating that it can stabilize the telomeric G-quadruplex (Smith et al., 2011). Furthermore, Phen-DC(3) can cause genomic instability by stimulating rearrangements of subtelomeric minisatellites (Piazza et al., 2012). In human cells, Phen-DC-(3) also causes genomic instability by impairing replication, specifically when the G-rich strand is the template of leading-strand replication (Lopes et al., 2011). The interesting properties to cause genome instability could potentially be deployed to target fast proliferating cells and it will be interesting to see further studies and developments of this class of G-quadruplex interacting small molecules.

CONCLUSION

Small molecules have been indispensable in the dissection of fundamental biological processes that involve G-quadruplexes. Some very useful chemical tools have been developed, which facilitated studies involving G-quadruplex nucleic acid isolation from cells, visualization of G-quadruplexes *in cellulo* and whole genome sequencing approaches, to reveal potential small molecule binding sites throughout the genome. G-quadruplexes have proposed functions in telomere biology, gene modulation and genetic control mechanisms, including replication. Small molecules have been developed to specifically target G-quadruplexes and a large body of literature available on the effects of G-quadruplex interacting small molecules highlights this. Some of these molecules have been developed further and tested in various xenograft models. Given some promising effects of a selection of these small molecules on inhibiting cancer cell growth both in culture and in xenograft models, they are promising drug candidates. Despite a great number of studies dissecting biological mechanisms with G-quadruplex interacting small molecules, it remains elusive how important G-quadruplexes are in many of the biological mechanisms described. The development of potential drugs has suffered from the absence of known biologically relevant G-quadruplexes as targets to allow suitable design. Despite a growing number of evidence for their existence *in vivo*, G-quadruplexes are highly dynamic and are challenged by changes in chromatin structure. As a result, non-validated targets have served as templates for the development of G-quadruplex binders, in particular for telomeric and promoter G-quadruplexes. There are two critical questions that need to be addressed in the context of the development of drugs targeting these nucleic acid structures: (1) What are all the *in vivo* targets of a G-quadruplex interacting small molecule and (2) how can their accessibility be controlled? Although telomeres, some promoters, rDNA and 5'UTRs of mRNAs have been described as potential targets, there are over 350,000 putative G-quadruplexes in the human genome (Huppert and Balasubramanian, 2005; Todd et al., 2005). Whether these molecules can target one specific target or cluster of G-quadruplexes is still under debate. The biological response of many of the small molecules is diverse, often including effects on telomere biology, initiation of the DNA damage response and effects on selected gene expression in one cell type. Indeed, this suggests that these molecules target a wide spectrum of G-quadruplex motifs. This was exemplified by pyridostatin, which causes DNA damage hotspots all over the genome and its correlation with Pif1 target sites, as revealed by deep sequencing (Rodriguez et al., 2012). In terms of accessibility, this is an underappreciated issue. DNA is packaged within chromatin and its accessibility controlled by histones and associated protein complexes (Chen and Dent, 2014; Filipescu et al., 2014; Müller and Almouzni, 2014; Müller and Rodriguez, 2014; Papamichos-Chronakis and Peterson, 2013; Swygert and Peterson, 2014; van Steensel, 2011). Dissecting G-quadruplex biology in the context of chromatin organisation and dynamics will be important for future investigations. Furthermore, interfering with chromatin remodelling processes with drugs in combination with G-quadruplex interacting small molecules will be an interesting angle for the development of combination therapies. The further development of G-quadruplex interacting small molecules, both in terms of new biological tools and for drug development, will unravel further biological functions and possibly unanticipated therapeutic targets for the treatment for human disease.

Acknowledgments

We thank the European Union for funding (ERC grant).

References

Ambrus, A., Chen, D., Dai, J., Jones, R. A., and Yang, D. (2005). Solution structure of the biologically relevant G-quadruplex element in the human c-MYC promoter. Implications for G-quadruplex stabilization. *Biochemistry 44*, 2048-2058.

Audry, J., Maestroni, L., Delagoutte, E., Gauthier, T., Nakamura, T. M., Gachet, Y., Saintome, C., Geli, V., and Coulon, S. (2015). RPA prevents G-rich structure formation at lagging-strand telomeres to allow maintenance of chromosome ends. *The EMBO journal 34*, 1942-1958.

Balasubramanian, S., Hurley, L. H., and Neidle, S. (2011). Targeting G-quadruplexes in gene promoters: a novel anticancer strategy? *Nature reviews Drug discovery 10*, 261-275.

Balasubramanian, S., and Neidle, S. (2009). G-quadruplex nucleic acids as therapeutic targets. *Current opinion in chemical biology 13*, 345-353.

Baral, A., Kumar, P., Pathak, R., and Chowdhury, S. (2013). Emerging trends in G-quadruplex biology--role in epigenetic and evolutionary events. *Molecular bioSystems 9*, 1568-1575.

Bejugam, M., Gunaratnam, M., Müller S, Sanders, D. A., Sewitz, S., Fletcher, J. A., Neidle, S., and Balasubramanian, S. (2010). Targeting the c-Kit promoter G-quadruplexes with 6-substituted indenoisoquinolines,. *ACS medicinal chemistry letters 1*, 306-310.

Bejugam, M., Sewitz, S., Shirude, P. S., Rodriguez, R., Shahid, R., and Balasubramanian, S. (2007). Trisubstituted isoalloxazines as a new class of G-quadruplex binding ligands: small molecule regulation of c-kit oncogene expression. *Journal of the American Chemical Society 129*, 12926-12927.

Berardinelli, F., Siteni, S., Tanzarella, C., Stevens, M. F., Sgura, A., and Antoccia, A. (2015). The G-quadruplex-stabilising agent RHPS4 induces telomeric dysfunction and enhances radiosensitivity in glioblastoma cells. *DNA repair 25*, 104-115.

Bharti, S. K., Sommers, J. A., George, F., Kuper, J., Hamon, F., Shin-ya, K., Teulade-Fichou, M. P., Kisker, C., and Brosh, R. M., Jr. (2013). Specialization among iron-sulfur cluster helicases to resolve G-quadruplex DNA structures that threaten genomic stability. *The Journal of biological chemistry 288*, 28217-28229.

Biffi, G., Di Antonio, M., Tannahill, D., and Balasubramanian, S. (2014). Visualization and selective chemical targeting of RNA G-quadruplex structures in the cytoplasm of human cells. *Nature chemistry 6*, 75-80.

Biffi, G., Tannahill, D., McCafferty, J., and Balasubramanian, S. (2013). Quantitative visualization of DNA G-quadruplex structures in human cells. *Nature chemistry 5*, 182-186.

Bochman, M. L., Paeschke, K., and Zakian, V. A. (2012). DNA secondary structures: stability and function of G-quadruplex structures. Nature reviews *Genetics 13*, 770-780.

Bochman, M. L., Sabouri, N., and Zakian, V. A. (2010). Unwinding the functions of the Pif1 family helicases. *DNA repair 9*, 237-249.

Boddupally, P. V., Hahn, S., Beman, C., De, B., Brooks, T. A., Gokhale, V., and Hurley, L. H. (2012). Anticancer activity and cellular repression of c-MYC by the G-quadruplex-stabilizing 11-piperazinylquindoline is not dependent on direct targeting of the G-quadruplex in the c-MYC promoter. *Journal of medicinal chemistry 55*, 6076-6086.

Booy, E. P., McRae, E. K., and McKenna, S. A. (2015). Biochemical characterization of G4 quadruplex telomerase RNA unwinding by the RNA helicase RHAU. *Methods Mol Biol 1259*, 125-135.

Brooks, T. A., and Hurley, L. H. (2010). Targeting MYC Expression through G-Quadruplexes. *Genes and cancer 1*, 641-649.

Brosh, R. M., Jr., and Bohr, V. A. (2007). Human premature aging, DNA repair and RecQ helicases. *Nucleic acids research 35*, 7527-7544.

Budd, M. E., Reis, C. C., Smith, S., Myung, K., and Campbell, J. L. (2006). Evidence suggesting that Pif1 helicase functions in DNA replication with the Dna2 helicase/nuclease and DNA polymerase delta. *Molecular and cellular biology 26*, 2490-2500.

Budhathoki, J. B., Stafford, E. J., Yodh, J. G., and Balci, H. (2015). ATP-dependent G-quadruplex unfolding by Bloom helicase exhibits low processivity. *Nucleic acids research 43*, 5961-5970.

Bugaut, A., Rodriguez, R., Kumari, S., Hsu, S. T., and Balasubramanian, S. (2010). Small molecule-mediated inhibition of translation by targeting a native RNA G-quadruplex. *Organic and biomolecular chemistry 8*, 2771-2776.

Burger, A. M., Dai, F., Schultes, C. M., Reszka, A. P., Moore, M. J., Double, J. A., and Neidle, S. (2005). The G-quadruplex-interactive molecule BRACO-19 inhibits tumor growth, consistent with telomere targeting and interference with telomerase function. *Cancer research 65*, 1489-1496.

Cahoon, L. A., and Seifert, H. S. (2009). An alternative DNA structure is necessary for pilin antigenic variation in Neisseria gonorrhoeae. *Science 325*, 764-767.

Cali, F., Bharti, S. K., Perna, R. D., Brosh, R. M., Jr., and Pisani, F. M. (2016). Tim/Timeless, a member of the replication fork protection complex, operates with the Warsaw breakage syndrome DNA helicase DDX11 in the same fork recovery pathway. *Nucleic acids research 44*, 705-717.

Castillo Bosch, P., Segura-Bayona, S., Koole, W., van Heteren, J. T., Dewar, J. M., Tijsterman, M., and Knipscheer, P. (2014). FANCJ promotes DNA synthesis through G-quadruplex structures. *The EMBO journal 33*, 2521-2533.

Cayrou, C., Ballester, B., Peiffer, I., Fenouil, R., Coulombe, P., Andrau, J. C., van Helden, J., and Mechali, M. (2015). The chromatin environment shapes DNA replication origin organization and defines origin classes. *Genome research 25*, 1873-1885.

Chai, W., Zheng, L., and Shen, B. (2013). DNA2, a new player in telomere maintenance and tumor suppression. *Cell Cycle 12*, 1985-1986.

Chatterjee, S., Zagelbaum, J., Savitsky, P., Sturzenegger, A., Huttner, D., Janscak, P., Hickson, I. D., Gileadi, O., and Rothenberg, E. (2014). Mechanistic insight into the interaction of BLM helicase with intra-strand G-quadruplex structures. *Nature communications 5*, 5556.

Chen, T., and Dent, S. Y. (2014). Chromatin modifiers and remodellers: regulators of cellular differentiation. *Nature reviews Genetics 15*, 93-106.

Chung, W. J., Heddi, B., Hamon, F., Teulade-Fichou, M.P., and Phan, A.T. (2014a). Solution structure of a G-quadruplex bound to the bisquinolinium compound Phen-DC(3). *Angew Chem Int Ed Engl 53*, 999-1002.

Chung, Y. L., Pan, C. H., Liou, W. H., Sheu, M. J., Lin, W. H., Chen, T. C., Huang, H. S., and Wu, C. H. (2014b). NSC746364, a G-quadruplex-stabilizing agent, suppresses cell growth of A549 human lung cancer cells through activation of the ATR/Chk1-dependent pathway. *Journal of pharmacological sciences 124*, 7-17.

Clynes, D., Jelinska, C., Xella, B., Ayyub, H., Scott, C., Mitson, M., Taylor, S., Higgs, D. R., and Gibbons, R. J. (2015). Suppression of the alternative lengthening of telomere pathway by the chromatin remodelling factor ATRX. *Nature communications 6*, 7538.

Cogoi, S., and Xodo, L. E. (2006). G-quadruplex formation within the promoter of the KRAS proto-oncogene and its effect on transcription. *Nucleic acids research 34*, 2536-2549.

Comoglio, F., Schlumpf, T., Schmid, V., Rohs, R., Beisel, C., and Paro, R. (2015). High-resolution profiling of Drosophila replication start sites reveals a DNA shape and chromatin signature of metazoan origins. *Cell* reports *11*, 821-834.

Cookson, J. C., Dai, F., Smith, V., Heald, R. A., Laughton, C. A., Stevens, M. F., and Burger, A. M. (2005). Pharmacodynamics of the G-quadruplex-stabilizing telomerase inhibitor 3,11-difluoro-6,8,13-trimethyl-8H-quino[4,3,2-kl]acridinium methosulfate (RHPS4) in vitro: activity in human tumor cells correlates with telomere length and can be enhanced, or antagonized, with cytotoxic agents. *Molecular pharmacology 68*, 1551-1558.

Cosconati, S., Rizzo, A., Trotta, R., Pagano, B., Iachettini, S., De Tito, S., Lauri, I., Fotticchia, I., Giustiniano, M., Marinelli, L., et al. (2012). Shooting for selective druglike G-quadruplex binders: evidence for telomeric DNA damage and tumor cell death. *Journal of medicinal chemistry 55*, 9785-9792.

Creacy, S. D., Routh, E. D., Iwamoto, F., Nagamine, Y., Akman, S. A., and Vaughn, J. P. (2008). G4 resolvase 1 binds both DNA and RNA tetramolecular quadruplex with high affinity and is the major source of tetramolecular quadruplex G4-DNA and G4-RNA resolving activity in HeLa cell lysates. *The Journal of biological chemistry 283*, 34626-34634.

d'Adda di Fagagna, F., Reaper, P. M., Clay-Farrace, L., Fiegler, H., Carr, P., Von Zglinicki, T., Saretzki, G., Carter, N. P., and Jackson, S. P. (2003). A DNA damage checkpoint response in telomere-initiated senescence. *Nature 426*, 194-198.

Dai, J., Carver, M., Punchihewa, C., Jones, R. A., and Yang, D. (2007). Structure of the Hybrid-2 type intramolecular human telomeric G-quadruplex in K+ solution: insights into structure polymorphism of the human telomeric sequence. *Nucleic acids research 35*, 4927-4940.

Damerla, R. R., Knickelbein, K. E., Strutt, S., Liu, F. J., Wang, H., and Opresko, P. L. (2012). Werner syndrome protein suppresses the formation of large deletions during the replication of human telomeric sequences. *Cell Cycle 11*, 3036-3044.

David, A. P., Margarit, E., Domizi, P., Banchio, C., Armas, P., and Calcaterra, N. B. (2016). G-quadruplexes as novel cis-elements controlling transcription during embryonic development. *Nucleic acids research.*

Davis, J. T. (2004). G-quartets 40 years later: from 5'-GMP to molecular biology and supramolecular chemistry. *Angew Chem Int Ed Engl 43*, 668-698.

De Cian, A., Delemos, E., Mergny, J. L., Teulade-Fichou, M. P., and Monchaud, D. (2007). Highly efficient G-quadruplex recognition by bisquinolinium compounds. *Journal of the American Chemical Society 129*, 1856-1857.

de Lange, T. (2005). Shelterin: the protein complex that shapes and safeguards human telomeres. *Genes and development 19*, 2100-2110.

Debray, J., Zeghida, W., Jourdan, M., Monchaud, D., Dheu-Andries, M. L., Dumy, P., Teulade-Fichou, M. P., and Demeunynck, M. (2009). Synthesis and evaluation of fused bispyrimidinoacridines as novel pentacyclic analogues of quadruplex-binder BRACO-19. *Organic and biomolecular chemistry 7*, 5219-5228.

Dexheimer, T. S., Carey, S. S., Zuohe, S., Gokhale, V. M., Hu, X., Murata, L. B., Maes, E. M., Weichsel, A., Sun, D., Meuillet, E. J., et al. (2009). NM23-H2 may play an indirect role in transcriptional activation of c-myc gene expression but does not cleave the nuclease hypersensitive element III(1). *Molecular cancer therapeutics 8*, 1363-1377.

Dixon, I. M., Lopez, F., Tejera, A. M., Esteve, J. P., Blasco, M. A., Pratviel, G., and Meunier, B. (2007). A G-quadruplex ligand with 10000-fold selectivity over duplex DNA. *Journal of the American Chemical Society 129*, 1502-1503.

Doi, T., Yoshida, M., Shin-ya, K., and Takahashi, T. (2006). Total synthesis of (R)-telomestatin. *Organic letters 8*, 4165-4167.

Douarre, C., Mergui, X., Sidibe, A., Gomez, D., Alberti, P., Mailliet, P., Trentesaux, C., and Riou, J. F. (2013). DNA damage signaling induced by the G-quadruplex ligand 12459 is modulated by PPM1D/WIP1 phosphatase. *Nucleic acids research 41*, 3588-3599.

Drosopoulos, W. C., Kosiyatrakul, S. T., and Schildkraut, C. L. (2015). BLM helicase facilitates telomere replication during leading strand synthesis of telomeres. *The Journal of cell biology 210*, 191-208.

Drygin, D., Siddiqui-Jain, A., O'Brien, S., Schwaebe, M., Lin, A., Bliesath, J., Ho, C. B., Proffitt, C., Trent, K., Whitten, J. P., et al. (2009). Anticancer activity of CX-3543: a direct inhibitor of rRNA biogenesis. *Cancer research 69*, 7653-7661.

Duan, W., Rangan, A., Vankayalapati, H., Kim, M. Y., Zeng, Q., Sun, D., Han, H., Fedoroff, O. Y., Nishioka, D., Rha, S. Y., et al. (2001). Design and synthesis of fluoroquinophenoxazines that interact with human telomeric G-quadruplexes and their biological effects. *Molecular cancer therapeutics 1*, 103-120.

Eddy, S., Ketkar, A., Zafar, M. K., Maddukuri, L., Choi, J. Y., and Eoff, R. L. (2014). Human Rev1 polymerase disrupts G-quadruplex DNA. *Nucleic acids research 42*, 3272-3285.

Erlitzki, R., and Fry, M. (1997). Sequence-specific binding protein of single-stranded and unimolecular quadruplex telomeric DNA from rat hepatocytes. *The Journal of biological chemistry 272*, 15881-15890.

Fernando, H., Reszka, A. P., Huppert, J., Ladame, S., Rankin, S., Venkitaraman, A. R., Neidle, S., and Balasubramanian, S. (2006). A conserved quadruplex motif located in a transcription activation site of the human c-kit oncogene. *Biochemistry 45*, 7854-7860.

Filipescu, D., Müller, S., and Almouzni, G. (2014). Histone H3 variants and their chaperones during development and disease: contributing to epigenetic control. *Annual review of cell and developmental biology 30*, 615-646.

Fry, M., and Loeb, L. A. (1999). Human werner syndrome DNA helicase unwinds tetrahelical structures of the fragile X syndrome repeat sequence d(CGG)n. *The Journal of biological chemistry 274*, 12797-12802.

Fujimori, J., Matsuo, T., Shimose, S., Kubo, T., Ishikawa, M., Yasunaga, Y., and Ochi, M. (2011). Antitumor effects of telomerase inhibitor TMPyP4 in osteosarcoma cell lines. *Journal of orthopaedic research : official publication of the Orthopaedic Research Society 29*, 1707-1711.

Gao, X., Ma, W., Nie, J., Zhang, C., Zhang, J., Yao, G., Han, J., Xu, J., Hu, B., Du, Y., et al. (2015). A G-quadruplex DNA structure resolvase, RHAU, is essential for spermatogonia differentiation. *Cell Death Dis 6*, e1610.

Gellert, M., Lipsett, M. N., and Davies, D. R. (1962). Helix formation by guanylic acid. *Proceedings of the National Academy of Sciences of the United States of America 48*, 2013-2018.

Ghosh, A., and Bansal, M. (2003). A glossary of DNA structures from A to Z. Acta crystallographica Section D, *Biological crystallography 59*, 620-626.

Gilbert, D. E., and Feigon, J. (1999). Multistranded DNA structures. *Current opinion in structural biology 9*, 305-314.

Giri, B., Smaldino, P. J., Thys, R. G., Creacy, S. D., Routh, E. D., Hantgan, R. R., Lattmann, S., Nagamine, Y., Akman, S. A., and Vaughn, J. P. (2011). G4 resolvase 1 tightly binds and unwinds unimolecular G4-DNA. *Nucleic acids research 39*, 7161-7178.

Gomez, D., Aouali, N., Londono-Vallejo, A., Lacroix, L., Megnin-Chanet, F., Lemarteleur, T., Douarre, C., Shin-ya, K., Mailliet, P., Trentesaux, C., et al. (2003). Resistance to the short term antiproliferative activity of the G-quadruplex ligand 12459 is associated with telomerase overexpression and telomere capping alteration. *The Journal of biological chemistry 278*, 50554-50562.

Gomez, D., Guedin, A., Mergny, J. L., Salles, B., Riou, J. F., Teulade-Fichou, M. P., and Calsou, P. (2010). A G-quadruplex structure within the 5'-UTR of TRF2 mRNA represses translation in human cells. *Nucleic acids research 38*, 7187-7198.

Gomez, D., Lemarteleur, T., Lacroix, L., Mailliet, P., Mergny, J. L., and Riou, J. F. (2004). Telomerase downregulation induced by the G-quadruplex ligand 12459 in A549 cells is mediated by hTERT RNA alternative splicing. *Nucleic acids research 32*, 371-379.

Gomez, D., O'Donohue, M. F., Wenner, T., Douarre, C., Macadre, J., Koebel, P., Giraud-Panis, M. J., Kaplan, H., Kolkes, A., Shin-ya, K., et al. (2006). The G-quadruplex ligand telomestatin inhibits POT1 binding to telomeric sequences in vitro and induces GFP-POT1 dissociation from telomeres in human cells. *Cancer research 66*, 6908-6912.

Gonzalez, V., Guo, K., Hurley, L., and Sun, D. (2009). Identification and characterization of nucleolin as a c-myc G-quadruplex-binding protein. *The Journal of biological chemistry 284*, 23622-23635.

Gowan, S. M., Harrison, J. R., Patterson, L., Valenti, M., Read, M. A., Neidle, S., and Kelland, L. R. (2002). A G-quadruplex-interactive potent small-molecule inhibitor of telomerase exhibiting in vitro and in vivo antitumor activity. *Molecular pharmacology 61*, 1154-1162.

Grand, C. L., Han, H., Munoz, R. M., Weitman, S., Von Hoff, D. D., Hurley, L. H., and Bearss, D. J. (2002). The cationic porphyrin TMPyP4 down-regulates c-MYC and human telomerase reverse transcriptase expression and inhibits tumor growth in vivo. *Molecular cancer therapeutics 1*, 565-573.

Granotier, C., Pennarun, G., Riou, L., Hoffschir, F., Gauthier, L. R., De Cian, A., Gomez, D., Mandine, E., Riou, J. F., Mergny, J. L., et al. (2005). Preferential binding of a G-quadruplex ligand to human chromosome ends. *Nucleic acids research 33*, 4182-4190.

Guliaev, A. B., and Leontis, N. B. (1999). Cationic 5,10,15,20-tetrakis(N-methylpyridinium-4-yl)porphyrin fully intercalates at 5'-CG-3' steps of duplex DNA in solution. *Biochemistry 38*, 15425-15437.

Gunaratnam, M., Greciano, O., Martins, C., Reszka, A. P., Schultes, C. M., Morjani, H., Riou, J. F., and Neidle, S. (2007). Mechanism of acridine-based telomerase inhibition and telomere shortening. *Biochemical pharmacology 74*, 679-689.

Gunaratnam, M., Swank, S., Haider, S. M., Galesa, K., Reszka, A. P., Beltran, M., Cuenca, F., Fletcher, J. A., and Neidle, S. (2009). Targeting human gastrointestinal stromal tumor cells with a quadruplex-binding small molecule. *Journal of medicinal chemistry 52*, 3774-3783.

Halder, R., Riou, J. F., Teulade-Fichou, M. P., Frickey, T., and Hartig, J. S. (2012). Bisquinolinium compounds induce quadruplex-specific transcriptome changes in HeLa S3 cell lines. *BMC research notes 5*, 138.

Hanahan, D., and Weinberg, R. A. (2011). Hallmarks of cancer: the next generation. *Cell 144*, 646-674.

Harrison, R. J., Cuesta, J., Chessari, G., Read, M. A., Basra, S. K., Reszka, A. P., Morrell, J., Gowan, S. M., Incles, C. M., Tanious, F.A., et al. (2003). Trisubstituted acridine derivatives as potent and selective telomerase inhibitors. *Journal of medicinal chemistry 46*, 4463-4476.

Hasegawa, D., Okabe, S., Okamoto, K., Nakano, I., Shin-Ya, K., and Seimiya, H. (2016). G-quadruplex ligand-induced DNA damage response coupled with telomere dysfunction and replication stress in glioma stem cells. *Biochemical and biophysical research communications 471*, 75-81.

Heddi, B., Cheong, V. V., Martadinata, H., and Phan, A. T. (2015). Insights into G-quadruplex specific recognition by the DEAH-box helicase RHAU: Solution structure of a peptide-quadruplex complex. *Proceedings of the National Academy of Sciences of the United States of America 112*, 9608-9613.

Heinrich, M. C., Corless, C. L., Demetri, G. D., Blanke, C. D., von Mehren, M., Joensuu, H., McGreevey, L. S., Chen, C. J., Van den Abbeele, A. D., Druker, B. J., et al. (2003). Kinase mutations and imatinib response in patients with metastatic gastrointestinal stromal tumor. *Journal of clinical oncology : official journal of the American Society of Clinical Oncology 21*, 4342-4349.

Hemann, M.T., and Greider, C.W. (1999). G-strand overhangs on telomeres in telomerase-deficient mouse cells. *Nucleic acids research 27*, 3964-3969.

Henderson, A., Wu, Y., Huang, Y. C., Chavez, E. A., Platt, J., Johnson, F. B., Brosh, R. M., Jr., Sen, D., and Lansdorp, P. M. (2014). Detection of G-quadruplex DNA in mammalian cells. *Nucleic acids research 42*, 860-869.

Huppert, J. L., and Balasubramanian, S. (2005). Prevalence of quadruplexes in the human genome. *Nucleic acids research 33*, 2908-2916.

Huppert, J. L., and Balasubramanian, S. (2007). G-quadruplexes in promoters throughout the human genome. *Nucleic acids research 35*, 406-413.

Huppert, J. L., Bugaut, A., Kumari, S., and Balasubramanian, S. (2008). G-quadruplexes: the beginning and end of UTRs. *Nucleic acids research 36*, 6260-6268.

Iida, K., Tera, M., Hirokawa, T., Shin-ya, K., and Nagasawa, K. (2009a). G-quadruplex recognition by macrocyclic hexaoxazole (6OTD) dimer: greater selectivity than monomer. *Chem Commun (Camb),* 6481-6483.

Iida, K., Tera, M., Shin-Ya, K., and Nagasawa, K. (2009b). G-quadruplex recognition by macrocyclic hexaoxazole (6OTD) dimer. *Nucleic Acids Symp Ser (Oxf),* 233-234.

Incles, C. M., Schultes, C. M., Kempski, H., Koehler, H., Kelland, L. R., and Neidle, S. (2004). A G-quadruplex telomere targeting agent produces p16-associated senescence and chromosomal fusions in human prostate cancer cells. *Molecular cancer therapeutics 3*, 1201-1206.

Ishiguro, A., Kimura, N., Watanabe, Y., Watanabe, S., and Ishihama, A. (2016). TDP-43 binds and transports G-quadruplex-containing mRNAs into neurites for local translation. *Genes to cells : devoted to molecular and cellular mechanisms*.

Islam, M. A., Thomas, S. D., Murty, V. V., Sedoris, K. J., and Miller, D. M. (2014). c-Myc Quadruplex-forming Sequence Pu-27 Induces Extensive Damage in Both Telomeric and Nontelomeric Regions of DNA. *The Journal of biological chemistry 289*, 8521-8531.

Jantos, K., Rodriguez, R., Ladame, S., Shirude, P. S., and Balasubramanian, S. (2006). Oxazole-based peptide macrocycles: a new class of G-quadruplex binding ligands. *Journal of the American Chemical Society 128*, 13662-13663.

Kang, H. J., and Park, H. J. (2009). Novel molecular mechanism for actinomycin D activity as an oncogenic promoter G-quadruplex binder. *Biochemistry 48*, 7392-7398.

Kanoh, Y., Matsumoto, S., Fukatsu, R., Kakusho, N., Kono, N., Renard-Guillet, C., Masuda, K., Iida, K., Nagasawa, K., Shirahige, K., et al. (2015). Rif1 binds to G quadruplexes and suppresses replication over long distances. *Nature structural and molecular biology 22*, 889-897.

Kelland, L. R. (2005). Overcoming the immortality of tumour cells by telomere and telomerase based cancer therapeutics--current status and future prospects. *Eur J Cancer 41*, 971-979.

Kim, M. Y., Vankayalapati, H., Shin-Ya, K., Wierzba, K., and Hurley, L.H. (2002). Telomestatin, a potent telomerase inhibitor that interacts quite specifically with the human telomeric intramolecular g-quadruplex. *Journal of the American Chemical Society 124*, 2098-2099.

Kim, N. W., Piatyszek, M. A., Prowse, K. R., Harley, C. B., West, M. D., Ho, P. L., Coviello, G. M., Wright, W. E., Weinrich, S. L., and Shay, J. W. (1994). Specific association of human telomerase activity with immortal cells and cancer. *Science 266*, 2011-2015.

Koole, W., van Schendel, R., Karambelas, A. E., van Heteren, J. T., Okihara, K. L., and Tijsterman, M. (2014). A Polymerase Theta-dependent repair pathway suppresses extensive genomic instability at endogenous G4 DNA sites. *Nature communications 5*, 3216.

Kumari, S., Bugaut, A., Huppert, J. L., and Balasubramanian, S. (2007). An RNA G-quadruplex in the 5' UTR of the NRAS proto-oncogene modulates translation. *Nature chemical biology 3*, 218-221.

Lagah, S., Tan, I. L., Radhakrishnan, P., Hirst, R. A., Ward, J. H., O'Callaghan, C., Smith, S. J., Stevens, M. F., Grundy, R. G., and Rahman, R. (2014). RHPS4 G-quadruplex ligand induces anti-proliferative effects in brain tumor cells. *PloS one 9*, e86187.

Lam, E. Y., Beraldi, D., Tannahill, D., and Balasubramanian, S. (2013). G-quadruplex structures are stable and detectable in human genomic DNA. *Nature communications 4*, 1796.

Lattmann, S., Stadler, M. B., Vaughn, J. P., Akman, S. A., and Nagamine, Y. (2011). The DEAH-box RNA helicase RHAU binds an intramolecular RNA G-quadruplex in TERC and associates with telomerase holoenzyme. *Nucleic acids research 39*, 9390-9404.

Law, M. J., Lower, K. M., Voon, H. P., Hughes, J. R., Garrick, D., Viprakasit, V., Mitson, M., De Gobbi, M., Marra, M., Morris, A., et al. (2010). ATR-X syndrome protein targets tandem repeats and influences allele-specific expression in a size-dependent manner. *Cell 143*, 367-378.

Lee, Y. A., Kim, J. O., Cho, T. S., Song, R., and Kim, S. K. (2003). Binding of meso-tetrakis(N-methylpyridium-4-yl)porphyrin to triplex oligonucleotides: evidence for the porphyrin stacking in the major groove. *Journal of the American Chemical Society 125*, 8106-8107.

Lee, Y. R., Yu, D. S., Liang, Y. C., Huang, K. F., Chou, S. J., Chen, T. C., Lee, C. C., Chen, C. L., Chiou, S. H., and Huang, H. S. (2013). New approaches of PARP-1 inhibitors in human lung cancer cells and cancer stem-like cells by some selected anthraquinone-derived small molecules. *PloS one 8*, e56284.

Lemarteleur, T., Gomez, D., Paterski, R., Mandine, E., Mailliet, P., and Riou, J. F. (2004). Stabilization of the c-myc gene promoter quadruplex by specific ligands' inhibitors of telomerase. *Biochemical and biophysical research communications 323*, 802-808.

Lennartsson, A., and Ekwall, K. (2009). Histone modification patterns and epigenetic codes. *Biochimica et biophysica acta 1790*, 863-868.

Leonetti, C., Amodei, S., D'Angelo, C., Rizzo, A., Benassi, B., Antonelli, A., Elli, R., Stevens, M. F., D'Incalci, M., Zupi, G., et al. (2004). Biological activity of the G-quadruplex ligand RHPS4 (3,11-difluoro-6,8,13-trimethyl-8H-quino[4,3,2-kl]acridinium methosulfate) is associated with telomere capping alteration. *Molecular pharmacology 66*, 1138-1146.

Li, J. L., Harrison, R. J., Reszka, A. P., Brosh, R. M., Jr., Bohr, V. A., Neidle, S., and Hickson, I. D. (2001). Inhibition of the Bloom's and Werner's syndrome helicases by G-quadruplex interacting ligands. *Biochemistry 40*, 15194-15202.

Lin, W., Sampathi, S., Dai, H., Liu, C., Zhou, M., Hu, J., Huang, Q., Campbell, J., Shin-Ya, K., Zheng, L., et al. (2013). Mammalian DNA2 helicase/nuclease cleaves G-quadruplex DNA and is required for telomere integrity. *The EMBO journal 32*, 1425-1439.

Linder, J., Garner, T. P., Williams, H. E., Searle, M. S., and Moody, C. J. (2011). Telomestatin: formal total synthesis and cation-mediated interaction of its seco-derivatives with G-quadruplexes. *Journal of the American Chemical Society 133*, 1044-1051.

Lipps, H. J., and Rhodes, D. (2009). G-quadruplex structures: in vivo evidence and function. *Trends in cell biology 19*, 414-422.

Liu, A. H., Sun, X., Wei, X. Q., and Zhang, Y. Z. (2013). Efficacy of multiple low-dose photodynamic TMPYP4 therapy on cervical cancer tumour growth in nude mice. *Asian Pacific journal of cancer prevention : APJCP 14*, 5371-5374.

Liu, J. N., Deng, R., Guo, J. F., Zhou, J. M., Feng, G. K., Huang, Z. S., Gu, L.Q., Zeng, Y. X., and Zhu, X. F. (2007). Inhibition of myc promoter and telomerase activity and induction of delayed apoptosis by SYUIQ-5, a novel G-quadruplex interactive agent in leukemia cells. *Leukemia 21*, 1300-1302.

London, T. B., Barber, L. J., Mosedale, G., Kelly, G. P., Balasubramanian, S., Hickson, I. D., Boulton, S. J., and Hiom, K. (2008). FANCJ is a structure-specific DNA helicase associated with the maintenance of genomic G/C tracts. *The Journal of biological chemistry 283*, 36132-36139.

Lopes, J., Piazza, A., Bermejo, R., Kriegsman, B., Colosio, A., Teulade-Fichou, M. P., Foiani, M., and Nicolas, A. (2011). G-quadruplex-induced instability during leading-strand replication. *The EMBO journal 30*, 4033-4046.

Maizels, N., and Gray, L. T. (2013). The G4 genome. *PLoS genetics 9*, e1003468.

McLuckie, K. I., Waller, Z. A., Sanders, D. A., Alves, D., Rodriguez, R., Dash, J., McKenzie, G. J., Venkitaraman, A. R., and Balasubramanian, S. (2011). G-quadruplex-binding benzo[a]phenoxazines down-regulate c-KIT expression in human gastric carcinoma cells. *Journal of the American Chemical Society 133*, 2658-2663.

Meier, M., Patel, T. R., Booy, E. P., Marushchak, O., Okun, N., Deo, S., Howard, R., McEleney, K., Harding, S. E., Stetefeld, J., et al. (2013). Binding of G-quadruplexes to the N-terminal recognition domain of the RNA helicase associated with AU-rich element (RHAU). *The Journal of biological chemistry 288*, 35014-35027.

Mergny, J. L., Lacroix, L., Teulade-Fichou, M. P., Hounsou, C., Guittat, L., Hoarau, M., Arimondo, P. B., Vigneron, J. P., Lehn, J. M., Riou, J. F., et al. (2001). Telomerase inhibitors based on quadruplex ligands selected by a fluorescence assay. *Proceedings of the National Academy of Sciences of the United States of America 98*, 3062-3067.

Mikami-Terao, Y., Akiyama, M., Yuza, Y., Yanagisawa, T., Yamada, O., Kawano, T., Agawa, M., Ida, H., and Yamada, H. (2009). Antitumor activity of TMPyP4 interacting G-quadruplex in retinoblastoma cell lines. *Experimental eye research 89*, 200-208.

Mikami-Terao, Y., Akiyama, M., Yuza, Y., Yanagisawa, T., Yamada, O., and Yamada, H. (2008). Antitumor activity of G-quadruplex-interactive agent TMPyP4 in K562 leukemic cells. *Cancer letters 261*, 226-234.

Miller, K. M., and Rodriguez, R. (2011). G-quadruplexes: selective DNA targeting for cancer therapeutics? *Expert review of clinical pharmacology 4*, 139-142.

Miyazaki, T., Pan, Y., Joshi, K., Purohit, D., Hu, B., Demir, H., Mazumder, S., Okabe, S., Yamori, T., Viapiano, M., et al. (2012). Telomestatin impairs glioma stem cell survival and growth through the disruption of telomeric G-quadruplex and inhibition of the proto-oncogene, c-Myb. *Clinical cancer research : an official journal of the American Association for Cancer Research 18*, 1268-1280.

Mohaghegh, P., Karow, J. K., Brosh, R. M., Jr., Bohr, V. A., and Hickson, I. D. (2001). The Bloom's and Werner's syndrome proteins are DNA structure-specific helicases. *Nucleic acids research 29*, 2843-2849.

Monchaud, D., Allain, C., and Teulade-Fichou, M. P. (2007). Thiazole orange: a useful probe for fluorescence sensing of G-quadruplex-ligand interactions. *Nucleosides, nucleotides and nucleic acids 26*, 1585-1588.

Monchaud, D., and Teulade-Fichou, M. P. (2008). A hitchhiker's guide to G-quadruplex ligands. *Organic and biomolecular chemistry 6*, 627-636.

Moye, A. L., Porter, K. C., Cohen, S. B., Phan, T., Zyner, K. G., Sasaki, N., Lovrecz, G. O., Beck, J. L., and Bryan, T. M. (2015). Telomeric G-quadruplexes are a substrate and site of localization for human telomerase. *Nature communications 6*, 7643.

Müller, S., and Almouzni, G. (2014). A network of players in H3 histone variant deposition and maintenance at centromeres. *Biochimica et biophysica acta 1839*, 241-250.

Müller, S., Kumari, S., Rodriguez, R., and Balasubramanian, S. (2010). Small-molecule-mediated G-quadruplex isolation from human cells. *Nature* chemistry *2*, 1095-1098.

Müller, S., Laxmi-Reddy, K., Rodriguez, R., Jena, P. V., Baptiste, B., Dong, Z., Godde, F., Ha, T., Balasubramanian, S., and Huc, I. (2014). Targeting DNA G-quadruplexes with helical small molecules. *Chembiochem: a European journal of chemical biology 15*, 2563-2570.

Müller, S., Pantos, G. D., Rodriguez, R., and Balasubramanian, S. (2009). Controlled-folding of a small molecule modulates DNA G-quadruplex recognition. *Chem Commun (Camb),* 80-82.

Müller, S., and Rodriguez, R. (2014). G-quadruplex interacting small molecules and drugs: from bench toward bedside. *Expert review of clinical pharmacology 7*, 663-679.

Müller, S., Sanders, D. A., Di Antonio, M., Matsis, S., Riou, J. F., Rodriguez, R., and Balasubramanian, S. (2012). Pyridostatin analogues promote telomere dysfunction and long-term growth inhibition in human cancer cells. *Organic and biomolecular chemistry 10*, 6537-6546.

Murat, P., and Balasubramanian, S. (2013). Existence and consequences of G-quadruplex structures in DNA. *Current opinion in genetics and development 25C*, 22-29.

Murat, P., Zhong, J., Lekieffre, L., Cowieson, N. P., Clancy, J. L., Preiss, T., Balasubramanian, S., Khanna, R., and Tellam, J. (2014). G-quadruplexes regulate Epstein-Barr virus-encoded nuclear antigen 1 mRNA translation. *Nature chemical biology*.

Nakajima, A., Tauchi, T., Sashida, G., Sumi, M., Abe, K., Yamamoto, K., Ohyashiki, J. H., and Ohyashiki, K. (2003). Telomerase inhibition enhances apoptosis in human acute leukemia cells: possibility of antitelomerase therapy. *Leukemia 17*, 560-567.

Neidle, S. (2009). The structures of quadruplex nucleic acids and their drug complexes. *Current opinion in structural biology 19*, 239-250.

Neidle, S. (2010). Human telomeric G-quadruplex: the current status of telomeric G-quadruplexes as therapeutic targets in human cancer. *The FEBS journal 277*, 1118-1125.

Nguyen, G. H., Tang, W., Robles, A. I., Beyer, R. P., Gray, L. T., Welsh, J. A., Schetter, A. J., Kumamoto, K., Wang, X. W., Hickson, I. D., et al. (2014). Regulation of gene expression by the BLM helicase correlates with the presence of G-quadruplex DNA motifs. *Proceedings of the National Academy of Sciences of the United States of America 111*, 9905-9910.

Ou, T. M., Lin, J., Lu, Y. J., Hou, J. Q., Tan, J. H., Chen, S. H., Li, Z., Li, Y. P., Li, D., Gu, L. Q., et al. (2011). Inhibition of cell proliferation by quindoline derivative (SYUIQ-05) through its preferential interaction with c-myc promoter G-quadruplex. *Journal of medicinal chemistry 54*, 5671-5679.

Ou, T. M., Lu, Y. J., Zhang, C., Huang, Z. S., Wang, X.D., Tan, J. H., Chen, Y., Ma, D. L., Wong, K. Y., Tang, J. C., et al. (2007). Stabilization of G-quadruplex DNA and down-regulation of oncogene c-myc by quindoline derivatives. *Journal of medicinal chemistry 50*, 1465-1474.

Paeschke, K., Bochman, M. L., Garcia, P. D., Cejka, P., Friedman, K. L., Kowalczykowski, S. C., and Zakian, V. A. (2013). Pif1 family helicases suppress genome instability at G-quadruplex motifs. *Nature 497*, 458-462.

Paeschke, K., Capra, J. A., and Zakian, V. A. (2011). DNA replication through G-quadruplex motifs is promoted by the Saccharomyces cerevisiae Pif1 DNA helicase. *Cell 145*, 678-691.

Papadopoulou, C., Guilbaud, G., Schiavone, D., and Sale, J.E. (2015). Nucleotide Pool Depletion Induces G-Quadruplex-Dependent Perturbation of Gene Expression. *Cell reports 13*, 2491-2503.

Papamichos-Chronakis, M., and Peterson, C. L. (2013). Chromatin and the genome integrity network. *Nature reviews Genetics 14*, 62-75.

Parkinson, G. N., Lee, M.P., and Neidle, S. (2002). Crystal structure of parallel quadruplexes from human telomeric DNA. *Nature 417*, 876-880.

Pennarun, G., Granotier, C., Gauthier, L. R., Gomez, D., Hoffschir, F., Mandine, E., Riou, J. F., Mergny, J. L., Mailliet, P., and Boussin, F. D. (2005). Apoptosis related to telomere instability and cell cycle alterations in human glioma cells treated by new highly selective G-quadruplex ligands. *Oncogene 24*, 2917-2928.

Pennarun, G., Granotier, C., Hoffschir, F., Mandine, E., Biard, D., Gauthier, L. R., and Boussin, F. D. (2008). Role of ATM in the telomere response to the G-quadruplex ligand 360A. *Nucleic acids research 36*, 1741-1754.

Perrone, R., Butovskaya, E., Daelemans, D., Palu, G., Pannecouque, C., and Richter, S. N. (2014). Anti-HIV-1 activity of the G-quadruplex ligand BRACO-19. *J Antimicrob Chemother 69*, 3248-3258.

Perry, P. J., Read, M. A., Davies, R. T., Gowan, S. M., Reszka, A. P., Wood, A. A., Kelland, L. R., and Neidle, S. (1999). 2,7-Disubstituted amidofluorenone derivatives as inhibitors of human telomerase. *Journal of medicinal chemistry 42*, 2679-2684.

Petenzi, M., Verga, D., Largy, E., Hamon, F., Doria, F., Teulade-Fichou, M.P., Guedin, A., Mergny, J. L., Mella, M., and Freccero, M. (2012). Cationic pentaheteroaryls as selective G-quadruplex ligands by solvent-free microwave-assisted synthesis. *Chemistry 18*, 14487-14496.

Phan, A. T., Kuryavyi, V., Luu, K. N., and Patel, D. J. (2007). Structure of two intramolecular G-quadruplexes formed by natural human telomere sequences in K+ solution. *Nucleic acids research 35*, 6517-6525.

Phan, A. T., Modi, Y. S., and Patel, D. J. (2004a). Propeller-type parallel-stranded G-quadruplexes in the human c-myc promoter. *Journal of the American Chemical Society 126*, 8710-8716.

Phan, A. T., Modi, Y. S., and Patel, D. J. (2004b). Two-repeat Tetrahymena telomeric d(TGGGGTTGGGGT) Sequence interconverts between asymmetric dimeric G-quadruplexes in solution. *Journal of molecular biology 338*, 93-102.

Phatak, P., Cookson, J. C., Dai, F., Smith, V., Gartenhaus, R. B., Stevens, M. F., and Burger, A. M. (2007). Telomere uncapping by the G-quadruplex ligand RHPS4 inhibits clonogenic tumour cell growth in vitro and in vivo consistent with a cancer stem cell targeting mechanism. *British journal of cancer 96*, 1223-1233.

Piazza, A., Boule, J. B., Lopes, J., Mingo, K., Largy, E., Teulade-Fichou, M. P., and Nicolas, A. (2010). Genetic instability triggered by G-quadruplex interacting Phen-DC compounds in Saccharomyces cerevisiae. *Nucleic acids research 38*, 4337-4348.

Piazza, A., Serero, A., Boule, J. B., Legoix-Ne, P., Lopes, J., and Nicolas, A. (2012). Stimulation of gross chromosomal rearrangements by the human CEB1 and CEB25 minisatellites in Saccharomyces cerevisiae depends on G-quadruplexes or Cdc13. *PLoS genetics 8*, e1003033.

Raiber, E. A., Kranaster, R., Lam, E., Nikan, M., and Balasubramanian, S. (2012). A non-canonical DNA structure is a binding motif for the transcription factor SP1 in vitro. *Nucleic acids research 40*, 1499-1508.

Rangan, A., Fedoroff, O. Y., and Hurley, L. H. (2001). Induction of duplex to G-quadruplex transition in the c-myc promoter region by a small molecule. *The Journal of biological chemistry 276*, 4640-4646.

Rankin, S., Reszka, A. P., Huppert, J., Zloh, M., Parkinson, G. N., Todd, A. K., Ladame, S., Balasubramanian, S., and Neidle, S. (2005). Putative DNA quadruplex formation within the human c-kit oncogene. *Journal of the American Chemical Society 127*, 10584-10589.

Rapozzi, V., Zorzet, S., Zacchigna, M., Della Pietra, E., Cogoi, S., and Xodo, L. E. (2014). Anticancer activity of cationic porphyrins in melanoma tumour-bearing mice and mechanistic in vitro studies. *Molecular cancer 13*, 75.

Ray, S., Bandaria, J. N., Qureshi, M. H., Yildiz, A., and Balci, H. (2014). G-quadruplex formation in telomeres enhances POT1/TPP1 protection against RPA binding. *Proceedings of the National Academy of Sciences of the United States of America 111*, 2990-2995.

Read, M., Harrison, R. J., Romagnoli, B., Tanious, F. A., Gowan, S. H., Reszka, A. P., Wilson, W. D., Kelland, L. R., and Neidle, S. (2001). Structure-based design of selective and potent G quadruplex-mediated telomerase inhibitors. *Proceedings of the National Academy of Sciences of the United States of America 98*, 4844-4849.

Ribeyre, C., Lopes, J., Boule, J. B., Piazza, A., Guedin, A., Zakian, V. A., Mergny, J. L., and Nicolas, A. (2009). The yeast Pif1 helicase prevents genomic instability caused by G-quadruplex-forming CEB1 sequences in vivo. *PLoS genetics 5*, e1000475.

Riou, J. F., Guittat, L., Mailliet, P., Laoui, A., Renou, E., Petitgenet, O., Megnin-Chanet, F., Helene, C., and Mergny, J. L. (2002). Cell senescence and telomere shortening induced by a new series of specific G-quadruplex DNA ligands. *Proceedings of the National Academy of Sciences of the United States of America 99*, 2672-2677.

Rodriguez, R., Miller, K. M., Forment, J. V., Bradshaw, C. R., Nikan, M., Britton, S., Oelschlaegel, T., Xhemalce, B., Balasubramanian, S., and Jackson, S. P. (2012). Small-molecule-induced DNA damage identifies alternative DNA structures in human genes. *Nature chemical biology 8*, 301-310.

Rodriguez, R., Müller, S., Yeoman, J. A., Trentesaux, C., Riou, J. F., and Balasubramanian, S. (2008). A novel small molecule that alters shelterin integrity and triggers a DNA-damage response at telomeres. *Journal of the American Chemical Society 130*, 15758-15759.

Salvati, E., Leonetti, C., Rizzo, A., Scarsella, M., Mottolese, M., Galati, R., Sperduti, I., Stevens, M. F., D'Incalci, M., Blasco, M., et al. (2007). Telomere damage induced by the G-quadruplex ligand RHPS4 has an antitumor effect. *The Journal of clinical investigation 117*, 3236-3247.

Salvati, E., Scarsella, M., Porru, M., Rizzo, A., Iachettini, S., Tentori, L., Graziani, G., D'Incalci, M., Stevens, M. F., Orlandi, A., et al. (2010). PARP1 is activated at telomeres upon G4 stabilization: possible target for telomere-based therapy. *Oncogene 29*, 6280-6293.

Sarkies, P., Murat, P., Phillips, L. G., Patel, K. J., Balasubramanian, S., and Sale, J. E. (2012). FANCJ coordinates two pathways that maintain epigenetic stability at G-quadruplex DNA. *Nucleic acids research 40*, 1485-1498.

Sarkies, P., Reams, C., Simpson, L. J., and Sale, J. E. (2010). Epigenetic instability due to defective replication of structured DNA. *Molecular cell 40*, 703-713.

Schaffitzel, C., Postberg, J., Paeschke, K., and Lipps, H. J. (2010). Probing telomeric G-quadruplex DNA structures in cells with in vitro generated single-chain antibody fragments. *Methods Mol Biol 608*, 159-181.

Schwab, R. A., Nieminuszczy, J., Shin-ya, K., and Niedzwiedz, W. (2013). FANCJ couples replication past natural fork barriers with maintenance of chromatin structure. *The Journal of cell biology 201*, 33-48.

Sedoris, K. C., Thomas, S. D., Clarkson, C. R., Muench, D., Islam, A., Singh, R., and Miller, D. M. (2012). Genomic c-Myc quadruplex DNA selectively kills leukemia. *Molecular cancer therapeutics 11*, 66-76.

Seenisamy, J., Rezler, E. M., Powell, T. J., Tye, D., Gokhale, V., Joshi, C. S., Siddiqui-Jain, A., and Hurley, L. H. (2004). The dynamic character of the G-quadruplex element in the c-MYC promoter and modification by TMPyP4. *Journal of the American Chemical Society 126*, 8702-8709.

Shalaby, T., Fiaschetti, G., Nagasawa, K., Shin-ya, K., Baumgartner, M., and Grotzer, M. (2013). G-quadruplexes as potential therapeutic targets for embryonal tumors. *Molecules 18*, 12500-12537.

Shalaby, T., von Bueren, A., Hurlimann, M. L., Fiaschetti, G., Castelletti, D., Masayuki, T., Nagasawa, K., Arcaro, A., Jelesarov, I., Shin-ya, K., et al. (2010). Disabling c-Myc in childhood medulloblastoma and atypical teratoid/rhabdoid tumor cells by the potent G-quadruplex interactive agent S2T1-6OTD. *Molecular cancer therapeutics 9*, 167-179.

Shammas, M. A., Shmookler Reis, R. J., Li, C., Koley, H., Hurley, L. H., Anderson, K. C., and Munshi, N. C. (2004). Telomerase inhibition and cell growth arrest after telomestatin treatment in multiple myeloma. *Clinical cancer research : an official journal of the American Association for Cancer Research 10*, 770-776.

Shay, J. W., and Wright, W. E. (2006). Telomerase therapeutics for cancer: challenges and new directions. *Nature reviews Drug discovery 5*, 577-584.

Shen, F. H., Jin, J., Li, J., Wang, Y., Zhu, S. H., Lu, Y. J., Ou, T. M., Huang, Z. S., Huang, M., and Huang, Z. Y. (2013). The G-quadruplex ligand, SYUIQ-FM05, targets proto-oncogene c-kit transcription and induces apoptosis in K562 cells. *Pharmaceutical biology 51*, 447-454.

Shin-ya, K., Wierzba, K., Matsuo, K., Ohtani, T., Yamada, Y., Furihata, K., Hayakawa, Y., and Seto, H. (2001). Telomestatin, a novel telomerase inhibitor from Streptomyces anulatus. *Journal of the American Chemical Society 123*, 1262-1263.

Shivalingam, A., Izquierdo, M. A., Le Marois, A., Vysniauskas, A., Suhling, K., Kuimova, M. K., and Vilar, R. (2015). The interactions between a small molecule and G-quadruplexes are visualized by fluorescence lifetime imaging microscopy. *Nature communications 6*, 8178.

Shklover, J., Weisman-Shomer, P., Yafe, A., and Fry, M. (2010). Quadruplex structures of muscle gene promoter sequences enhance in vivo MyoD-dependent gene expression. *Nucleic acids research 38*, 2369-2377.

Siddiqui-Jain, A., Grand, C. L., Bearss, D. J., and Hurley, L. H. (2002). Direct evidence for a G-quadruplex in a promoter region and its targeting with a small molecule to repress c-MYC transcription. *Proceedings of the National Academy of Sciences of the United States of America 99*, 11593-11598.

Smith, J. S., Chen, Q., Yatsunyk, L. A., Nicoludis, J. M., Garcia, M. S., Kranaster, R., Balasubramanian, S., Monchaud, D., Teulade-Fichou, M. P., Abramowitz, L., et al. (2011). Rudimentary G-quadruplex-based telomere capping in Saccharomyces cerevisiae. *Nature structural and molecular* biology *18*, 478-485.

Stewart, S. A., and Weinberg, R. A. (2006). Telomeres: cancer to human aging. *Annual review of cell and developmental biology 22*, 531-557.

Sumi, M., Tauchi, T., Sashida, G., Nakajima, A., Gotoh, A., Shin-Ya, K., Ohyashiki, J. H., and Ohyashiki, K. (2004). A G-quadruplex-interactive agent, telomestatin (SOT-095), induces telomere shortening with apoptosis and enhances chemosensitivity in acute myeloid leukemia. *International journal of oncology 24*, 1481-1487.

Sun, D., Thompson, B., Cathers, B. E., Salazar, M., Kerwin, S. M., Trent, J. O., Jenkins, T. C., Neidle, S., and Hurley, L. H. (1997). Inhibition of human telomerase by a G-quadruplex-interactive compound. *Journal of medicinal chemistry 40*, 2113-2116.

Sun, H., Karow, J. K., Hickson, I. D., and Maizels, N. (1998). The Bloom's syndrome helicase unwinds G4 DNA. *The Journal of biological chemistry 273*, 27587-27592.

Swygert, S. G., and Peterson, C. L. (2014). Chromatin dynamics: Interplay between remodeling enzymes and histone modifications. *Biochimica et biophysica acta.*

Taetz, S., Baldes, C., Murdter, T. E., Kleideiter, E., Piotrowska, K., Bock, U., Haltner-Ukomadu, E., Mueller, J., Huwer, H., Schaefer, U. F., et al. (2006). Biopharmaceutical characterization of the telomerase inhibitor BRACO19. *Pharmaceutical research 23*, 1031-1037.

Taetz, S., Murdter, T. E., Zapp, J., Boettcher, S., Baldes, C., Kleideiter, E., Piotrowska, K., Schaefer, U. F., Klotz, U., and Lehr, C. M. (2008). Decomposition of the telomere-targeting agent BRACO19 in physiological media results in products with decreased inhibitory potential. *International journal of pharmaceutics 357*, 6-14.

Tauchi, T., Shin-ya, K., Sashida, G., Sumi, M., Okabe, S., Ohyashiki, J. H., and Ohyashiki, K. (2006). Telomerase inhibition with a novel G-quadruplex-interactive agent, telomestatin: in vitro and in vivo studies in acute leukemia. *Oncogene 25*, 5719-5725.

Temime-Smaali, N., Guittat, L., Sidibe, A., Shin-ya, K., Trentesaux, C., and Riou, J. F. (2009). The G-quadruplex ligand telomestatin impairs binding of topoisomerase IIIalpha to G-quadruplex-forming oligonucleotides and uncaps telomeres in ALT cells. *PloS one 4*, e6919.

Thakur, R. K., Kumar, P., Halder, K., Verma, A., Kar, A., Parent, J. L., Basundra, R., Kumar, A., and Chowdhury, S. (2009). Metastases suppressor NM23-H2 interaction with G-quadruplex DNA within c-MYC promoter nuclease hypersensitive element induces c-MYC expression. *Nucleic acids research 37*, 172-183.

Todd, A. K., Johnston, M., and Neidle, S. (2005). Highly prevalent putative quadruplex sequence motifs in human DNA. *Nucleic acids research 33*, 2901-2907.

Tong, X., Lan, W., Zhang, X., Wu, H., Liu, M., and Cao, C. (2011). Solution structure of all parallel G-quadruplex formed by the oncogene RET promoter sequence. *Nucleic acids research 39*, 6753-6763.

Tosoni, E., Frasson, I., Scalabrin, M., Perrone, R., Butovskaya, E., Nadai, M., Palu, G., Fabris, D., and Richter, S. N. (2015). Nucleolin stabilizes G-quadruplex structures folded by the LTR promoter and silences HIV-1 viral transcription. *Nucleic acids research 43*, 8884-8897.

Tran, H., Schilling, M., Wirbelauer, C., Hess, D., and Nagamine, Y. (2004). Facilitation of mRNA deadenylation and decay by the exosome-bound, DExH protein RHAU. *Molecular cell 13*, 101-111.

van Steensel, B. (2011). Chromatin: constructing the big picture. *The EMBO journal 30*, 1885-1895.

Vasilyev, N., Polonskaia, A., Darnell, J. C., Darnell, R. B., Patel, D. J., and Serganov, A. (2015). Crystal structure reveals specific recognition of a G-quadruplex RNA by a beta-turn in the RGG motif of FMRP. *Proceedings of the National Academy of Sciences of the United States of America 112*, E5391-5400.

Vaughn, J. P., Creacy, S. D., Routh, E. D., Joyner-Butt, C., Jenkins, G. S., Pauli, S., Nagamine, Y., and Akman, S. A. (2005). The DEXH protein product of the DHX36 gene is the major source of tetramolecular quadruplex G4-DNA resolving activity in HeLa cell lysates. *The Journal of biological chemistry 280*, 38117-38120.

Waddington, C. H., and Kacser, H. (1957). *The Strategy of the Genes, a Discussion of Some Aspects of Theoretical Biology* (Allen and Unwin).

Wang, P., Leung, C. H., Ma, D. L., Yan, S. C., and Che, C. M. (2010). Structure-based design of platinum(II) complexes as c-myc oncogene down-regulators and luminescent probes for G-quadruplex DNA. *Chemistry 16*, 6900-6911.

Wang, Y., and Patel, D. J. (1993). Solution structure of a parallel-stranded G-quadruplex DNA. *Journal of molecular biology 234*, 1171-1183.

Wei, D., Parkinson, G. N., Reszka, A. P., and Neidle, S. (2012). Crystal structure of a c-kit promoter quadruplex reveals the structural role of metal ions and water molecules in maintaining loop conformation. *Nucleic acids research 40*, 4691-4700.

Wei, D., Todd, A. K., Zloh, M., Gunaratnam, M., Parkinson, G. N., and Neidle, S. (2013). Crystal structure of a promoter sequence in the B-raf gene reveals an intertwined dimer quadruplex. *Journal of the American Chemical Society 135*, 19319-19329.

Williamson, J. R., Raghuraman, M. K., and Cech, T. R. (1989). Monovalent cation-induced structure of telomeric DNA: the G-quartet model. *Cell 59*, 871-880.

Wu, X., and Maizels, N. (2001). Substrate-specific inhibition of RecQ helicase. *Nucleic acids research 29*, 1765-1771.

Wu, Y., and Brosh, R. M., Jr. (2010). G-quadruplex nucleic acids and human disease. *The FEBS journal 277*, 3470-3488.

Wu, Y., Shin-ya, K., and Brosh, R. M., Jr. (2008). FANCJ helicase defective in Fanconia anemia and breast cancer unwinds G-quadruplex DNA to defend genomic stability. *Molecular and cellular biology 28*, 4116-4128.

Wu, Y., Suhasini, A. N., and Brosh, R. M., Jr. (2009). Welcome the family of FANCJ-like helicases to the block of genome stability maintenance proteins. *Cellular and molecular life sciences : CMLS 66*, 1209-1222.

Xu, C. X., Zheng, Y. X., Zheng, X. H., Hu, Q., Zhao, Y., Ji, L. N., and Mao, Z. W. (2013). V-shaped dinuclear Pt(II) complexes: selective interaction with human telomeric G-quadruplex and significant inhibition towards telomerase. *Scientific reports 3*, 2060.

Yang, P., De Cian, A., Teulade-Fichou, M. P., Mergny, J. L., and Monchaud, D. (2009). Engineering bisquinolinium/thiazole orange conjugates for fluorescent sensing of G-quadruplex DNA. *Angew Chem Int Ed Engl 48*, 2188-2191.

Zahler, A. M., Williamson, J. R., Cech, T. R., and Prescott, D. M. (1991). Inhibition of telomerase by G-quartet DNA structures. *Nature 350*, 718-720.

Zakian, V. A. (1995). Telomeres: beginning to understand the end. *Science 270*, 1601-1607.

Zhang, D. H., Zhou, B., Huang, Y., Xu, L. X., and Zhou, J. Q. (2006). The human Pif1 helicase, a potential Escherichia coli RecD homologue, inhibits telomerase activity. *Nucleic acids research 34*, 1393-1404.

Zhang, Z., Dai, J., Veliath, E., Jones, R. A., and Yang, D. (2010). Structure of a two-G-tetrad intramolecular G-quadruplex formed by a variant human telomeric sequence in K+ solution: insights into the interconversion of human telomeric G-quadruplex structures. *Nucleic acids research 38*, 1009-1021.

Zhao, J., Bacolla, A., Wang, G., and Vasquez, K. M. (2010). Non-B DNA structure-induced genetic instability and evolution. *Cellular and molecular life sciences : CMLS 67*, 43-62.

Zimmer, J., Tacconi, E. M., Folio, C., Badie, S., Porru, M., Klare, K., Tumiati, M., Markkanen, E., Halder, S., Ryan, A., et al. (2016). Targeting BRCA1 and BRCA2 Deficiencies with G-Quadruplex-Interacting Compounds. *Molecular cell 61*, 449-460.

BIOGRAPHICAL SKETCH

Sebastian Müller
Institut Curie/ Paris, France, European Union

Education: Baccalaureate, Evangelisches Schulzentrum, Leipzig, Germany
MSci in Biology and Chemistry, University of Durham, England
PhD in Chemistry, University of Cambridge, England

Research and Professional Experience:

2016 – present – Senior Research Scientist, Institut Curie, Paris/ France, Group of Dr. Raphaël Rodriguez

2011-2015 – Postdoctoral Researcher, Institut Curie, Paris/ France, Group of Dr. Geneviève Almouzni

2006-2010 – Doctoral Researcher, Cambridge University/ England, Group of Prof. Shankar Balasubramanian

2003-2006 – Research Assistant at the same time as undergraduate studies, Durham University/ England, Group of Dr. Rumaisa Bashir

Professional Appointments:

Senior Research Scientist, Institut Curie, France 2016
Postdoctoral Researcher, Institut Curie, France, 2011
Doctoral Researcher, University of Cambridge, 2006

Honors: MSci, PhD

Publications Last 3 Years:

# equal contribution; * corresponding author

Research Articles

(4) L. Prendergast, S. Müller, Y. Liu, H. Huang, F. Dingli, D. Loew, I. Vassias, R. Amignacco, D.J. Patel, K.F. Sullivan, G. Almouzni. The CENP-T-CENP-W complex is a binding partner of the H2A-H2B chaperone FACT. *Genes and Dev.*

(3) H. Tachiwana[#], S. Müller[#], J. Blümer, K. Klare, A. Musacchio G. Almouzni. HJURP involvement in *de novo* CenH3^{CENP-A} and CENP-C recruitment. *Cell Rep*, 2015, *11*, 1-11.

(2) S. Müller, R. Montes de Oca, N. Lacoste, F. Dingli, D. Loew, G. Almouzni. Phosphorylation and DNA-binding of HJURP determine its centromeric recruitment and function in CenH3$^{\text{CENP-A}}$ loading. *Cell Rep*, 2014, *8*, 190-203.

(1) S. Müller, K. Laxmi-Reddy, R. Rodriguez, P. V. Jena, B. Baptiste, Z. Dong, F. Godde, T. Ha, S. Balasubramanian, I. Huc. Targeting DNA G-quadruplexes with helical small molecules. *ChemBioChem*, 2014, *15*, 2563-2570.

Review Articles

(5) S. Müller, G. Almouzni. Centromeric chromatin during the cell cycle, *Nature Reviews Genetics*.

(4) S. Müller, D. Filipescu, G. Almouzni. Developmental roles of histone H3 variants and their chaperones. Book chapter in *"The Functional Nucleus"*, 2016, *Springer Verlag*.

(3) D. Filipescu[#], S. Müller[#], G. Almouzni. Histone H3 variants and their chaperones in development and disease: contributing to epigenetic control. *Ann Rev Cell Dev Biol*, 2014, *30*, 615-646.

(2) S. Müller*, R. Rodriguez. G-quadruplex interacting small molecules and drugs: from bench toward bedside. *Expert Rev Clin Pharmacol*, 2014, *7*, 663-679.

(1) S. Müller, G. Almouzni. A network of players in H3 histone variant deposition and maintenance at centromeres. *BBA Gene Reg Mech*, 2014, 1839, pp. 241-250.

In: G-Quadruplex Structures, Formation and Role in Biology ISBN: 978-1-63485-512-9
Editor: Hannah Santos

Chapter 6

ASSAYS FOR TELOMERASE ACTIVITY TOWARD APPLICATIONS OF CANCER DIAGNOSIS

Hidenobu Yaku[1,*] and Daisuke Miyoshi[2,†]

[1] Advanced Technology Division, Panasonic Corporation, Kyoto, Japan

[2] Faculty of Frontiers of Innovative Research in Science and Technology (FIRST), Konan University, Kobe, Japan

ABSTRACT

Telomeres at the ends of eukaryotic chromosomes contain accessory proteins and telomeric DNA. Telomeric DNA is composed of double-stranded tandem repeats of a G-rich motif (5'-TTAGGG-3' in humans) and a complementary C-rich motif, and a single-stranded 3' overhang of the G-rich motif, termed a G-tail. In normal somatic cells, telomeric DNA shortens with every cell division due to the end replication problem. The shortening of telomeric DNA finally leads to apoptosis. In contrast, in most tumor cells, the length of telomeric DNA is maintained by highly activated telomerase, which is a ribonucleoprotein that adds new six nucleotides, 5'-GGTTAG-3' at the 3'-end of the G-rich motif. The maintenance of telomeric DNA is a critical biological event for carcinogenesis. The specificity and universality of the telomerase activity for tumor cells demonstrates that the detection and quantification of the telomerase activity should be greatly helpful for a precise cancer diagnosis. The telomeric repeat amplification protocol (TRAP) assay, which is the most widely used telomerase assay, can detect telomerase activity with high sensitivity because it amplifies the telomerase reaction products by polymerase chain reaction (PCR). However, this assay is not appropriate for clinical applications because PCR inhibitors in clinical samples cause false negative results. Thus, researchers have proposed other telomerase assays to solve the problem. Those assays are classified into three main groups: (i) Using non enzymatic signal amplification processes. (ii) Using enzymatic signal amplification processes other than PCR. (iii) Combining processes that remove the inhibitors from the reaction mixture and enzymatic signal amplification processes including PCR. In this review, we focus on the latest

* Correspondence should be addressed: yaku.hidenobu@jp.panasonic.com; Tel.: +81-774-98-2556; Fax: +81-774-98-2585.

† Correspondence should be addressed: miyoshi@center.konan-u.ac.jp; Tel.: +81-78-303-1426; Fax: +81-78-303-1495.

advances in the development of the telomerase assays and further discuss the application potentiality of the assays for clinical samples.

1. INTRODUCTION

Telomeres at ends of eukaryotic chromosomes contain telomeric DNA with tandem repeats of a G-rich motif (Figure 1) [1-4]. The sequence of the G-rich motif and the number of repeats varies greatly among species. In ciliate *Oxytricha nova*, telomeric DNA is a fixed 4.5 repeats of 5'-TTTTGGGG-3', whereas human telomeric DNA is composed of 10-15 kbp-long 5'-TTAGGG-3' repeats (Table 1) [2-4]. Telomeric DNA further has a structural feature at the 3'-end with a single-stranded G-rich overhang called a G-tail [3-9]. In humans, the length of the G-tail, 5'-TTAGGG-3', is 150-250 bases. Intrinsically, linear DNA is recognized as damaged DNA and the ends are rapidly joined via catalysis by various nucleases and DNA ligases. However, chromosomal DNA is protected from illicit DNA end-joining events because at least six telomeric proteins including telomeric repeat-binding factor 1 (TRF1), TRF2, repressor and activator protein 1 (RAP1), TRF1-interacting nuclear protein 2 (TIN2), protection of telomeres 1 (POT1) and TPP1 (*abbreviation* derived from its former names TINT1, PTOP or PIP1) bind to telomeric DNA to form a capped complex known as shelterin (Figure 2) [3, 4, 10, 11]. TRF1 and TRF2 directly bind to telomeric double-stranded DNA (dsDNA) after the formation of their respective homodimers [3, 4, 10, 11]. On the other hand, POT1 directly binds to the G-tail. TIN2 forms a bridge between TRF1, TRF2 and TPP1, and TPP1 forms another bridge between TIN2 and POT1 [3, 4, 10, 11]. The RAP1 homologue in humans associates with telomeric DNA via its interaction with TRF2 [3, 4, 10, 11].

Table 1. Telomeric G-rich sequences

Sequences (5'→3')	Species
TTGGGG	*Tetrahymena thermophila*
TTAGGG	*Human*
TTTAGGG	*Arabidopsis thaliana*
TTTTGGGG	*Oxytricha nova*
TTTTGGGG	*Stylonychia lemnae*

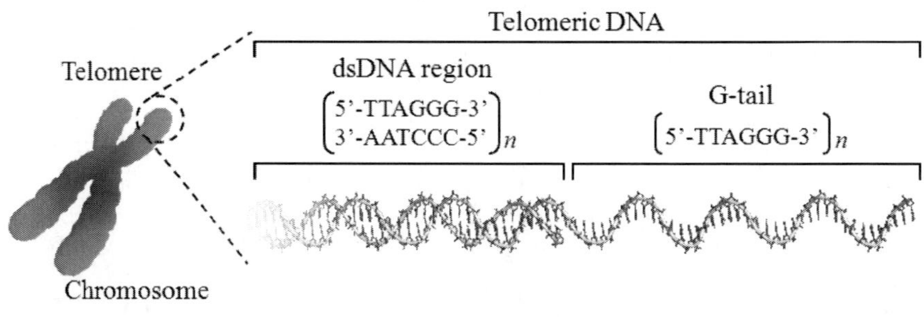

Figure 1. Human telomeric DNA.

Telomeric DNA can form an alternative higher-order structure termed a telomere loop (T-loop) (Figure 3) [3, 4, 10-12]. A T-loop is a lariat-like configuration in which the G-tail invades the dsDNA region to form a displacement loop or D-loop. Because a T-loop is not thermodynamically favorable, the telomeric accessory proteins should be involved in T-loop formation. This structure may also be another option to protect chromosomal DNA. Telomeric DNA is responsible for not only the protection of chromosomal ends but also the control of cell life-span. During S-phase of the cell cycle, chromosomal DNA is replicated from RNA primers by DNA polymerase, and then RNA primers are removed by nucleases (Figure 4A) [13, 14]. Thus, daughter strands are shorter than parent strands. This end replication problem triggers a loss of 100-200 bases of telomeric DNA with every cell division in normal somatic cells [15-18]. The shortening of telomeric DNA finally leads to inefficiency of the protection of chromosome ends by telomeric DNA (Figure 4B) [14, 18]. Furthermore, this deprotection causes cellular senescence or apoptosis [14, 18]. A series of these biological events serves as a huge barrier to carcinogenesis. Even though oncogenic mutations occur, carcinogenesis should be prevented as long as telomeric DNA shortens in the normal way.

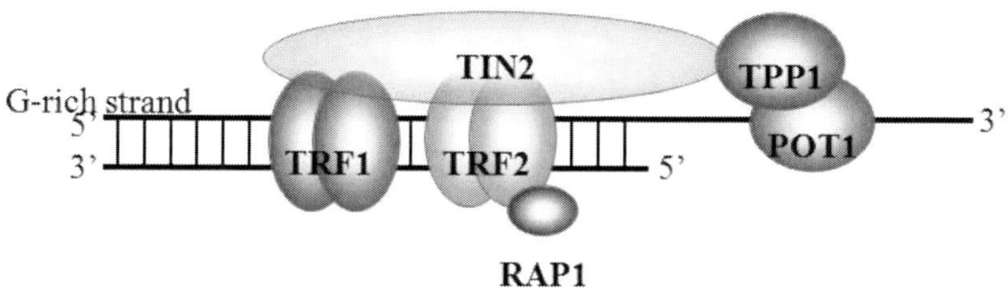

Figure 2. Shelterin composed of telomeric DNA and accessory proteins.

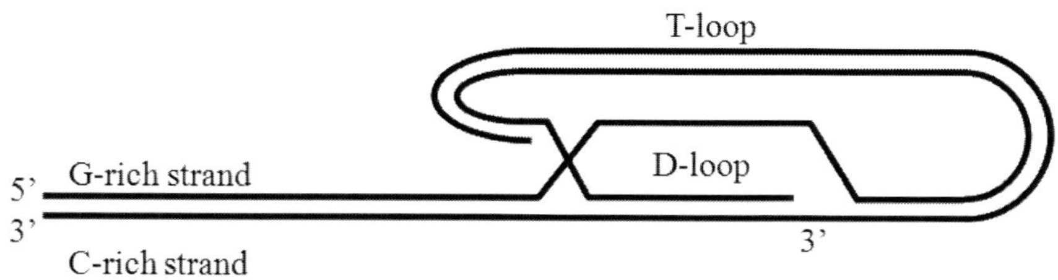

Figure 3. Structure of a T-loop.

Tumor cells can proliferate indefinitely [15, 19]. Importantly, 85-90% of human tumor cells solve the end replication problem by telomeric DNA extension using highly activated telomerase (Figure 5) [20-23]. Telomerase has a complex structure composed of many subunits, although not all of them are well understood. Previous studies have shown that two subunits, which are found in the central part of a human telomerase holoenzyme, have critical functions in telomeric DNA extension [24-27]. One of the subunits is telomerase reverse transcriptase (abbreviated to TERT, or ***hTERT*** in humans), which can synthesize DNA using an RNA template (Figure 5) [24, 25]. Another subunit is a telomerase RNA component (abbreviated to TR, or hTR in humans), which includes a template sequence for telomeric

DNA extension (Figure 5) [26, 27]. Antisense RNA of hTR was shown to competitively prevent human telomerase from binding to telomeric DNA by binding to hTR [28]. This competitive prevention caused a reduction in telomerase activity, and then tumor cells stopped proliferating in 23 to 26 days. A dominant-negative mutant of hTERT also inhibited the proliferation of tumor cells [28]. These results indicate that constitutive telomerase activation is essential for carcinogenesis as well as activation and inactivation of oncogenes and tumor suppressor genes, respectively. More importantly, most normal cells excepting germ cells, stem cells and activated lymphocytes show little or no telomerase activity [15, 19, 29]. The specificity and universality of the telomerase activity for tumor cells demonstrates that a measurement technology of the telomerase activity should be great help for a precise cancer diagnosis.

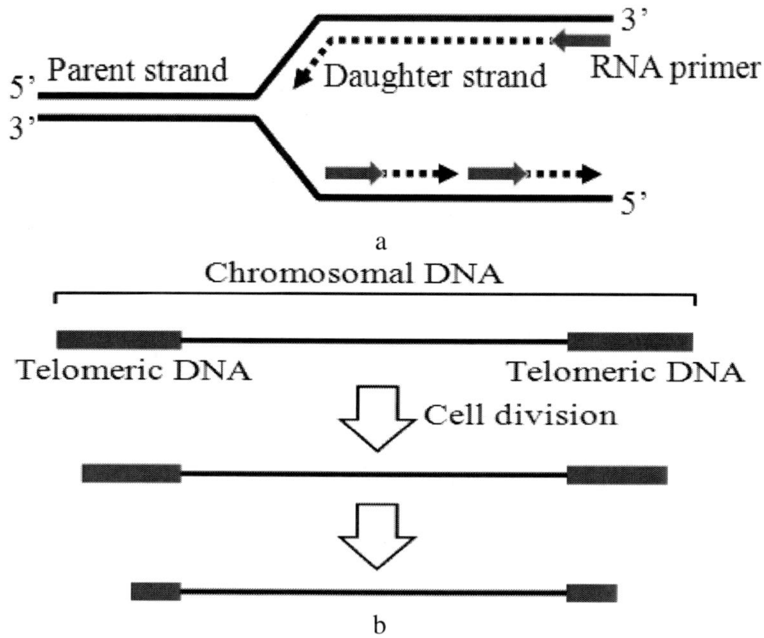

Figure 4. (A) The end replication problem and (B) loss of telomeric DNA with every cell division.

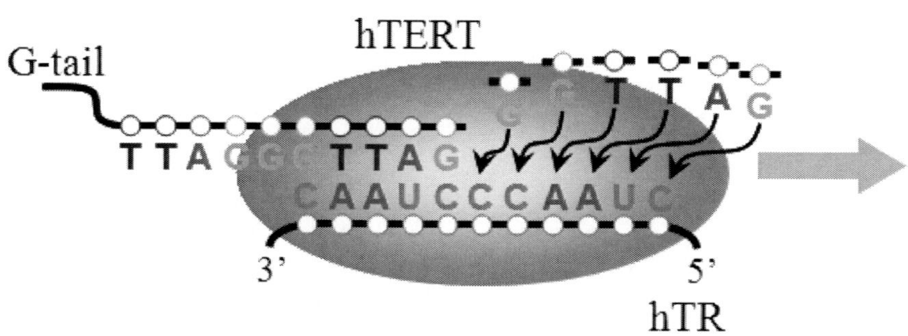

Figure 5. Telomeric DNA extension by telomerase.

2. Telomeric Repeat Amplification Protocol (TRAP) Assay

Telomerase has been suggested as a promising marker for cancer diagnosis. Many studies have shown the association between telomerase activity and a number of prognostic and clinicopathological features in colorectal cancer [30-36]. For the diagnosis with telomerase activity as a marker, an assay for telomerase activity that is precise and sensitive is required. A conventional procedure for telomerase activity is TRAP assay (Figure 6) [23]. In the assay, a telomerase reaction is carried out using a telomerase substrate primer and biological samples such as cell lysate including telomerase. Following the telomerase reaction, the reaction products are amplified by PCR using the telomerase substrate primer as a forward primer. Finally, the amplification products are analyzed electrophoretically. Because telomerase adds six nucleotides with every extension reaction, a ladder of PCR products with six base pair (bp) increments is observed. Given the simple principle of TRAP assay, the telomerase substrate primer and the reverse primer in PCR should be a telomeric repetitive oligonucleotide and its complementary oligonucleotide, respectively. However, such primers generate primer dimers of different length because of their repetitive complementary sequences (Figure 7). The primer dimers further bind to the primers, leading to longer primer artifacts with six bp increments. To avoid a false positive result, TS primer (telomerase substrate primer), which is a non-telomeric oligonucleotide, is used as the primer for telomerase reaction (Table 2) [23]. In addition, several reverse primers, which have non-complementary sequences to the telomeric sequence, have been designed (Table 2) [23, 37-39]. These improvements drastically reduce the probability of a false positive result. However, TRAP assay still has a severe problem in that some polymerase inhibitors in clinical samples inhibit PCR, which leads to a false negative result [38]. Thus, a novel telomerase assay that can avoid a false negative result is required for practical cancer diagnostics.

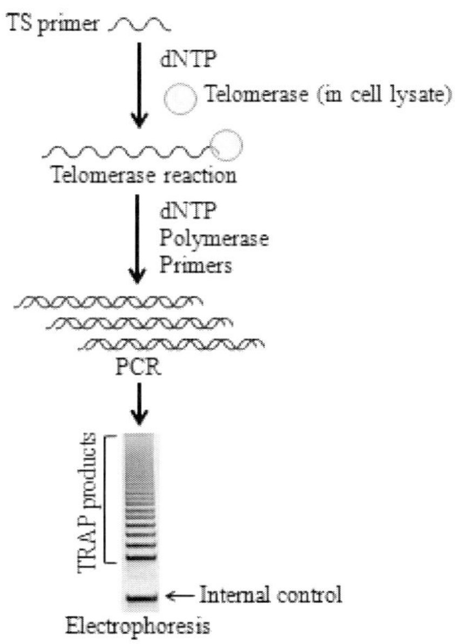

Figure 6. TRAP assay.

Table 2. Primers used in TRAP assay

Name	Function	Sequence [a]	Ref.
TS	Telomerase substrate primer and forward primer	5'-AATCCGTCGAGCAGAGTT-3'	[23]
CX	Reverse primer	5'-CCCTTACCCTTACCCTTACCCTAA-3'	[23]
CX-ext	Reverse primer	5'-GTGCCCTTACCCTTACCCTTACCCTAA-3'	[37]
ACX	Reverse primer	5'-GCGCGGCTTACCCTTACCCTTACCCTAACC-3'	[38]
ACT	Reverse primer	5'-GCGCGGCTAACCCTAACCCTAACC-3'	[38]
Cxa	Reverse primer	5'-GTGTAACCCTAACCCTAACCC-3'	[39]

[a] Underlined sequences of reverse primers are uncomplimentary with telomeric repetitive sequences.

Forward primer: 5'-TTAGGG TTAGGG TTAGGG TTAGGG-3'

+

Reverse primer: 3'-AATCCC AATCCC AATCCC AATCCC-5'

5'-TTAGGG TTAGGG TTAGGG TTAGGG-3'
3'-AATCCC AATCCC AATCCC AATCCC-5'

5'-TTAGGG TTAGGG TTAGGG TTAGGG TTAGGG-3'
3'-AATCCC AATCCC AATCCC AATCCC AATCCC-5'

5'-TTAGGG TTAGGG TTAGGG TTAGGG TTAGGG TTAGGG-3'
3'-AATCCC AATCCC AATCCC AATCCC AATCCC AATCCC-5'

5'-TTAGGG TTAGGG TTAGGG TTAGGG TTAGGG TTAGGG TTAGGG-3'
3'-AATCCC AATCCC AATCCC AATCCC AATCCC AATCCC AATCCC-5'

Figure 7. Example of primer dimers of different length caused by the telomeric repetitive forward primer and its complementary reverse primer. Underlined sequences are elongated products *via* PCR.

3. Telomerase Activity Assays without Any Enzymatic Reaction

3.1. Fluorescence-Based Assays

Fluorescence-labeled oligo DNA is widely-used for various biosensing methods. In the same way, some assays using fluorescence-labeled oligo DNA to detect the telomerase-elongated DNA have been reported [40-45]. Schmidt et al. developed an optical fiber immobilized with TS primer, which contained phosphorothioate (PS)-modification for strong binding to telomerase (Figure 8) [40]. In this assay, after the telomerase elongation reaction of the PS-modified TS primer, a fluorescence (FITC)-labeled DNA probe complementary to the telomere sequence is hybridized with the elongated DNA. Finally, the fluorescence intensity of the DNA probe bound to the elongated DNA is measured by the optical fiber. For example, the assay allowed successful detection of telomerase activity in HL-60 cell extract without PCR amplification.

Kawamura et al. also developed a very simple assay using a fluorescence resonance energy transfer (FRET) technique with a fluorescence (FAM)-labeled DNA probe to the telomere sequence (Figure 9) [41]. The assay included a telomerase reaction and hybridization of the DNA probe with the telomerase-elongated DNA in the presence of ethidium bromide. Although the fluorescence intensity of the DNA probe was high before the hybridization reaction, the fluorescence of the DNA probe bound to the telomerase-elongated DNA was quenched by ethidium bromide intercalated into the dsDNA of the DNA probe and the telomerase-elongated DNA. The detection limit was 400 HeLa cells μL^{-1} (Table 3).

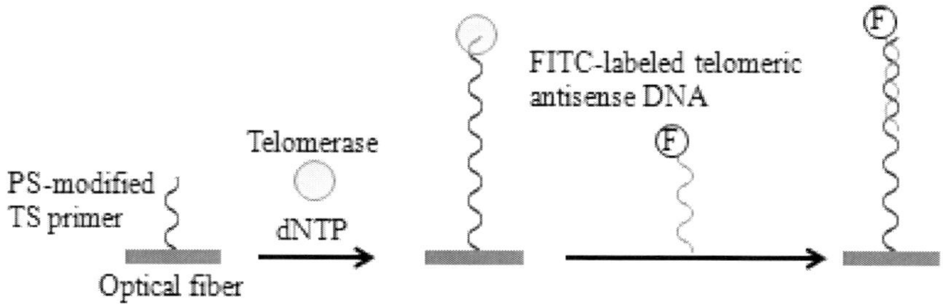

Figure 8. Principle of the fluorescence telomerase assay using an optical fiber [40].

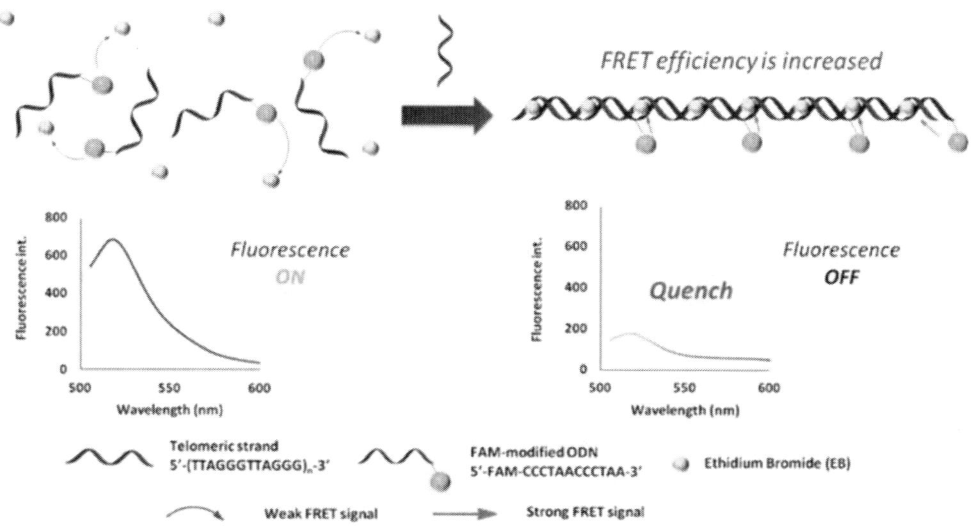

Figure 9. Principle of the fluorescence telomerase assay using FAM-labeled telomeric antisense DNA probe and ethidium bromide. Reused with permission from ref 41. Copyright 2014 Royal Society of Chemistry.

Table 3. Principle and sensitivity of each telomerase assay

Principle	Fig.	Sensitivity		Ref.
		Cell line	LOD (Limit of detection)	
Fluorescence enhancement by hybridization of FITC-labeled DNA probes with telomerase-elongated DNA on an optical fiber	8	HL-60	-	40
FRET between FAM-labeled DNA probes hybridized with telomerase-elongated DNA and ethidium bromide bound to the hybrid	9	HeLa	400 cells μL^{-1}	41
Fluorescence enhancement of opening of Cy5-labeled nicked molecular beacon DNA probes on an AuNP by telomerization	10	HeLa	-	42
Electron transfer from QD to hemin on G-quadruplexes of telomerase-elongated DNA on QDs	11	293T	270 cells μL^{-1}	43
Fluorescence enhancement of Zn (II) PPIX by binding to G-quadruplexes of telomerase-elongated DNA	12	293T	380 cells μL^{-1}	44
Fluorescence enhancement of SYBR Green I by binding to G-quadruplexes of telomerase-elongated DNA	13	HeLa	4 cells	45
Electrochemiluminescence of ECL nanoprobes tethered to a magnetic bead with telomerase-elongated DNA, which is gathered to an electrode by a magnet	14	HeLa	500 cells	76
Electrochemiluminescence of $Ru(bpy)_3^{2+}$, which binds to telomerase-elongated DNA adsorbed on a TAPP/CCG-modified glassy carbon electrode	15	HeLa	10 cells mL^{-1}	77
Color change via aggregation of AuNPs modified with TS primers	16	HeLa	1 cell μL^{-1}	80
Color change via aggregation of AuNPs comodified with TS primers and telomeric antisense DNA	17	MCF-7	-	81
Electrochemical signal enhancement via binding of ferrocenylnaphthalene diimide to G-quadruplexes of telomerase-elongated DNA	18	HeLa	40-140 cells μL^{-1}	82
Electrochemical signal enhancement by binding of $[Ru(NH_3)_6]^{3+}$ to DNA immobilized to AuNPs attached to an electrode	19	HeLa	10 cells	83
Electrochemical signal decrement by release of ferrocene-labeled DNA from DNA probe containing a TS primer on an Au electrode	20	HeLa	100 cells mL^{-1}	84
SPR signal enhancement via telomerization on a sensor chip	21	T47-D	100 cells with 1×10^5 normal fibroblast cells	89
SPR signal enhancement by binding of AuNPs to telomerase-elongated DNA via hybridization on a sensor chip	22	293T	18 cells μL^{-1}	43
SERS signal enhancement by aggregation of AuNPs comodified with DTNB and TS primers, which is prevented by telomerization	23	SKBR3	1 cell mL^{-1}	94
SERS signal enhancement and color change by accumulation of AuNPs modified with telomeric antisense DNA on magnetic beads modified with TS primers after telomerization	24	SKBR3	10 cells mL^{-1} (Color) 1 cell mL^{-1} (SERS)	95

Principle	Fig.	Sensitivity		Ref.
		Cell line	LOD	
Chemiluminescence signal generated by the pyrosequencing technique	25	A549 HeLA	5 cells 10 cells	98 99
Color change of ABTS oxidization by a G-quadruplex DNAzyme generated from a hairpin primer for telomerization	26	HeLa	500 cells	105
Color change of ABTS oxidization by a G-quadruplex DNAzyme formed by telomerase-elongated DNA with dGTP	27	HeLa	200 cells	106
Fluorescence enhancement by digestion of a molecular beacon by 8–17 DNAzyme generated from a hairpin primer for telomerization	28	HeLa	200 cells	107
Fluorescence enhancement by digestion of a TaqMan probe hybridized with telomerase-elongated DNA by T7 exonuclease	29	HeLa	5 cells	113
Electrochemical signal enhancement by digestion of a methylene blue-labeled hairpin DNA probe hybridized with telomerase-elongated DNA by T7 exonuclease	30	HeLa	1 cell	114
Chemiluminescence enhancement with luminol oxidization by a G-quadruplex DNAzyme exponentially generated by EXPAR	32A	HeLa	1 cell	117
Fluorescence enhancement with digestion of molecular beacons by DNAzyme exponentially generated by EXPAR	32B	HEK293T	1 cell	118
Fluorescence enhancement of digestion of reporter probes or SYBR Green I during exponential isothermal amplification of telomere repeat (EXPIATR) assay utilizing SD polymerase and NEase	33	HeLa	1 cell	119
Advanced TRAP assay containing the washing process of telomerase-elongated DNA on a magnetic bead to remove impurities	34	-	-	120
Chemiluminescence enhancement by the ALP reaction, in which anti-DIG-ALP is attached to a plate with binding to DIG-labeled telomeric antisense DNA hybridized with telomerase-elongated DNA	35	293T, HeLa, Raji, HT1080, and A431	-	125
Enhancement of electrode resistance or frequency change of Au-quartz crystal by precipitation of the insoluble product of the ALP reaction, in which avidin-ALP is bound to telomerase-elongated DNA with biotin on the electrode or the QCM crystal, respectively	36	HeLa	1000 to 5000 cells	126
Electrochemical signal enhancement with TMB oxidation by avidin-HRP bound to telomerase-elongated DNA with biotin from a spired DNA tetrahedron TS primer	37	HeLa	10 cells	127
Chemiluminescence enhancement with luminol oxidization by a G-quadruplex DNAzyme that is bound to telomerase-elongated DNA via hybridization	38A	HeLa	1000 cells	128
Chemiluminescence enhancement via luminol oxidization by G-quadruplex DNAzyme on AuNP that is bound to telomerase-elongated DNA via hybridization	38B	HeLa	1000 cells	129

Table 3. (Continued)

Principle	Fig.	Sensitivity		Ref.
		Cell line	LOD	
Electrochemical signal enhancement by adsorption of PtNPs on GCE after removing an unelongated TS primer hybridized with TS antisense DNA by exonuclease III	39A	A549	Less than 2 cells	130
Electrochemical signals enhancement via deposition of AgNPs to telomerase-elongated DNA on an Au electrode after removing unelongated TS primer hybridized with TS antisense DNA by exonuclease III	39B	MCF-7	1 cell	131
Fluorescence enhancement with digestion of probe RNA by CPT after A-PCR of telomerase-elongated DNA on a magnetic bead	40	HeLa	50 cells	133

Furthermore, a gold nanoparticle (AuNP) immobilized with a fluorescence (Cy5)-labeled nicked molecular beacon DNA probe was designed for the quantification of intracellular telomerase activity (Figure 10) [42]. The molecular beacon contained two segments: a shorter TS primer sequence as a part of the 5'end stem and a longer sequence to form a loop with the 3'end stem immobilized to the AuNP. When the molecular beacon closed, the fluorescence of Cy5 was quenched by AuNP. On the other hand, when the telomerase reaction proceeded, the telomerase reaction products opened the molecular beacon, leading to an increase in the fluorescence intensity. The telomerase activity in a HeLa cell was successfully monitored by using this technique.

Telomeric DNA (or telomerase-elongated DNA) that is a repetitive sequence of 5'-TTAGGG-3' can fold to form a four-stranded DNA called a G-quadruplex in the presence of metal ions such as K^+ and Na^+ [46-64]. Because G-quadruplex contains large π-planars termed G-quartets, in which four guanines are held together via eight Hoogsteen hydrogen bonds, it can bind to various π-conjugated molecules via π-π stacking interactions [65-68]. In particular, complexes of telomere G-quadruplex and protoporphyrins such as protoporphyrin IX (PPIX) and iron (III) protoporphyrin IX (hemin) are often used for detection of telomerase activity because they have two properties: the fluorescence increase of protoporphyrins upon binding to G-quadruplexes and the peroxidase-like DNAzyme activity of hemin-G-quadruplex complexes [69-74]. In order to detect telomerase activity without PCR, Willner's group conducted a telomerase reaction from TS primer on CdSe/ZnS quantum dots (QDs) and then added K^+ and hemin to form the telomere G-quadruplex / hemin complex (Figure 11) [43]. Because the fluorescence of the QDs was quenched by the complex, telomerase activity was measured with a detection limit of 270 293T cells μL^{-1} by analyzing the quenching ratio (Table 3). The same group developed another assay for telomerase activity in which K^+ and Zn (II) PPIX were added into the reaction solution after telomerization by 293T cells lysate (Figure 12) [44]. The fluorescence intensity of Zn (II) PPIX increased depending on the number of 293T cells and, the detection limit was 380 cells μL^{-1} (Table 3).

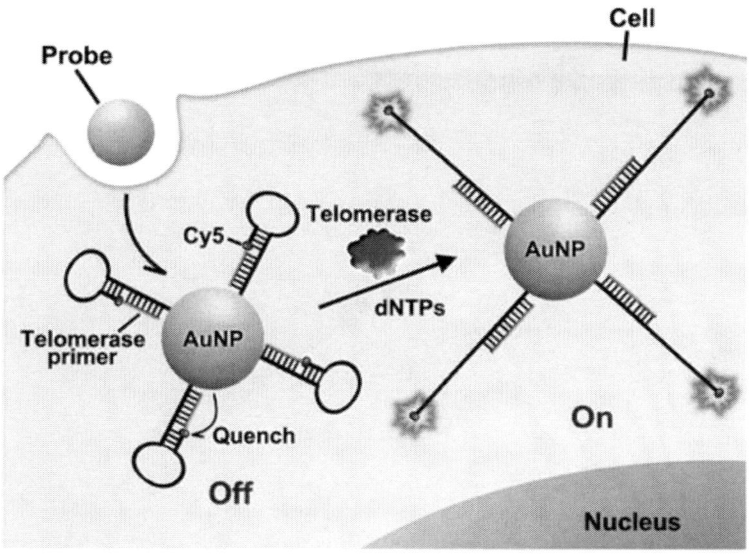

Figure 10. Principle of the fluorescence telomerase assay using AuNPs modified with Cy5-labeled molecular beacons. Reused with permission from ref 42. Copyright 2014 American Chemical Society.

The assays using PPIXs are very simple but the sensitivity is relatively low for clinical applications. Recently, it was reported that the fluorescence of SYBR Green I, which is known as a dsDNA specific ligand over ssDNA, increases by binding to G-quadruplex [75]. Thus, in the similar way as the assay using Zn (II) PPIX described above, K^+ and SYBR Green I were added into the telomerase reaction solution, which resulted in highly sensitive detection of telomerase activity with a detection limit of four HeLa cells (Figure 13 and Table 3) [45]. In the practical use of the assay, clinical samples obtained by biopsy generally contain a larger amount of normal cells. Thus, the fluorescence of SYBR Green I should not increase even in the presence of genomic dsDNA in cells for clinical applications. In this point of view, the authors further demonstrated that 0-400 normal cells showed very low fluorescence intensity in the assay.

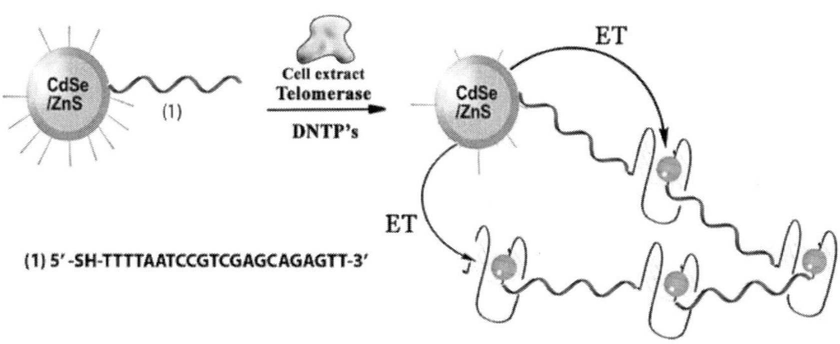

Figure 11. Principle of the fluorescence telomerase assay based on electron transfer from a QD to hemin on the G-quadruplexes of telomerase-elongated DNA. Reused with permission from ref 43. Copyright 2010 American Chemical Society.

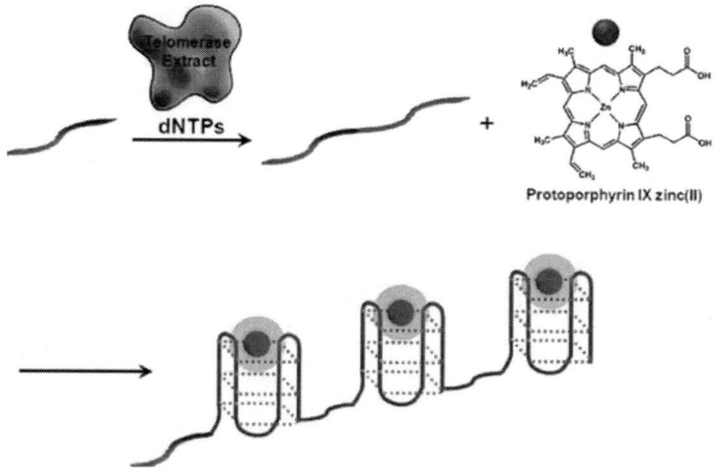

Figure 12. Principle of the telomerase assay using Zn (II) PPIX, which binds to the G-quadruplexes of telomerase-elongated DNA. Reused with permission from ref 44. Copyright 2012 American Chemical Society.

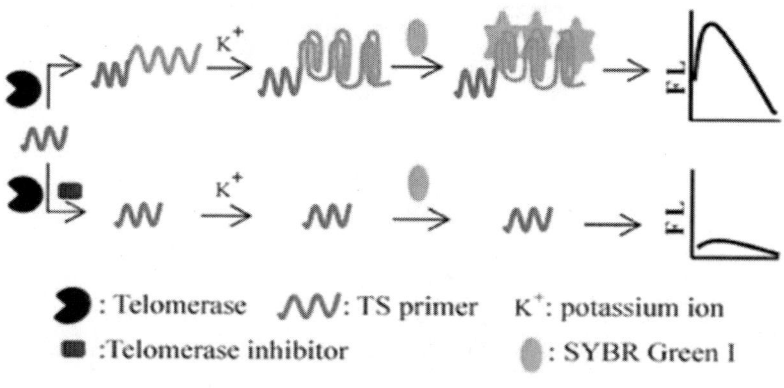

Figure 13. Principle of the telomerase assay using SYBR Green I, which binds to the G-quadruplexes of telomerase-elongated DNA. Reused with permission from ref 45. Copyright 2013 Royal Society of Chemistry.

3.2. Chemiluminescence-Based Assays

Chemiluminescence-based assays have an important advantage in low background noise compared to fluorescence-based assays. For the telomerase assay by electrochemiluminescence (ECL) method, Zhou et al. constructed an ECL nanoprobe, which was AuNP immobilized with two kinds of oligo DNAs: a capture DNA complementary to telomeric DNA and an ECL signal probe having a ruthenium (Ru) complex (Figure 14) [76]. Telomerase-elongated DNA was bound to the ECL nanoprobe and to a magnetic bead via hybridization and an avidin-biotin system, respectively. After the ECL nanoprobe-telomerase-elongated DNA-magnetic bead complex was captured on an electrode by a magnet, electrochemiluminescence of the Ru complex was measured. This assay allowed telomerase detection in HeLa cells with a detection limit of 500 cells (Table 3).

A label-free electrohemiluminescence assay for telomerase activity was also proposed. A cation porphyrin, meso-tetra(4-N,N,N-trimethylanilinium) porphyrin (TAPP) functionalized chemically converted graphene oxide (TAPP/CCG)-modified glassy carbon electrode was constructed (Figure 15) [77]. After telomerization, telomerase-elongated DNA treated by a dialysis tube to remove unreacted dNTPs was adsorbed to the TAPP/CCG via electrostatic and π-π stacking interactions because the TAPP/CCG has large π-planar and positive charges derived from graphene and TAPP, respectively. Furthermore, the positively charged $Ru(bpy)_3^{2+}$ was brought to the electrode surface via electrostatic attraction with the telomerase-elongated DNA, and consequently produced the ECL signal. Because the amount of $Ru(bpy)_3^{2+}$ on the electrode surface depended on that of telomerase-elongated DNA, telomerase activity was measured by analysis of the ECL signal. Although dialysis treatment was relatively cumbersome, this method could measure telomerase activity down to 10 HeLa cells mL^{-1} (Table 3).

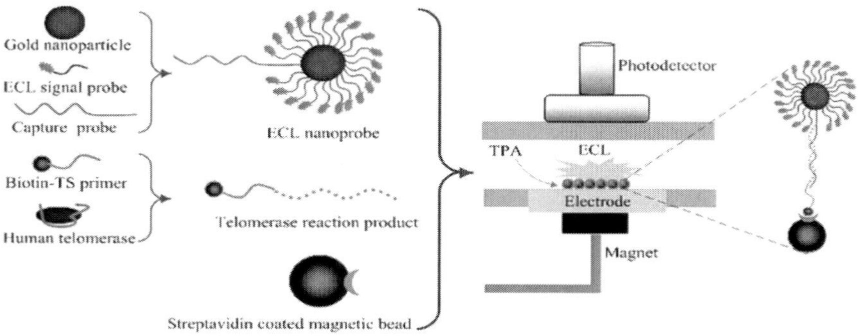

Figure 14. Principle of the telomerase assay using ECL nanoprobes, which are AuNPs modified with ECL signal probes and telomeric antisense probes. Reused with permission from ref 76. Copyright 2009 American Chemical Society.

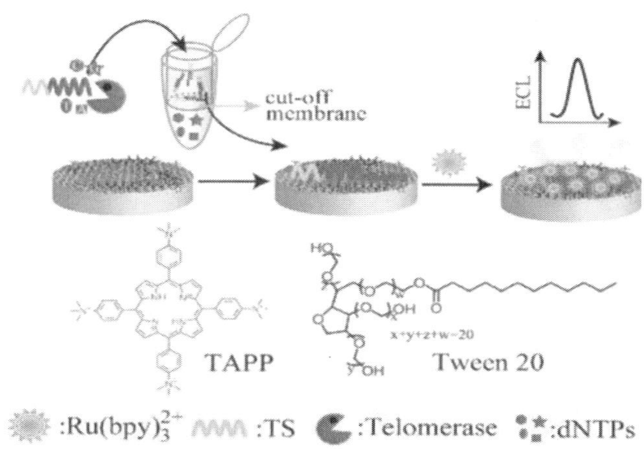

Figure 15. Principle of electrochemiluminescence telomerase assay using a TAPP/CCG-modified glassy carbon electrode. Reused with permission from ref 77. Copyright 2012 John Wiley and Sons, Inc.

3.3. Colorimetric Assay

Because detection of telomerase activity by naked eyes does not need any expensive equipment, it may allow cancer diagnosis even in small medical institutions. In a colorimetric biosensing field, AuNP has been widely employed as an indicator because absorption wavelength caused by the surface plasmon resonance (SPR) of AuNP varies depending on the size, shape and dispersion state of AuNP [78, 79]. A TS primer-modified AuNP was constructed for a colorimetric assay for telomerase activity (Figure 16) [80]. The addition of NaCl and $MgCl_2$ aggregated the functionalized AuNP, which was dispersed in a telomerase reaction solution without telomerase, resulting in a red color solution. The aggregation induced by the Na^+ and Mg^{2+} screening electrostatic repulsion of the citrate ions around the AuNP led to a rapid red-to-blue color change. On the other hand, when telomerization was performed on the AuNP, no color change was observed even in the presence of the cations. Because the telomerase-elongated DNA can fold into a G-quadruplex under the condition with Na^+ and Mg^{2+}, it is attributed that the telomerase-elongated DNA prevented the aggregation and the subsequent color change by forming the G-quadruplex. The assay enabled the detection of six cancer cell lines including HeLa, K562, MCF7, HepG2, 293T and 3T3, and the detection limit was 1 HeLa cell μL^{-1} (Table 3).

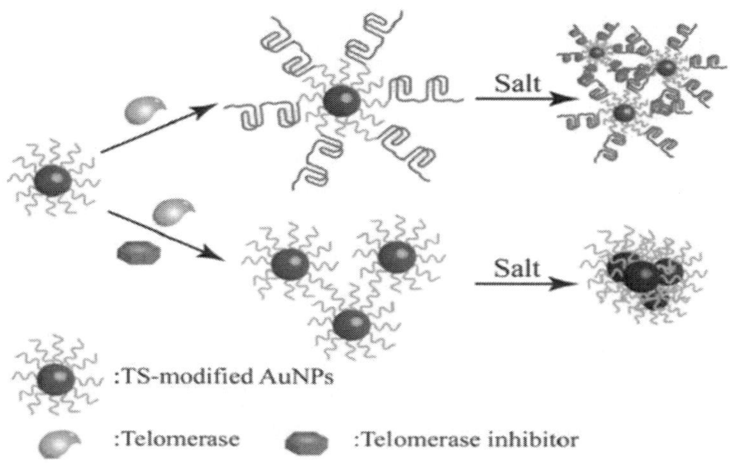

Figure 16. Principle of the colorimetric telomerase assay using AuNPs modified with TS primers. Reused with permission from ref 80. Copyright 2012 John Wiley and Sons, Inc.

In another colorimetric telomerase assay, an AuNP was coimmobilized with a TS primer and a telomeric antisense DNA probe (Figure 17) [81]. In the absence of telomerase, the telomerization solution containing the AuNP showed a blue color and displayed the red shift and broadened surface plasmon band in the UV-vis spectrum. The characteristic spectrum indicated the aggregation of the AuNP, which occurred mainly because ions in the cell lysis buffer screened the electrostatic repulsion of the AuNP. Likewise, for a small amount of telomerase from MCF-7, the red shift and broadened surface plasmon band was observed, but the degree was smaller than that without telomerase. This phenomenon indicated that the unorderly aggregation caused by ions of the lysis buffer and the orderly assembly via hybridization between the telomerase-elongated DNA and the DNA probe occurred simultaneously. On the other hand, highly concentrated telomerase caused precipitation of the AuNPs, and a remarkable increase in absorption and a weak red shift was observed. This absorption property was probably caused by a multi assembly network of the AuNP and the long distance between AuNPs in the network. Importantly, the authors quantitatively expressed the amount of telomerase as a value of A650/A520. They further succeeded in the diagnosis of 18 urine specimens from bladder cancer and inflammation patients and normal individuals by using the index of the absorbance value.

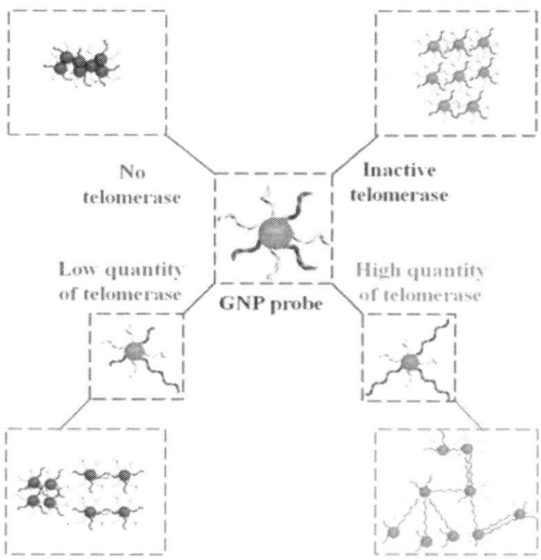

Figure 17. Principle of the colorimetric assay for urine telomerase using AuNPs comodified with TS primers and telomeric antisense DNA probes. Reused with permission from ref 81. Copyright 2014 American Chemical Society.

3.4. Electrochemical Assays

An electrochemical assay is attractive due to its simplicity, low instrumentation costs, and small equipment compared to an optical detection system. Sato et al. synthesized ferrocenylnaphthalene diimide, which is a G-quadruplex ligand having electrochemical properties, for a label-free electrochemical telomerase assay (Figure 18) [82]. After the telomerization reaction from TS primer immobilized on an Au electrode, the G-quadruplex ligand was added to the telomerase-elongated DNA in the presence of K^+, which caused the G-quadruplex ligand to bind the G-quadruplex formed by the elongated DNA. The electrochemical signal derived from the ligand bound to the G-quadruplex depended on the telomerase activity. A good correlation was observed between the current value derived from the ligand bound to the G-quadruplex and the HeLa cell concentration over the range of 40–140 cells μL^{-1} (Table 3). This was the first example of electrochemical detection of telomerase activity without relying on PCR or any enzymatic reactions.

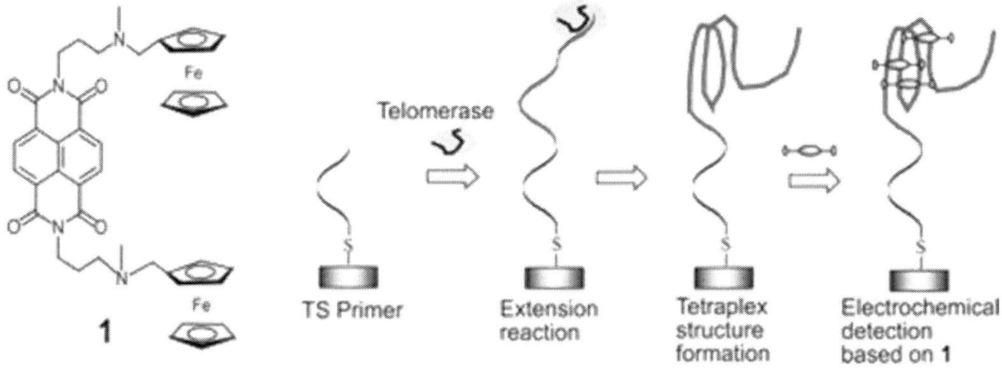

Figure 18. Principle of the electrochemical telomerase assay using ferrocenylnaphthalene diimide, which binds to the G-quadruplexes of telomerase-elongated DNA on an Au electrode. Reused with permission from ref 82. Copyright 2005 American Chemical Society.

Binding of $[Ru(NH_3)_6]^{3+}$ to DNA strands was also utilized for an electrochemical telomerase assay (Figure 19) [83]. In the assay, telomerase-elongated DNA from TS primer was hybridized with a TS antisense DNA on an Au electrode. The telomerase-elongated DNA attached to the electrode further bound to a signal DNA complementary to telomeric DNA on AuNP. $[Ru(NH_3)_6]^{3+}$ quantitatively bound to the DNA on AuNP via an electrostatic interaction, which caused electrochemical signals depending on the telomerase activity. The detection limit of the assay was 10 HeLa cells (Table 3).

Yi et al. constructed an Au electrode immobilized with a DNA probe, which contained a TS primer sequence at its 3' end and a recognition sequence complementary to a signal DNA with a ferrocene moiety (Figure 20) [84]. The DNA probe was also designed to form a hairpin structure with its 3′ end after the telomerase reaction, which resulted in the release of the signal DNA. Therefore, although an electrochemical signal was detected before telomerization, the signal intensity decreased depending on the degree of telomerization. This method showed a dynamic range from 10^2 to 6×10^4 HeLa cells mL^{-1} with a detection limit of 100 HeLa cells mL^{-1} (Table 3).

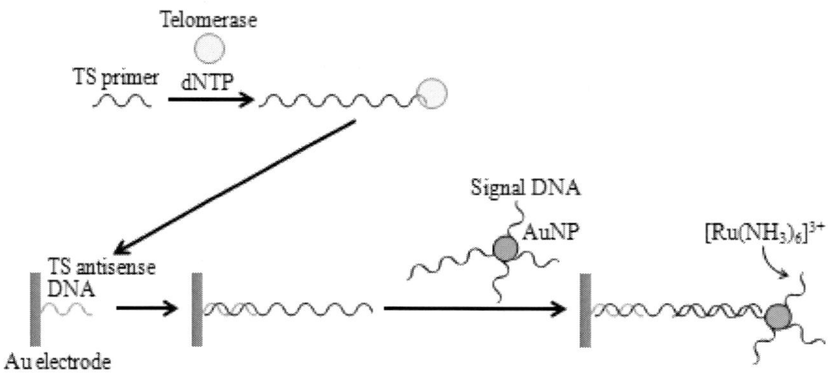

Figure 19. Principle of the electrochemical telomerase assay using $[Ru(NH_3)_6]^{3+}$, which binds to DNA on AuNPs attached to an Au electrode via hybridization with telomerase-elongated DNA.

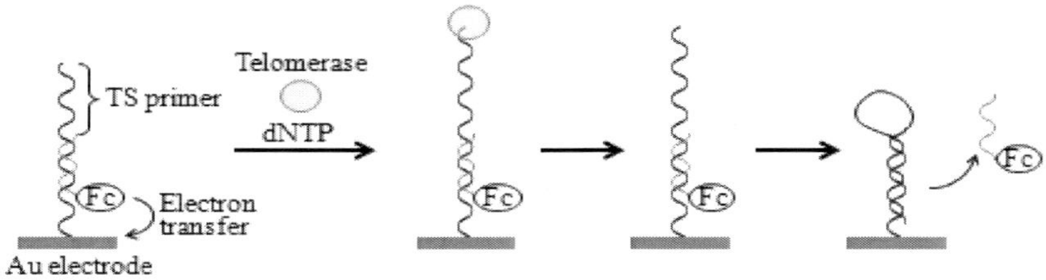

Figure 20. Principle of the electrochemical telomerase assay based on the release of ferrocene labeled signal DNA from DNA probe containing a TS primer.

3.5. SPR Technique

SPR technique has come into widespread use for measuring biological activity. In principle, SPR can detect the adsorption of molecules on a metal surface by measuring changes in refractive index at the surface, which enables the detection of unlabeled molecules' interactions in real-time [85]. Researchers have demonstrated that SPR is an effective tool for analyzing the DNA elongation reaction by DNA polymerase or reverse transcriptase [86-88]. Maesawa et al. applied cancer lysate with dNTP to a TS primer containing a telomere repeat DNA sequence immobilized on a SPR sensor chip to allow the telomerization reaction to occur on the chip (Figure 21) [89]. They analyzed the telomerase elongation rates of 18 cancer cell lines, three normal cell lines, 12 human primary carcinomas and matching normal tissues by measuring SPR signals. The elongation rates of cancer cell lines were two to 10 times higher than normal cell lines and normal tissues. They could not systematically determine the detection limit of cancer cells because of the weak telomerase activity of normal cells used in the study. However, 0.1 mg mL^{-1} extract of 100 T47-D breast cancers showed larger signal compared to extracts without cancer cells (Table 3).

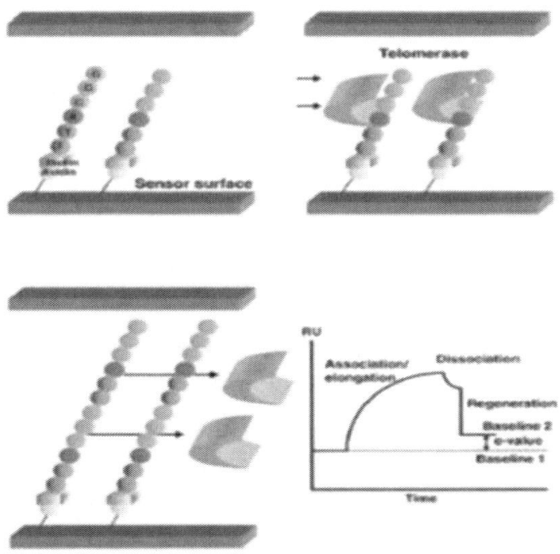

Figure 21. Principle of the telomerase assay using an SPR sensor anchoring primer DNA for telomerase. Reused with permission from ref 89. Copyright 2003 Oxford University Press.

In another SPR-based telomerase assay, after telomerization had occurred on the SPR sensor chip in similar manner to Maesawa et al., AuNP modified with a telomeric antisense DNA probe was applied for binding to the telomerization product via hybridization (Figure 22) [43]. The attachment of AuNP on the sensor chip amplified the SPR signal due to the coupling between the local plasmon of AuNP and the surface plasmon wave, which enabled the detection of 293T cells with a detection limit of 18 cells μL^{-1} despite the inclusion of 180 cells μL^{-1} without the AuNP (Table 3).

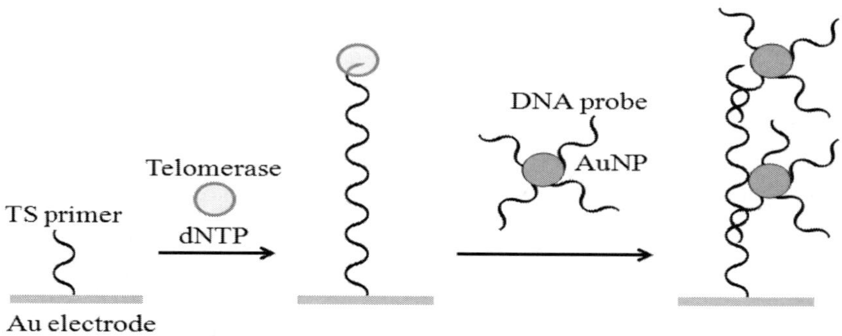

Figure 22. Principle of the telomerase assay using an SPR sensor anchoring TS primers, and AuNPs modified with telomeric antisense DNA probes.

3.6. Surface Enhanced Raman Scattering (SERS) Technique

SERS is a Raman spectroscopy technique to enhance the Raman scattering signal of a target molecule by placing the molecule on a SERS substrate covered with nanostructures, such as AuNP and silver nanoparticles [90-92]. Because the Raman signal strongly depends on the molecular structure and because the signal enhancement factor by SERS can reach 10^{10} to 10^{11}, SERS allows the detection of the target molecule with high selectivity and sensitivity [93]. Zong et al. developed a SERS-based telomerase assay using an AuNPs modified with a TS primer and a Raman-active molecule, 5,5′ dithiobis (2-nitrobenzoic acid) (DTNB) (Figure 23) [94]. The functionalized AuNPs were aggregated by adding the salts NaCl and $MgCl_2$. However, in the presence of telomerase, telomerase-elongated DNA on the AuNPs prevented the aggregation by folding the G-quadruplex with the salts. Finally, the reaction mixtures were pipetted onto a glass slide and subjected to the SERS measurements. Although the strong SERS signal of DTNB was observed without telomerase, the signal intensity was reduced depending on the cancer cell extract concentration. The authors demonstrated that three cancer cell lines, including HeLa, SKBR3, and MCF7 were detected by this assay, and the detection limit of SKBR3 was 1 cell mL^{-1} (Table 3). The same group also reported another SERS-based telomerase assay (Figure 24) [95]. In the assay, magnetic beads modified with a TS primer and AuNPs modified with DTNB and a telomeric antisense capture DNA were constructed. The telomerase-elongated DNA on the functionalized magnetic beads accumulated the functionalized AuNPs via hybridization with the capture DNA. The accumulation of the AuNPs caused a significant difference in the color and SERS intensity of the magnetically separated sediments. The detection limits of the colorimetric detection and SERS measurement were 10 cells mL^{-1} and 1 cell mL^{-1}, respectively (Table 3).

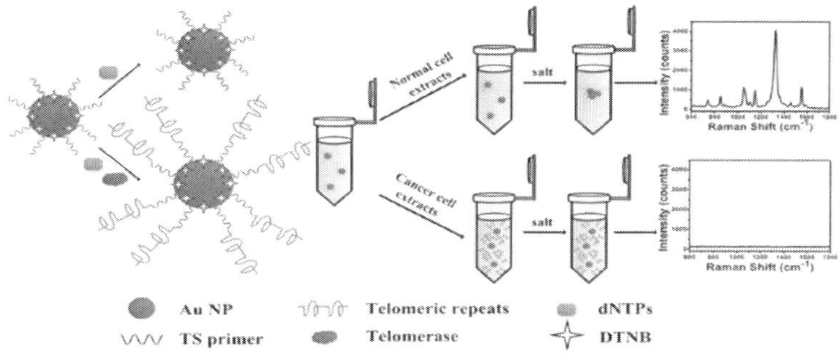

Figure 23. Principle of the SERS-based telomerase assay using AuNPs comodified with TS primers and DTNB. Reused with permission from ref 94. Copyright 2013 John Wiley and Sons, Inc.

4. Telomerase Activity Assays with Enzymatic Signal Amplification without PCR

4.1. Signal Amplification by Luciferase

DNA elongation reactions such as DNA polymerization, reverse transcription and telomerization are accompanied by the release of a single pyrophosphate (PPi) molecule following the incorporation of a single nucleotide. This means that measuring techniques of PPi enable the quantitative analysis of those elongation reactions. Pyrosequencing is a well-known sequencing technique, which includes a luciferin/luciferase reaction using adenosine triphosphate (ATP) converted from the released PPi by sulfurylase for identification of the nucleotide incorporated by DNA polymerase [96, 97]. Two groups reported biochemiluminescence-based telomerase assays in which PPi released during telomerization were detected based on the principle of pyrosequencing (Figure 25) [98, 99]. Both of them achieved highly sensitive detection of cancer cells with detection limits of five A549 cells [98] and 10 HeLa cells [99] without PCR (Table 3). However, those assays may be difficult to apply to clinical samples because a lot of ATP contained in both normal and cancer cells also serves as the substrate of luciferase, which will result in a high background signal. Even if the intracellular ATP is removed by some enzymes, those enzymes should be inactivated before telomerization because they probably remove dATP. It would be difficult to specifically inactivate those enzymes but not telomerase.

4.2. DNAzyme-Aided Signal Amplification

In the late 1990s, Sen's group discovered that some G-quadruplexes bind to hemin, which is an anionic porphyrin utilized as a cofactor of peroxidase in nature, despite its electrostatic repulsion [69]. They also demonstrated that the complexes show peroxidase like activity [69]. Since then, extensive studies have been conducted to elucidate the mechanism and to demonstrate that hemin acts as the cofactor for G-quadruplex-based peroxidase activity by end-stacking to G-quartets [100-104]. On the other hand, hemin cannot bind to ssDNA and dsDNA, which results in no peroxidase activity without G-quadruplexes [69-74]. It was also demonstrated that a parallel G-quadruplex in the presence of K^+ showed higher peroxidase like activity than an antiparallel G-quadruplex in the presence of Na^+. This was attributed to the antiparallel G-quadruplex protrusion of the loops over the top and bottom G-quartets inhibiting the end-stacking of hemin [100-104]. These findings opened a new avenue for biosensing techniques, including telomerase assays. Xiao et al. constructed a hairpin primer with an overhanging TS primer at its 3' end (Figure 26) [105]. Although the hairpin primer also contains two segments that can form a G-quadruplex structure together, in the absence of telomerase one of the two segments forms a stem duplex to prevent the G-quadruplex formation. The telomerase-elongated DNA from the TS primer binds to the other sequence of the stem, which opens the hairpin structure. Thus, the addition of K^+, hemin and hydrogen peroxide allows the G-quadruplex formed by the two segments to oxidize 2,2'-azinobis(3-ethylbenzothiazoline)-6-sulfonic acid (ABTS). The detection limit of the HeLa cells was ca. 500 cells (Table 3).

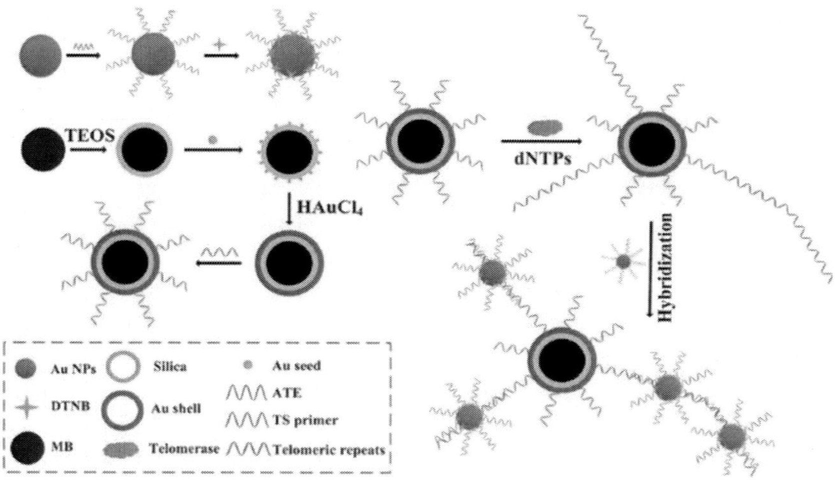

Figure 24. Principle of the telomerase assay based on SERS and colorimetric signals using MBs modified with TS primers and AuNPs comodified with telomeric antisense DNA and DTNB. Reused with permission from ref 95. Copyright 2014 Royal Society of Chemistry.

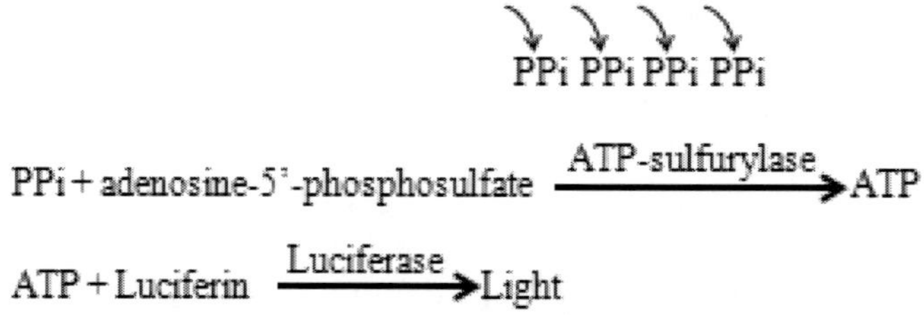

Figure 25. Principle of the telomerase assay using the pyrosequencing technique.

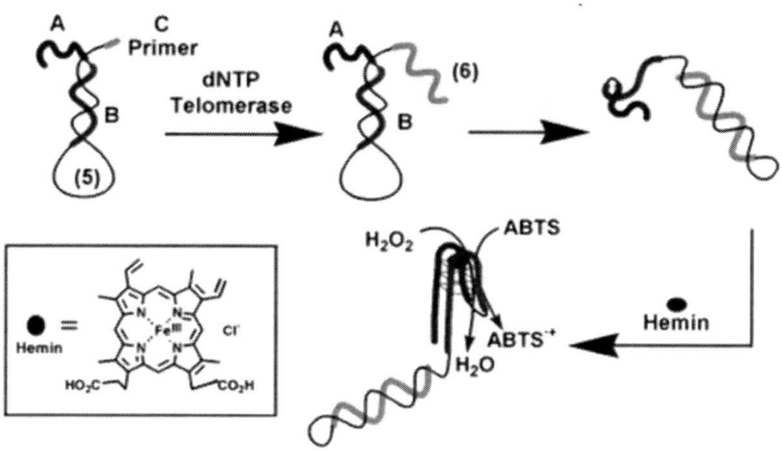

Figure 26. Principle of the colorimetric telomerase assay using a hairpin TS primer that generates a G-quadruplex DNAzyme after telomerization. Reused with permission from ref 105. Copyright 2004 American Chemical Society.

The peroxidase-like activity by G-quadruplexes formed by the telomerase-elongated DNA in the presence of dGTP but not dATP, dTTP and dCTP was also monitored for detection of telomerase activity (Figure 27) [106]. In the assay, the sequence of telomerase substrate primer was 5'-TGGGTAGGGCGGGTTA-3' because the primer itself had little peroxidase-like activity, but the model sequence of the telomerase-elongated DNA, 5'-TGGGTAGGGCGGGTTAGGG-3', in the presence of dGTP only showed a significant activity to oxidize ABTS. In the assay, when using heat-inactivated extract from 10^4 HeLa cells as the negative control, an obvious detection signal change above this negative control was observed for 200 HeLa cells, indicating that the detection limit of the HeLa cells was ca. 200 cells (Table 3).

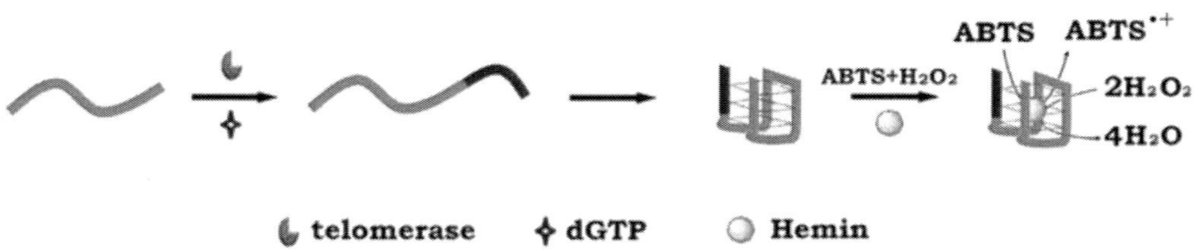

Figure 27. Principle of the colorimetric telomerase assay using a primer DNA that generates a G-quadruplex DNAzyme after telomerization using dGTP. Reused with permission from ref 106. Copyright 2015 Royal Society of Chemistry.

Besides the G-quadruplex-based DNAzyme, the 8–17 DNAzyme, which could cleave a complementary RNA strand, was applied for the telomerase assay. Tian et al. designed a hairpin primer that contained an overhanging TS primer at its 3' end and a 8–17 DNAzyme sequence in its loop region (Figure 28) [107]. When folded, the hairpin structure inhibited the 8–17 DNAzyme activity. However, in the presence of telomerase, the telomerase-elongated DNA unfolded the hairpin structure by binding to the strand including 5'-AATCCC-3' repeats of the stem. The unfolding of the primer subsequently activated the 8–17 DNAzyme to cleave a molecular beacon substrate modified with a fluorophore and a quencher for FRET. The enhancement of the fluorescence depending on cleavage of the molecular beacon led to the detection limit of 200 HeLa cells (Table 3). Furthermore, the system enabled the in situ detection of telomerase at the cellular level in HeLa cancer cells.

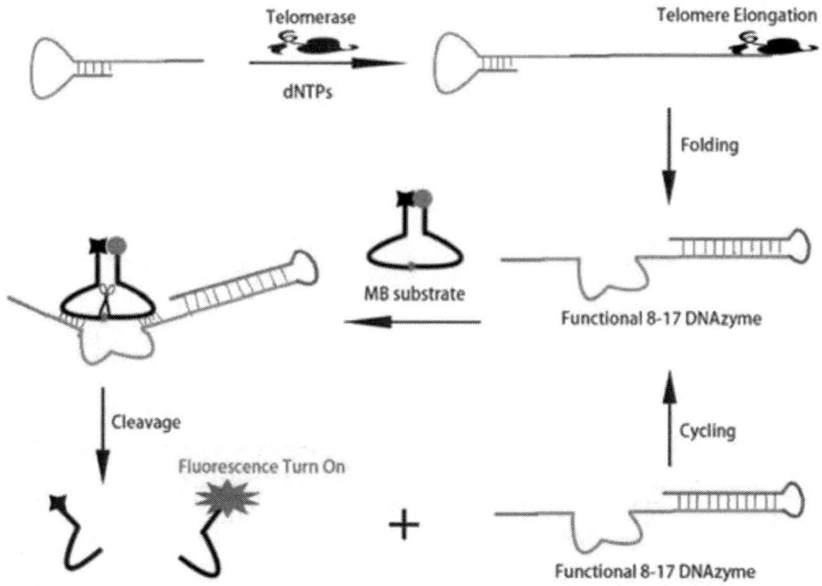

Figure 28. Principle of the fluorescence telomerase assay using a hairpin TS primer that generates a 8–17 DNAzyme after telomerization. Reused with permission from ref 107. Copyright 2013 Royal Society of Chemistry.

4.3. T7 Exonuclease-Aided Target Recycling

T7 exonuclease is a dsDNA-specific exonuclease, which catalyzes the removal of 5' mononucleotide in the 5' to 3' direction. T7 exonuclease can initiate the removal at blunt ends, sticky ends with 3' overhang, gaps and nicks, but not at sticky ends with a 5' overhang on dsDNA. This unique property of T7 exonuclease allowed researchers to develop sensitive DNA or RNA detection assays without PCR [108-112]. In the assays, the DNA probe and the target DNA was designed to form a dsDNA with a 5' overhang of the target DNA. T7 exonuclease digested only the DNA probe but not the target DNA strand in the dsDNA. Importantly, because these reactions including the hybridization and the digestion of the DNA probe occurred iteratively, the target DNA was recycled for the hybridization, leading to catalytic signal amplification. For the telomerase assay using T7 exonuclease, Wang et al. designed a TaqMan probe modified with a fluorophore (6-carboxy-fluorescein, FAM) and a quencher (tetramethyl-6-carboxyrhodamine, TAMRA), which was complementary to the telomere sequence (Figure 29) [113]. After the telomerization reaction, the TaqMan probe and the T7 exonuclease were added to the reaction mixture in order to hybridize the TaqMan probe with the telomerase-elongated DNA and to digest the hybridized TaqMan probe by T7 exonuclease. The digestion of the TaqMan probe separated FAM (donor) from TAMRA (acceptor), which resulted in enhancement of the fluorescence signal. Additionally, as described above, these repetitive processes including the hybridization and the digestion of the TaqMan probe ultimately led to signal amplification. The signal amplification resulted in the detection limit of 5 HeLa cells, although negligible signal was observed for even 1000 normal human fibroblast cells (Table 3).

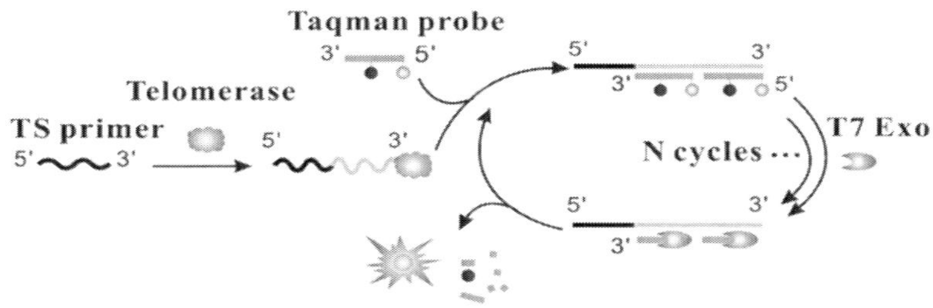

Figure 29. Principle of the fluorescence telomerase assay using a TaqMan probe for telomeric DNA and T7 exonuclease. Reused with permission from ref 113. Copyright 2013 Royal Society of Chemistry.

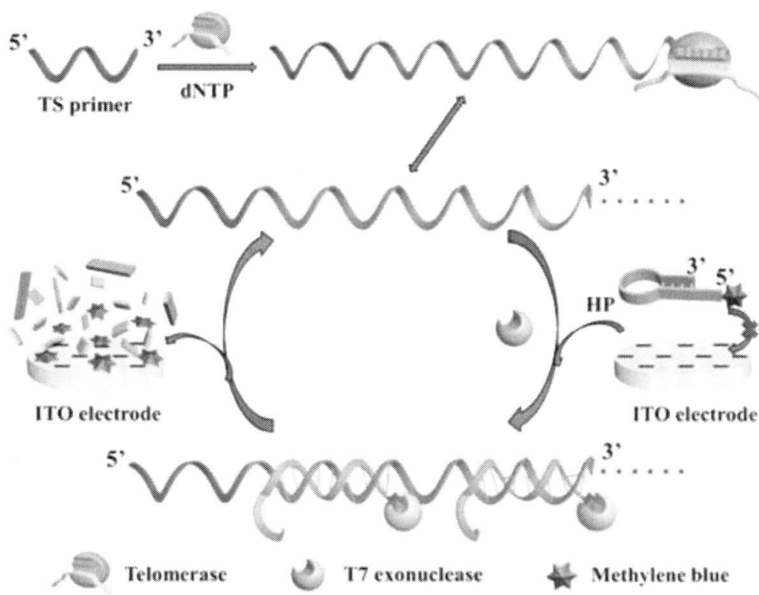

Figure 30. Principle of the electrochemical telomerase assay using a methylene blue-labeled hairpin DNA probe for telomeric DNA and T7 exonuclease. Reused with permission from ref 114. Copyright 2015 American Chemical Society.

T7 exonuclease-aided target recycling was also applied to the electrochemical detection of telomerase activity. Liu et al. designed a hairpin DNA probe modified with methylene blue as an electrochemical reporter at the 5' end (Figure 30) [114]. The hairpin probe contained a telomeric antisense sequence in its loop and 5' stem region. Furthermore, the stem of the hairpin DNA probe had a sticky end with a 5' overhang to prevent the digestion by T7 exonuclease. In the assay, after telomerization, the telomerase-elongated DNA, the hairpin DNA probe and the T7 exonuclease were added to a negatively charged indium tin oxide (ITO) electrode. Thus, in the absence of telomerase, the hairpin DNA probe kept the hairpin structure, which resulted in a weak electrochemical signal from methylene blue due to electrostatic repulsion between the DNA probe and the ITO electrode. On the other hand, in the presence of telomerase, the telomerase-elongated DNA formed dsDNA with the loop and 5' stem region of the hairpin DNA probe. Subsequently, T7 exonuclease released methylene blue-labeled mononucleotide by digestion. These repetitive processes accumulated methylene blue-labeled mononucleotide with much higher diffusivity toward the ITO electrode due to its less negative charge and smaller size, which resulted in an increase of the electrochemical signal. The assay enabled the sensitive detection of a single HeLa cell (Table 3). On the other hand, 100 normal cells also showed a weak electrochemical signal despite the lack of telomerase activity. Therefore, the elucidation of this mechanism must be essential for the clinical application of the assay.

4.4. Isothermal DNA Amplification Using DNA Polymerase and Nicking Endonuclease

Isothermal DNA amplification is also an effective method for the sensitive detection of telomerase-elongated DNA without PCR. The exponential amplification reaction (EXPAR) is a widely used isothermal DNA amplification technique using DNA polymerase and nicking endonuclease (NEase) (Figure 31) [115, 116]. In EXPAR, a 5' phosphate modified template DNA contains two identical sequences at its 5' and 3' ends that are respectively complementary to the target DNA, and the two sequences are separated by the nicking endonuclease recognition site. Because the reaction temperature is more than the T_m value of dsDNA between the target DNA and the complementary sequence in the template DNA, the target DNA cannot form a thermostable duplex with the complementary sequences. On the other hand, DNA polymerase can add nucleotides to the target DNA forming a transient dsDNA at the 5' end of the template, which results in the thermostable template duplex. Subsequently, the NEase recognizes the nicking site and nicks the elongated DNA of the duplex, which causes the release of the created target DNA. This primed template continues to create and release new target DNA via a linear amplification cycle because of the repetitive reactions of DNA polymerase and NEase. Furthermore, the created target DNA acts as a trigger on another template DNA, which allows an exponential amplification of the target DNA. Based on the principle of EXPAR, Zhang's group constructed a functional TS primer containing the template DNA at its 5' end. This exponentially created a G-quadruplex-based DNAzyme depending on the telomerase-elongated DNA primed DNA polymerization (Figure 32A) [117]. The authors successfully detected telomerase activity with a detection limit of a single HeLa cell via luminol chemiluminescence catalyzed by the exponentially-created G-quadruplex (Table 3). In a similar way, they also constructed another functional TS primer that exponentially created DNAzyme to cleave a molecular beacon. This strategy allowed the detection of a single HEK293T cell (Figure 32B and Table 3) [118].

Tian and Weizmann reported a novel exponential isothermal amplification of telomere repeat (EXPIATR) assay utilizing a strand-displacement (SD) polymerase and a NEase (Figure 33) [119]. In the assay, two primers were designed: (1) A nicking telomerase substrate (NTS) primer containing a TS primer at the 3' end and a nicking site. (2) A nicking fluorescent reporter probe (NFRP) primer in the hairpin structure that contained an ssDNA tail at its 3' end complementary to the telomere sequence and a nicking site in the loop. Telomerase added telomere sequences to the NTS primer, which allowed the NFRP primer to hybridize to the telomerase-elongated NTS primer. The hybridization of the NFRP primer initiated the DNA polymerization from the NFRP primer, forming a nick on the NTS template duplex by the NEase and a SD polymerization from the nick by SD polymerase. The SD polymerization opened the NFRP hairpin structure and completed the nicking site on the NFRP template duplex, which initiated further nicking and SD polymerization downstream. These reactions resulted in the repetitive release of the elongated ssDNA from the nicking sites on each primer, and the released each ssDNA acted as the template again. The exponential DNA amplification by EXPIATR was detected by monitoring the fluorescence of the FRET system on the NFRP (Figure 33A) or SYBR Green I (Figure 33B). The detection limit was a single HeLa cell in 25 mins (Table 3).

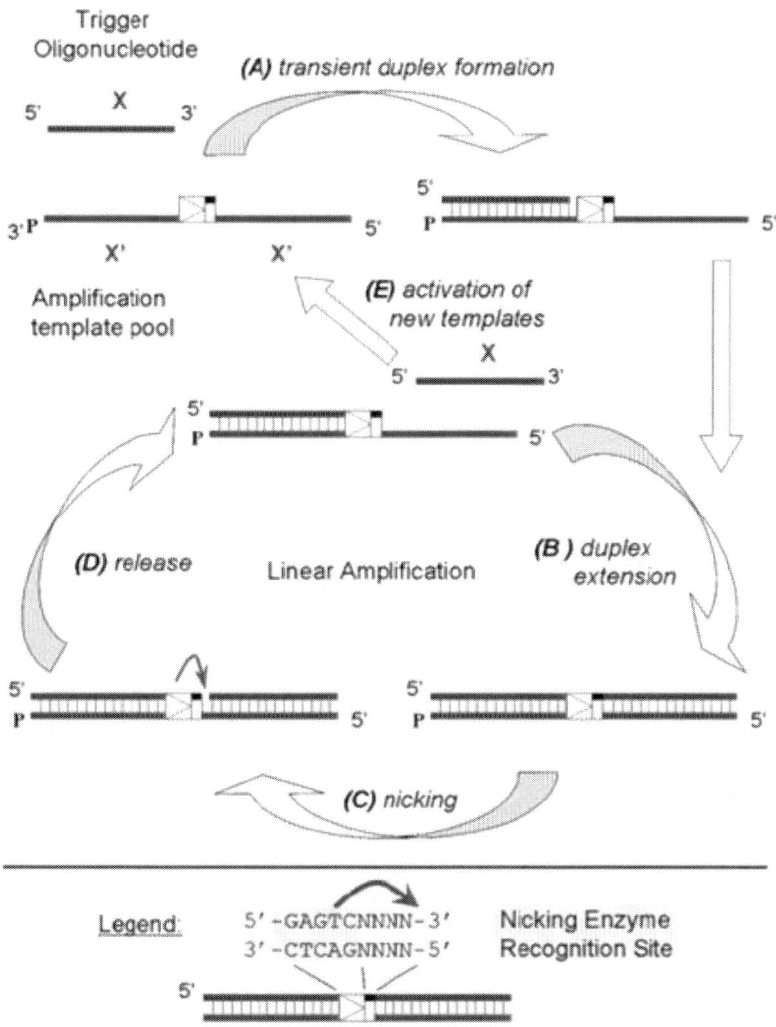

Figure 31. Principle of EXPAR. Reused with permission from ref 116. Copyright 2005 American Chemical Society.

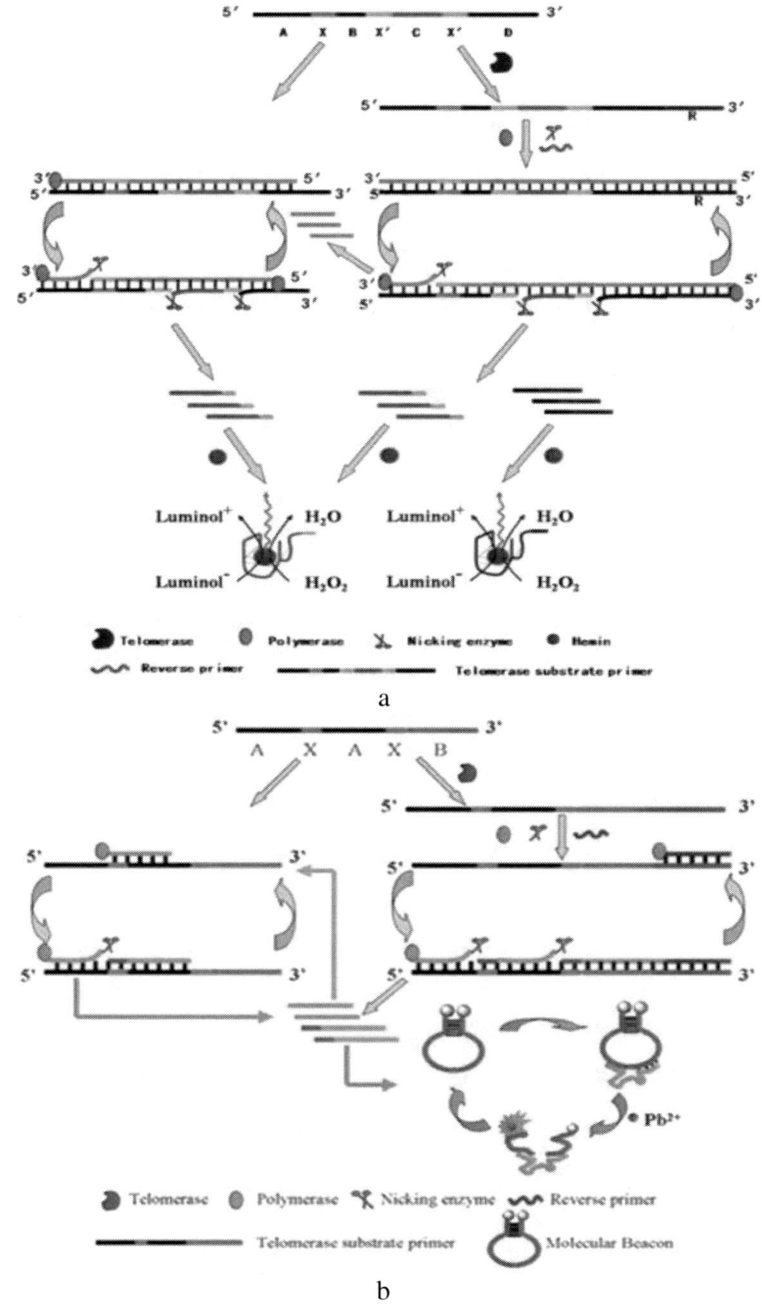

Figure 32. (A) Principle of the luminescence telomerase assay using EXPAR, which generates (A) G-quadruplex DNAzyme, or (B) DNAzyme for the digestion of the molecular beacon. Reused with permissions from refs 117 and 118. Copyright 2013 American Chemical Society for panel (A) and 2014 Royal Society of Chemistry for panel (B).

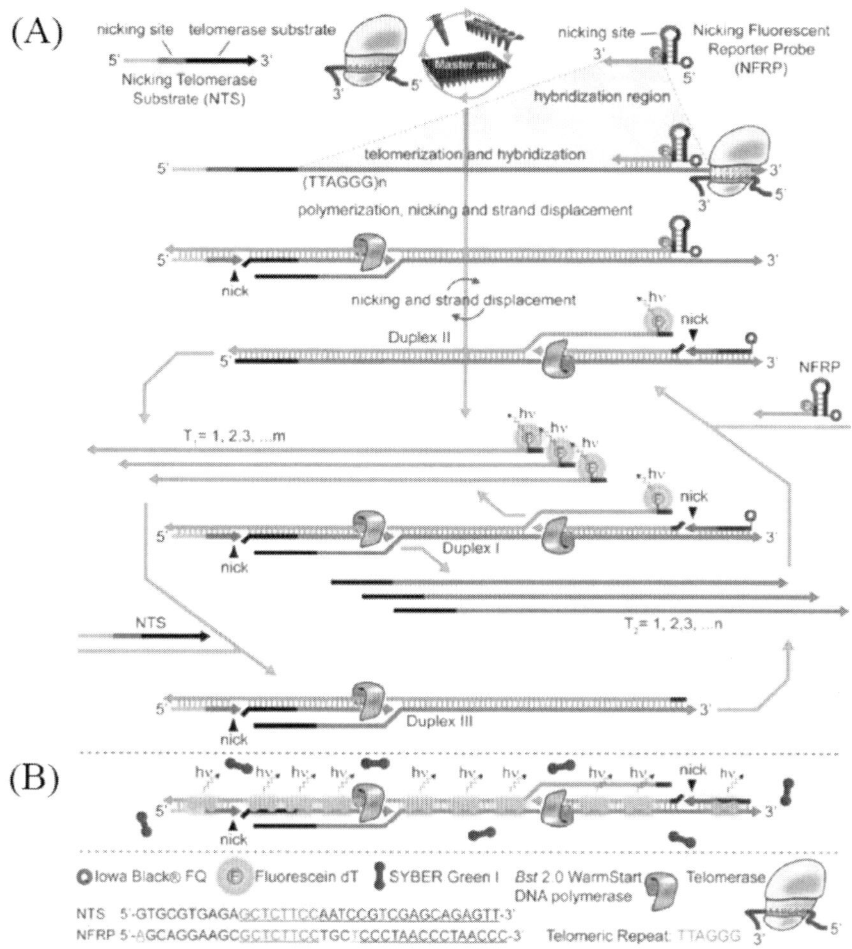

Figure 33. Principle of EXPIATR. Reused with permission from ref 119. Copyright 2013 American Chemical Society.

5. Telomerase Activity Assays with Processes for Avoidance of Inhibitors Prior to Enzymatic Signal Amplification

5.1. Advanced TRAP Assay

In a conventional TRAP assay, the reaction mixture containing all reagents for the telomerase reaction and PCR are mixed with the clinical sample, which sometimes causes a false-negative result by PCR inhibitors in the sample. Therefore, removal of PCR inhibitors prior to PCR can overcome this problem. Gollahon and Holt isolated the telomerase-elongated DNA from PCR inhibitors by the washing process (Figure 34) [120]. The washing process included hybridization between the elongated DNA and a biotinylated telomeric antisense DNA, immobilization of the biotinylated duplex on a streptavidin-coated magnetic bead and washing of the duplex-magnetic bead complex on a magnet. After the washing process, the telomerase-elongated DNA was released from the magnetic bead by heating and was applied to PCR. This advanced telomerase assay drastically increased the detection rate of telomerase activity in fine needle aspirates of breast tumor lesions and colonic effluents compared to the conventional one.

Commercially available spin columns for nucleic acids purification were also utilized to purify the telomerase-elongated DNA prior to PCR [121-123]. These methods have been conducted for accurate evaluation of the telomerase inhibition effects of a zinc-finger protein and small molecules including TMPyP4, BRACO-19, telomestatin, AA8, PIPER and BMVC4. According to both reports of Neidle's group [122] and Lin's group [123], the IC_{50} value (the concentration of inhibitor at 50% relative telomerase activity) of the small molecules in their advanced TRAP assay using the columns to remove the small molecules before PCR were larger than those in the conventional TRAP assay. These results indicate two important points: (i) the small molecules inhibit not only the telomerase reaction but also PCR and (ii) the conventional TRAP assay cannot accurately evaluate the telomerase inhibition effect. Telomerase inhibitors have been expected to be anticancer drugs with few side-effects because of the high specificity of telomerase for tumor cells over the normal somatic cells [68, 72, 124]. Thus, telomerase assays unsusceptible to PCR inhibitors are essential not only for accurate cancer diagnosis but also for screening of telomerase inhibitors as anticancer drugs.

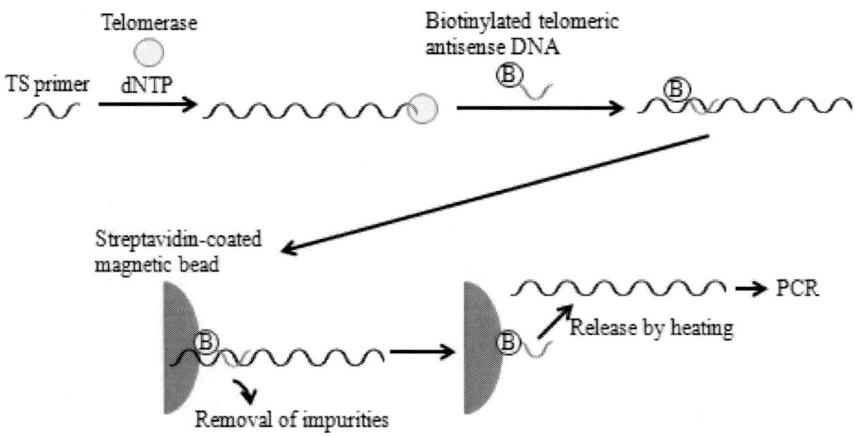

Figure 34. Principle of the advanced TRAP assay containing the washing process of telomerase-elongated DNA on a magnetic bead.

5.2. ELISA-Based Signal Amplification

The enzyme-linked immunosorbent assay (ELISA) is a popular solid-phase technique to detect target molecules such as proteins, peptides, hormones, and antibodies utilizing enzymatic signal amplification. There are several variations of ELISA, but commonly the antibody labeled with an enzyme such as alkaline phosphatase (ALP) and horse radish peroxidase (HRP) is immobilized on the plate via an antigen-antibody reaction and, finally, the enzymatic reaction converts the amounts of the target molecule into a detectable signal.

Because the enzymatic reaction allows for highly sensitive detection, similar principles were applied to the telomerase assay [125-127]. Kha et al. hybridized a digoxigenin (DIG)-labeled telomere antisense DNA probe with the telomerase-elongated DNA after the telomerization of a biotinylated TS primer (Figure 35) [125]. Then, the hybrid was immobilized on a streptavidin-coated plate via a biotin/streptavidin interaction and washed. Finally, after an anti-DIG-ALP conjugate was bound to the hybrid on the plate via an antigen/antibody interaction and washed, the ALP reaction with chemiluminescent substrate was performed. The ELISA-based assay showed the higher telomerase activity in 293T than HeLa, Raji, HT1080 or A431. The authors further successfully constructed an automated robotic system for their assay.

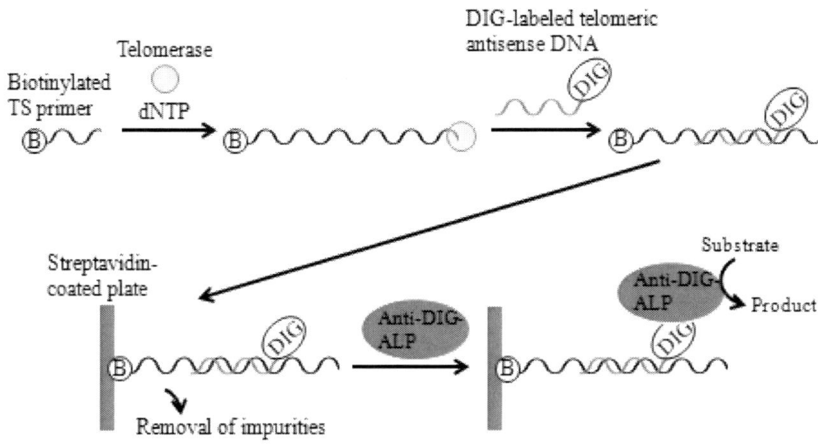

Figure 35. Principle of the ELISA-based telomerase assay, in which anti-DIG-ALP is attached to a plate via binding to DIG-labeled telomeric antisense DNA hybridized with telomerase-elongated DNA.

Pavlov et al. also developed a solid-phase telomerase assay utilizing the ALP reaction (Figure 36) [126]. In their assay, biotin groups were attached to a TS primer-modified Au plate by conducting the telomerization reaction on the plate in the presence of biotinylated dUTP (Figure 36A) or by hybridizing a biotinylated telomeric antisense DNA with the telomerase-elongated DNA after telomerization on the Au plate (Figure 36B). After washing the plate containing the biotin groups, an avidin-tagged ALP was bound to the plate via an avidin/biotin interaction. ALP bound to the plate catalyzed the oxidative hydrolysis of 5-bromo-4-chloro-3-indolyl phosphate to the insoluble indigo derivative, which was quantified based on the changes in electrode resistance using chronopotentiometry or the frequency changes of piezoelectric quartz crystals. The methods successfully detected telomerase activity with a detection limit of 1000 to 5000 HeLa cells (Table 3).

Li et al. constructed an Au electrode anchoring a spired DNA tetrahedron TS primer, which had a rigid dsDNA spire for regulation of the distance between the TS primer and the electrode surface (Figure 37) [127]. Telomerization of the tetrahedron TS primer in the presence of biotinylated dATP introduced biotin groups on the electrode, and the resulting electrode was washed. Then, avidin-tagged HRP was bound to the electrode via an avidin/biotin interaction. HRP attached to the electrode oxidized 3,3′,5,5′-tetramethylbenzidine (TMB) in the presence of hydrogen peroxide and the oxidized TMB was electrochemically detected. The signal-to-noise ratio was two times higher than the spired

DNA tetrahedron TS primer without the rigid dsDNA spire. The limit of detection was lower than 10 HeLa cells (Table 3).

ELISA-based telomerase assays using G-quadruplex DNAzyme instead of ALP or HRP have also been reported. Willner's group conducted the telomerization on a TS primer modified Au plate (Figure 38) [128, 129]. After washing the plate, the G-quadruplex DNAzyme was attached to the plate by hybridizing a telomere antisense DNA probe containing a G-quadruplex DNAzyme with the telomerase-elongated DNA (Figure 38A) [128] or by binding the AuNP anchoring the functionalized DNA probe to the telomerase-elongated DNA (Figure 38B) [129]. Finally, the G-quadruplex attached to the Au plate oxidized luminol in the presence of hemin and hydrogen peroxide, which resulted in a chemiluminescence signal. The detection limit of both assays was 1000 HeLa cells (Table 3).

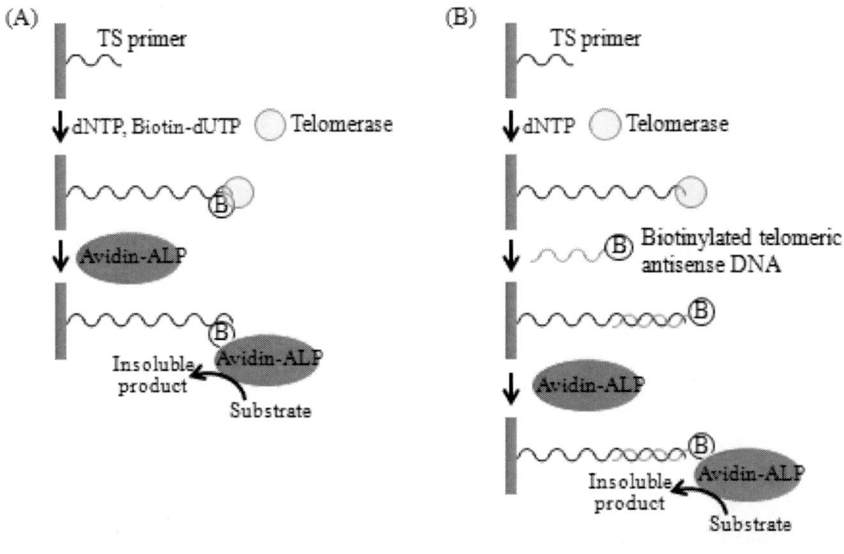

Figure 36. Principle of the ELISA-based telomerase assay, in which avidin-ALP is attached to a plate via avidin/biotin interaction after telomerization in the presence of biotinylated dUTP (A) or by hybridiziation between biotinylated telomeric antisense DNA and telomerase-elongated DNA (B).

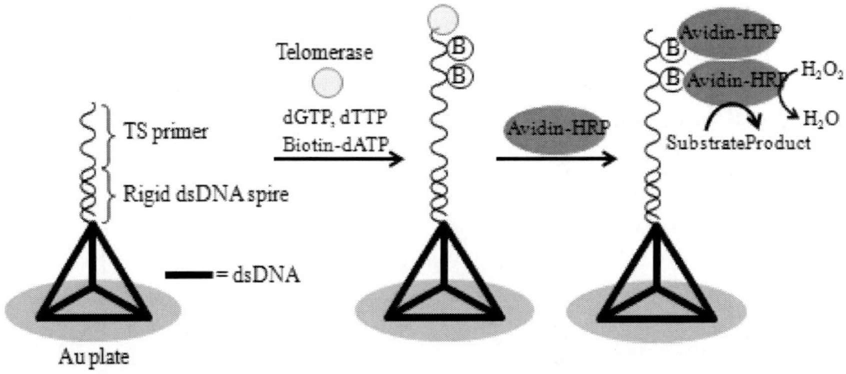

Figure 37. Principle of the ELISA-based telomerase assay using a spired DNA tetrahedron TS primer.

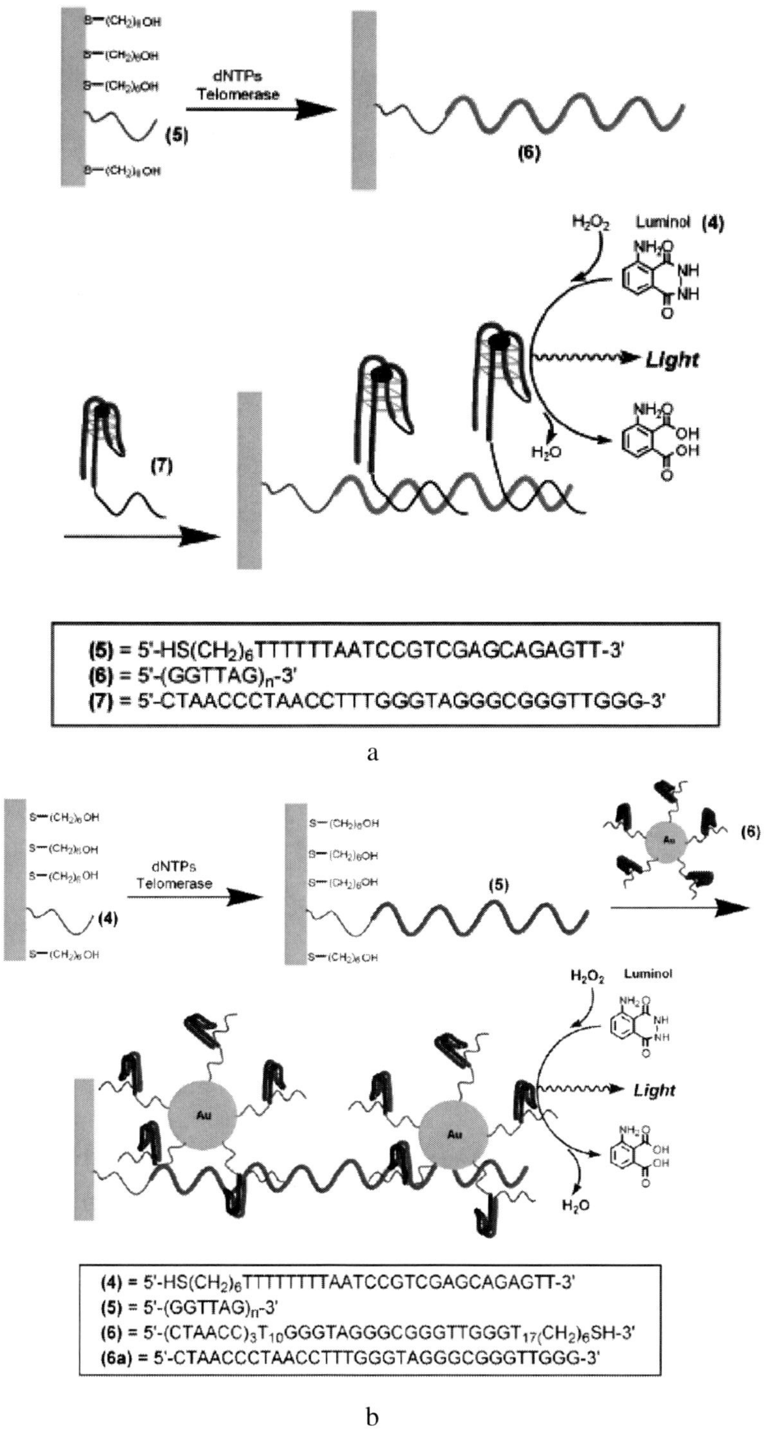

Figure 38. Principle of the ELISA-based telomerase assay using (A) a telomeric antisense DNA probe containing a G-quadruplex DNAzyme or (B) an AuNP anchoring telomeric antisense DNA probes containing G-quadruplex DNAzyme. Reused with permissions from refs 128 and 129. Copyright 2004 American Chemical Society. Copyright 2004 American Chemical Society.

5.3. Combination of the Washing Process and Nuclease-Aided Signal Amplification

As described above, nuclease digestions specific for DNA or RNA structures are very useful for the detection of DNA. Qu's group developed a solid-phase electrochemical telomerase assay utilizing exonuclease III (ExoIII), which specifically cleaves duplex DNA from blunt or recessed 3' termini (Figure 39A) [130]. In the assay, telomerization was performed on a glassy carbon electrode (GCE) immobilized with a TS primer. Subsequently, TS antisense DNA and Exo III were added to cleave the dsDNA of the unelongated TS primer and the TS antisense DNA by ExoIII. Note that Exo III could not digest dsDNA containing the elongated TS primer because it had a 3' overhang of ssDNA. After washing the electrode, negatively charged Pt nanoparticles (PtNPs) were added to adsorb on the electrode, which resulted in a hydrazine electrocatalytic oxidation signal. On the other hand, the uncleaved dsDNA prevented the PtNPs from adsorbing on the electrode due to electrostatic repulsion, thus reducing the signal. This assay was highly sensitive for telomerase with a detection limit of less than two A549 cells (Table 3). The same group also reported a similar assay performing telomerization on an Au electrode immobilized with TS primer (Figure 39B) [131]. After removing the unelongated TS primer based on the same principle described above, Ag^+ was added to bind to the uncleaved dsDNA with the elongated TS primer via electrostatic attraction. The bound Ag^+ allowed the DNA-template deposition of Ag nanoparticles as electroactive labels through a highly characteristic solid-state Ag/AgCl reaction. The detection limit of this assay was a single MCF-7 cell (Table 3).

Cycling probe technology (CPT) using ribonuclease H (RNase H), which recognizes an RNA/DNA duplex and hydrolyses the RNA in the hybrid, is also a highly sensitive detection technique for a target DNA [132]. In CPT, a probe RNA labeled with a fluorophore and a quencher is designed to hybridize with the target DNA, which allows specific hydrolysis of the hybridized probe RNA by RNase H. The hydrolysis of the probe RNA separates the fluorophore from the quencher, leading to enhancement of the fluorescent signal. These successive reactions in CPT occur repeatedly, which results in signal amplification. Based on this concept, CPT was applied to a highly sensitive assay for telomerase activity (Figure 40) [133]. The assay included the following steps: (i) The telomerase reaction was conducted from a 5' biotinylated TS primer; (ii) The biotinylated telomerase reaction products were bound to streptavidin-coated magnetic beads via a streptavidin-biotin interaction; (iii) The telomerase reaction products on magnetic beads were washed to remove sample contaminants including PCR inhibitors; (iv) The telomerase-elongated DNA sequences of the telomerase products were preferentially amplified by asymmetric PCR; (v) Amplified telomerase-elongated DNA sequences were then detected via CPT. The assay successfully detected HeLa cells with a detection limit of 50 cells and showed a specific detection of HeLa cells among cell populations comprising predominantly normal human dermal fibroblast (NHDF) cells (Table 3). It was also demonstrated that the assay eliminated false-negative results caused by PCR inhibitors such as bile salt, hemoglobin and heparin, which are often included in clinical samples. Another important finding of the report was that HeLa cell lysate with Mg^{2+} degraded a model oligo DNA of telomerase-elongated DNA, indicating that intracellular DNase caused a false-negative result by degrading the telomerase-elongated DNA. Moreover, the problem was solved by addition of λ DNA into the telomerase reaction mixture in order that λ DNA efficiently inhibited the DNase from degrading the telomerase-elongated DNA by serving as a competitive substrate for DNase.

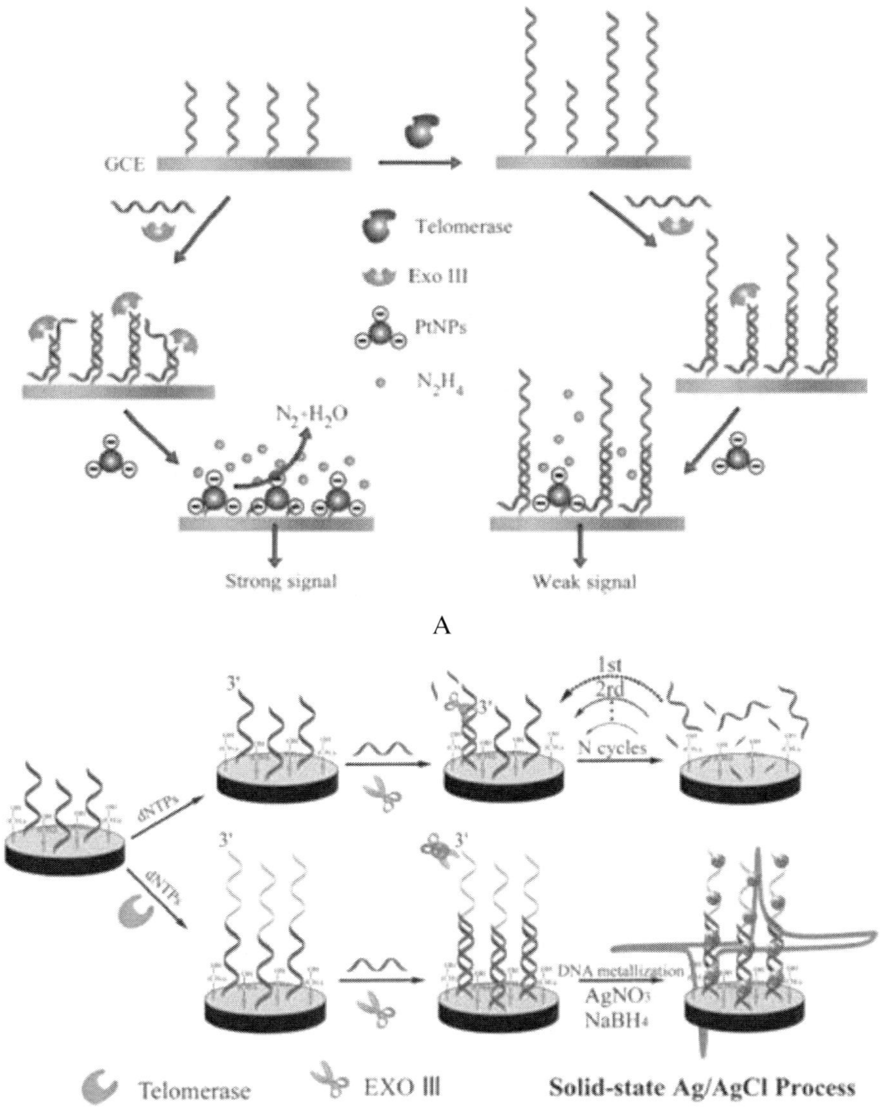

A

B

Figure 39. Principle of the telomerase assay using (A) PtNPs or (B) AgNPs as an electrochemical signal, in which the unelongated TS primer is hybridized with TS antisense DNA and is digested by exonuclease III. Reused with permissions from refs 130 and 131, respectively. Copyright 2013 Royal Society of Chemistry and copyright 2014 John Wiley and Sons, Inc.

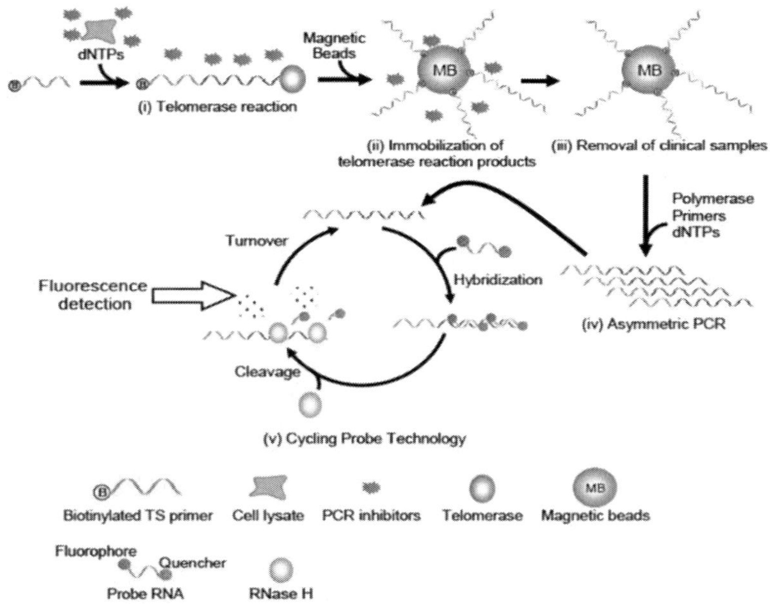

Figure 40. Principle of the telomerase assay based on A-PCR amplification of telomerase-elongated DNA on magnetic beads and subsequent application of CPT. Reused with permission from ref. 133.

6. Toward a Clinical Application

6.1. Application Potentiality of Telomerase Assays

With the primary purpose to eliminate false-negative results caused by PCR inhibitors in the conventional TRAP assay, researchers have developed many novel telomerase assays, which are classified into three main groups as described above. Some of them showed extremely high sensitivity of less than 10 cultured cancer cells, which means that these assays are equal or superior to TRAP assay in terms of sensitivity. However, the application potentiality of these assays to clinical samples is still questioned, because their capabilities were verified only for cell lines with a few exceptions. In particular, the telomerase assays described in Chapter 4 should face challenges for clinical applications, because in the assays the enzymatic signal amplifications in place of PCR will be conducted in the presence of impurities from clinical samples including a lot of normal cells, blood and other body fluids. The impurities probably inhibit the signal amplification reactions. In fact, an EXPIATR assay based on the isothermal amplification of telomerase-elongated DNA allowed the detection of a single HeLa cell in the absence any impurities (Figure 33) [119], but 1000 and 4000 normal cell lysates reduced the amplification efficiency, resulting in the signals of telomerase activity to be 20% and less than 5%, respectively [134]. Given that telomerase assays are often performed with clinical samples containing 1-10 μg of protein [135], which corresponds to 1400-14000 cells because of 7×10^{-10} g of total protein in a single cell, it is difficult to apply EXPIATR assay to clinical samples. Interestingly, however, the authors found that the addition of AuNPs remedied the problems: almost 100% and more than 20% of the signals of telomerase activity with 1000 and 4000 normal cell lysates, respectively, were observed compared to the signals without normal cell lysates. The effect of AuNPs is attributable to the fact local assembly of necessary components for the isothermal amplification reaction on AuNPs enhances the reaction efficiency within a lot of normal cell lysates. This solution is very simple and likely to be effective for other enzymatic signal amplification strategies, but is still not enough to completely eliminate the inhibition by normal cell lysates. Thus, further improvement is required. On the other hand, the telomerase assays described in Chapter 3 may be close to practical use, because they do not need any enzymatic signal amplifications. Takenaka's group evaluated the telomerase activity in the lysate of cancerous tissues of oral cancer patients using their electrochemical telomerase assay (Figure 18) [82, 136]. The positive rate was 85%, which was higher than the 50% rate evaluated by TRAP assay. Furthermore, the positive rate of exfoliated cells of patients was 95%, indicating that this minimally invasive test based on their assay could be promising for oral cancer diagnosis. The telomerase assays described in Chapter 6 are also suitable for practical use because impurities are removed before enzymatic signal amplifications. For example, in the assay using asymmetric PCR and CPT for signal amplification (Figure 40), removal of impurities from the telomerase-elongated DNA immobilized on magnetic beads could completely avoid the false-negative results caused by normal cell lysates and PCR inhibitors including bile salt, hemoglobin and heparin [133]. These results indicate that this assay could also be promising for clinical application. However, such removal strategies using magnetic beads, purification columns, or metal substrates tend to make the operation complicated, which should reduce the reproducibility of the results in the bed-side diagnosis. One key technology to solve the problem is a micro total analysis system (μ-TAS), which describes a small device to automatically conduct all necessary procedures for biochemical analysis [137-139]. Researchers have developed various μ-TAS chips to isolate biomolecules such as DNA, proteins, and cells using magnetic beads or filters [140-151]. Those μ-TAS technologies may play a critical role toward achieving a reliable and simple telomerase assay.

6.2. Needs in Clinical Practice

Pathological diagnosis, which involves the examination of tissues and cells in microscopic detail, remains the gold standard for cancer diagnosis. In general, cytological diagnosis examining cells from fluids, scrapings or aspirations is employed in cancer screening. Cytological diagnosis has advantages including easy sampling and little or no risk to patients when compared to histological diagnosis of tissues excised by biopsy or surgical resection. However, cytology reports are less conclusive than histology reports, because cytological examination cannot provide the architectural detail of whole tissues, and the collected cells are not always enough to examine. For example, the Papanicolaou (Pap) smear test, a cytological method for cervical cancer, shows low sensitivity for cancer and precancerous cells [152]. In addition, the test may give an ambiguous result referred to as atypical squamous cells of undetermined significance (ASC-US) meaning that some cells have a mildly unusual appearance—but are not so unusual as to appear precancerous [153]. DNA testing of human papillomavirus (HPV), of which infection is the first step of cervical cancer development, remedies the problems [152, 154-159]. However, the HPV test increases the probability of a false-positive result, because in most cases HPV infection is a transient event and does not result in neoplastic transformation [155, 157]. Thus, another biomarker with high sensitivity and specificity is required. Previous studies showed that E6 and E7 proteins of persistently infected high-risk HPV activate hTERT expression by interacting with transcriptional repressors and activators [154, 159-161]. Therefore, an accurate and sensitive telomerase assay would give a great contribution to the screening. Unlike cytological diagnosis, histological diagnosis dealing with whole tissues is regarded as a reliable enough method to render a definitive diagnosis. However, the diagnosis result of cervical intraepithelial neoplasia, grade II (CIN II) indicating moderate dysplasia for precancerous cervical cells does not offer clear information on whether the precancerous cells have become malignant or not [162-164]. Randomized controlled trials demonstrated that the HPV test caused a risk of overdiagnosis and overtreatment if no triage was implemented [165, 166]. Thus, also in this case, a telomerase assay would be useful. Similar problems of pathological diagnosis, which can be explained by taking cervical cancer as an example, occur in the pathological diagnosis of other cancers. Rather, there should be a greater need for a telomerase assay in cancer diagnosis other than cervical cancer diagnosis, because they do not use a HPV test. Thus, telomerase assay has a great potential to become a versatile cancer diagnosis technique in future, and it gives complementary information that can make the conventional histological diagnosis more accurate and quantitative.

CONCLUSION

The field of telomeres and telomerase is an extremely exciting research field, because they play key roles in cellular senescence and carcinogenesis. TRAP assay is a powerful tool for detecting telomerase activity and has made a significant contribution to elucidate the relationship between the activity and the various cells including tumor cells, stem cells, germ cells and normal somatic cells. However, TRAP assay cannot be applied for cancer diagnosis due to its susceptibility to inhibitors, especially PCR inhibitors, included in clinical samples. New technologies that we focused on here have been developed in order to solve the problems, and some of them are far superior in sensitivity and specificity for cancer cell lines to TRAP assay. Although there still seems to be a challenge for clinical application of the technologies, interdisciplinary approaches, such as cooperation with nanotechnology and μ-TAS technology, should overcome the challenge. Furthermore, telomerase assay may be useful for applications other than cancer diagnosis. For example, tumor development of cells derived from induced pluripotent stem cells (iPSCs) is a serious problem in regenerative medicine [167]. Although researchers have made extensive efforts to solve the problem, it would be difficult to completely prevent tumor development. Thus, when iPSCs are put into practical use, telomerase assay as a sampling inspection technology to detect tumor cells coexisting with cells differentiated from iPSCs may be required for quality assurance.

ACKNOWLEDGMENTS

Financial support by a Grant-in-Aid for Scientific Research on Innovative Areas "Nanomedicine Molecular Science" (No. 2306) from the Ministry of Education, Culture, Sports, Science and Technology (MEXT), Japan, by KAKENHI (no. 15H03840, 16K14042) from Japan Society for the Promotion of Science (JSPS) and by the Asahi Glass Foundation is gratefully acknowledged.

REFERENCES

[1] Szostak, J.W.; Blackburn, E.H. (1982). Cloning yeast telomeres on linear plasmid vectors. *Cell*, *29*, 245-255.

[2] Moyzis, R.K.; Buckingham, J.M.; Cram, L.S.; Dani, M.; Deaven, L.L.; Jones, M.D.; Meyne, J.; Ratliff, R.L.; Wu, J.R. (1988). A highly conserved repetitive DNA sequence, $(TTAGGG)_n$, present at the telomeres of human-chromosomes. *Proc. Natl. Acad. Sci. USA*, *85*, 6622-6626.

[3] Palm, W.; de Lange, T. (2008). How shelterin protects mammalian telomeres. *Annu. Rev. Genet.*, *42*, 301-334.

[4] Nandakumar, J.; Cech, T.R. (2013). Finding the end: Recruitment of telomerase to telomeres. *Nat. Rev. Mol. Cell Bio.*, *14*, 69-82.

[5] Henderson, E.R.; Blackburn, E.H. (1989). An overhanging 3' terminus is a conserved feature of telomeres. *Mol. Cell. Biol.*, *9*, 345-348.

[6] Wellinger, R.J.; Wolf, A.J.; Zakian, V.A. (1993). Saccharomyces telomeres acquire single-strand TG_{1-3} tails late in S phase. *Cell*, *72*, 51-60.

[7] Makarov, V.L.; Hirose, Y.; Langmore, J.P. (1997). Long G tails at both ends of human chromosomes suggest a C strand degradation mechanism for telomere shortening. *Cell*, *88*, 657-666.

[8] McElligott, R.; Wellinger, R.J. (1997). The terminal DNA structure of mammalian chromosomes. *EMBO J.*, *16*, 3705-3714.

[9] Wright, W.E.; Tesmer, V.M.; Huffman, K.E.; Levene, S.D.; Shay, J.W. (1997). Normal human chromosomes have long G-rich telomeric overhangs at one end. *Gene. Dev.*, *11*, 2801-2809.

[10] de Lange, T. (2004). T-loops and the origin of telomeres. *Nat. Rev. Mol. Cell Bio.*, *5*, 323-329.

[11] de Lange, T. (2005). Shelterin: The protein complex that shapes and safeguards human telomeres. *Gene. Dev.*, *19*, 2100-2110.

[12] Griffith, J.D.; Comeau, L.; Rosenfield, S.; Stansel, R.M.; Bianchi, A.; Moss, H.; de Lange, T. (1999). Mammalian telomeres end in a large duplex loop. *Cell*, *97*, 503-514.

[13] Rossi, M.L.; Purohit, V.; Brandt, P.D.; Bambara, R.A. (2006). Lagging strand replication proteins in genome stability and DNA repair. *Chem. Rev.*, *106*, 453-473.

[14] Gilson, E.; Géli, V. (2007). How telomeres are replicated. *Nat. Rev. Mol. Cell Bio.*, *8*, 825-838.

[15] Harley, C.B.; Futcher, A.B.; Greider, C.W. (1990). Telomeres shorten during ageing of human fibroblasts. *Nature*, *345*, 458-460.

[16] Counter, C.M.; Avilion, A.A.; LeFeuvre, C.E.; Stewart, N.G.; Greider, C.W.; Harley, C.B.; Bacchetti, S. (1992). Telomere shortening associated with chromosome instability is arrested in immortal cells which express telomerase activity. *EMBO J.*, *11*, 1921-1929.

[17] Shay, J.W.; Wright, W.E. (2000). Hayflick, his limit, and cellular ageing. *Nat. Rev. Mol.*

Cell Bio., *1*, 72-76.
[18] Verdun, R.E.; Karlseder, J. (2007). Replication and protection of telomeres. *Nature*, *447*, 924-931.
[19] Wright, W.E.; Piatyszek, M.A.; Rainey, W.E.; Byrd, W.; Shay, J.W. (1996). Telomerase activity in human germline and embryonic tissues and cells. *Dev. Genet.*, *18*, 173-179.
[20] Greider, C.W.; Blackburn, E.H. (1985). Identification of a specific telomere terminal transferase-activity in *Tetrahymena* extracts. *Cell*, *43*, 405-413.
[21] Greider, C.W.; Blackburn, E.H. (1989). A telomeric sequence in the RNA of *Tetrahymena* telomerase required for telomere repeat synthesis. *Nature*, *337*, 331-337.
[22] Morin, G.B. (1989). The human telomere terminal transferase enzyme is a ribonucleoprotein that synthesizes TTAGGG repeats. *Cell*, *59*, 521-529.
[23] Kim, N.W.; Piatyszek, M.A.; Prowse, K.R.; Harley, C.B.; West, M.D.; Ho, P.L.C.; Coviello, G.M.; Wright, W.E.; Weinrich, S.L.; Shay, J.W. (1994). Specific association of human telomerase activity with immortal cells and cancer. *Science*, *266*, 2011-2015.
[24] Counter, C.M.; Meyerson, M.; Eaton, E.N.; Weinberg, R.A. (1997). The catalytic subunit of yeast telomerase. *Proc. Natl. Acad. Sci. USA*, *94*, 9202-9207.
[25] Lingner, J.; Hughes, T.R.; Shevchenko, A.; Mann, M.; Lundblad, V.; Cech, T.R. (1997). Reverse transcriptase motifs in the catalytic subunit of telomerase. *Science*, *276*, 561-567.
[26] Chen, J.L.; Greider, C.W. (2004). An emerging consensus for telomerase RNA structure. *Proc. Natl. Acad. Sci. USA*, *101*, 14683-14684.
[27] Ly, H.; Blackburn, E.H.; Parslow, T.G. (2003). Comprehensive structure-function analysis of the core domain of human telomerase RNA. *Mol. Cell. Biol.*, *23*, 6849-6856.
[28] White, L.K.; Wright, W.E.; Shay, J.W. (2001). Telomerase inhibitors. *Trends Biotechnol.*, *19*, 114-120.
[29] Masutomi, K.; Yu, E.Y.; Khurts, S.; Ben-Porath, I.; Currier, J.L.; Metz, G.B.; Brooks, M.W.; Kaneko, S.; Murakami, S.; DeCaprio, J.A., et al. (2003). Telomerase maintains telomere structure in normal human cells. *Cell*, *114*, 241-253.
[30] Vidaurreta, M.; Maestro, M.; Rafael, S.; Veganzones, S.; Sanz-Casla, M.; Cerdan, J.; Arroyo, M. (2007). Telomerase activity in colorectal cancer, prognostic factor and implications in the microsatellite instability pathway. *World J. Gastroentero.*, *13*, 3868-3872.
[31] Uen, Y.H.; Lin, S.R.; Wu, D.C.; Su, Y.C.; Wu, J.Y.; Cheng, T.L.; Chi, C.W.; Wang, J.Y. (2007). Prognostic significance of multiple molecular markers for patients with stage II colorectal cancer undergoing curative resection. *Ann. Surg.*, *246*, 1040-1046.
[32] Saleh, S.; Lam, A.K.Y.; Ho, Y.H. (2008). Real-time PCR quantification of human telomerase reverse transcriptase (hTERT) in colorectal cancer. *Pathology*, *40*, 25-30.
[33] Lam, A.K.Y.; Ong, K.; Ho, Y.H. (2008). Aurora kinase expression in colorectal adenocarcinoma: correlations with clinicopathological features, p16 expression, and telomerase activity. *Hum. Pathol.*, *39*, 599-604.
[34] Lam, A.K.Y.; Saleh, S.; Smith, R.A.; Ho, Y.H. (2008). Quantitative analysis of survivin in colorectal adenocarcinoma: increased expression and correlation with telomerase activity. *Hum. Pathol.*, *39*, 1229-1233.
[35] Soreide, K.; Gudlaugsson, E.; Skaland, I.; Janssen, E.A.M.; Van Diermen, B.; Korner, H.; Baak, J.P.A. (2008). Metachronous cancer development in patients with sporadic colorectal adenomas - multivariate risk model with independent and combined value of

hTERT and survivin. *Int. J. Colorectal Dis.*, *23*, 389-400.

[36] Lam, A.K.Y.; Ong, K.; Ho, Y.H. (2008). hTERT expression in colorectal adenocarcinoma: correlations with p21, p53 expressions and clinicopathological features. *Int. J. Colorectal Dis.*, *23*, 587-594.

[37] Krupp, G.; Kuhne, K.; Tamm, S.; Klapper, W.; Heidorn, K.; Rott, A.; Parwaresch, R. (1997). Molecular basis of artifacts in the detection of telomerase activity and a modified primer for a more robust 'TRAP' assay. *Nucleic Acids Res.*, *25*, 919-921.

[38] Kim, N.W.; Wu, F. (1997). Advances in quantification and characterization of telomerase activity by the telomeric repeat amplification protocol (TRAP). *Nucleic Acids Res.*, *25*, 2595-2597.

[39] Falchetti, M.L.; Levi, A.; Molinari, P.; Verna, R.; D'Ambrosio, E. (1998). Increased sensitivity and reproducibility of TRAP assay by avoiding direct primers interaction. *Nucleic Acids Res.*, *26*, 862-863.

[40] Schmidt, P.M.; Lehmann, C.; Matthes, E.; Bier, F.F. (2002). Detection of activity of telomerase in tumor cells using fiber optical biosensors. *Biosens. Bioelectron.*, *17*, 1081-1087.

[41] Kawamura, K.; Yaku, H.; Miyoshi, D.; Murashima, T. (2014). A simple "Add and measure" FRET-based telomeric tandem repeat sequence detection and telomerase assay method. *Org. Biomol. Chem.*, *12*, 936-941.

[42] Qian, R.; Ding, L.; Yan, L.; Lin, M.; Ju, H. (2014). A robust probe for lighting up intracellular telomerase via primer extension to open a nicked molecular beacon. *J. Am. Chem. Soc.*, *136*, 8205-8208.

[43] Sharon, E.; Freeman, R.; Riskin, M.; Gil, N.; Tzfati, Y.; Willner, I. (2010). Optical, electrical and surface plasmon resonance methods for detecting telomerase activity. *Anal. Chem.*, *82*, 8390-8397.

[44] Zhang, Z.; Sharon, E.; Freeman, R.; Liu, X.; Willner, I. (2012). Fluorescence detection of DNA, adenosine-5'-triphosphate (ATP), and telomerase activity by zinc(II)-protoporphyrin IX/G-quadruplex labels. *Anal. Chem.*, *84*, 4789-4797.

[45] Quach, Q.H.; Jung, J.; Kim, H.; Chung, B.H. (2013). A simple, fast and highly sensitive assay for the detection of telomerase activity. *Chem. Commun.*, *49*, 6596-6598.

[46] Williamson, J.R.; Raghuraman, M.K.; Cech, T.R. (1989). Mono-valent cation induced structure of telomeric DNA: the G-quartet model. *Cell*, *59*, 871-880.

[47] Wang, Y.; Patel, D.J. (1993). Solution structure of the human telomeric repeat d [$AG_3(T_2AG_3)_3$] G-tetraplex. *Structure*, *1*, 263-282.

[48] Balagurumoorthy, P.; Brahmachari, S.K. (1994). Structure and stability of human telomeric sequence. *J. Biol. Chem.*, *269*, 21858-21869.

[49] Parkinson, G.N.; Lee, M.P.H.; Neidle, S. (2002). Crystal structure of parallel quadruplexes from human telomeric DNA. *Nature*, *417*, 876-880.

[50] Ying, L.; Green, J.J.; Li, H.; Klenerman, D.; Balasubramanian, S. (2003). Studies on the structure and dynamics of the human telomeric G quadruplex by single-molecule fluorescence resonance energy transfer. *Proc. Natl. Acad. Sci. USA*, *100*, 14629-14634.

[51] Redon, S.; Bombard, S.; Elizondo-Riojas, M.A.; Chottard, J.C. (2003). Platinum cross-linking of adenines and guanines on the quadruplex structures of the $AG_3(T_2AG_3)_3$ and $(T_2AG_3)_4$ human telomere sequences in Na^+ and K^+ solutions. *Nucleic Acids Res.*, *31*, 1605-1613.

[52] Phan, A.T.; Patel, D.J. (2003). Two-repeat human telomeric d(TAGGGTTAGGGT)

sequence forms interconverting parallel and antiparallel G-quadruplexes in solution: distinct topologies, thermodynamic properties, and folding/unfolding kinetics. *J. Am. Chem. Soc.*, *125*, 15021-15027.

[53] He, Y.; Neumann, R.D.; Panyutin, I.G. (2004). Intramolecular quadruplex conformation of human telomeric DNA assessed with ^{125}I-radioprobing. *Nucleic Acids Res.*, *32*, 5359-5367.

[54] Li, J.; Correia, J.J.; Wang, L.; Trent, J.O.; Chaires, J.B. (2005). Not so crystal clear: the structure of the human telomere G-quadruplex in solution differs from that present in a crystal. *Nucleic Acids Res.*, *33*, 4649-4659.

[55] Zhang, N.; Phan, A.T.; Patel, D.J. (2005). (3+1) assembly of three human telomeric repeats into an asymmetric dimeric G-quadruplex. *J. Am. Chem. Soc.*, *127*, 17277-17285.

[56] Włodarczyk, A.; Grzybowski, P.; Patkowski, A.; Dobek, A. (2005). Effect of ions on the polymorphism, effective charge, and stability of human telomeric DNA. Photon correlation spectroscopy and circular dichroism studies. *J. Phys. Chem. B*, *109*, 3594-3605.

[57] Qi, J.; Shafer, R.H. (2005). Covalent ligation studies on the human telomere quadruplex. *Nucleic Acids Res.*, *33*, 3185-3192.

[58] Ambrus, A.; Chen, D.; Dai, J.X.; Bialis, T.; Jones, R.A.; Yang, D.Z. (2006). Human telomeric sequence forms a hybrid-type intramolecular G-quadruplex structure with mixed parallel/antiparallel strands in potassium solution. *Nucleic Acids Res*, *34*, 2723-2735.

[59] Luu, K.N.; Phan, A.T.; Kuryavyi, V.; Lacroix, L.; Patel, D.J. (2006). Structure of the human telomere in K^+ solution: An intramolecular (3+1) G-quadruplex scaffold. *J. Am. Chem. Soc.*, *128*, 9963-9970.

[60] Xu, Y.; Noguchi, Y.; Sugiyama, H. (2006). The new models of the human telomere d[AGGG(TTAGGG)$_3$] in K^+ solution. *Bioorgan. Med. Chem.*, *14*, 5584-5591.

[61] Yu, H.Q.; Miyoshi, D.; Sugimoto, N. (2006). Characterization of structure and stability of long telomeric DNA G-quadruplexes. *J. Am. Chem. Soc.*, *128*, 15461-15468.

[62] Dai, J.X.; Carver, M.; Punchihewa, C.; Jones, R.A.; Yang, D.Z. (2007). Structure of the hybrid-2 type intramolecular human telomeric G-quadruplex in K^+ solution: insights into structure polymorphism of the human telomeric sequence. *Nucleic Acids Res.*, *35*, 4927-4940.

[63] Phan, A.T.; Kuryavyi, V.; Luu, K.N.; Patel, D.J. (2007). Structure of two intramolecular G-quadruplexes formed by natural human telomere sequences in K^+ solution. *Nucleic Acids Res.*, *35*, 6517-6525.

[64] Dai, J.X.; Carver, M.; Yang, D.Z. (2008). Polymorphism of human telomeric quadruplex structures. *Biochimie*, *90*, 1172-1183.

[65] Balasubramanian, S.; Neidle, S. (2009). G-quadruplex nucleic acids as therapeutic targets. *Curr. Opin. Chem. Biol.*, *13*, 345-353.

[66] Cuesta, J.; Read, M.A.; Neidle, S. (2003). The design of G-quadruplex ligands as telomerase inhibitors. *Mini. Rev. Med. Chem.*, *3*, 11-21.

[67] Tan, J.H.; Gu, L.Q.; Wu, J.Y. (2008). Design of selective G-quadruplex ligands as potential anticancer agents. *Mini Rev Med Chem*, *8*, 1163-1178.

[68] Yaku, H.; Fujimoto, T.; Murashima, T.; Miyoshi, D.; Sugimoto, N. (2012). Phthalocyanines: a new class of G-quadruplex-ligands with many potential

applications. *Chem. Commun.*, *48*, 6203-6216.

[69] Travascio, P.; Li, Y.F.; Sen, D. (1998). DNA-enhanced peroxidase activity of a DNA aptamer-hemin complex. *Chem. Biol.*, *5*, 505-517.

[70] Ma, D.L.; Chan, D.S.H.; Yang, H.; He, H.Z.; Leung, C.H. (2012). Luminescent G-quadruplex probes. *Curr. Pharm. Design*, *18*, 2058-2075.

[71] Neo, J.L.; Kamaladasan, K.; Uttamchandani, M. (2012). G-quadruplex based probes for visual detection and sensing. *Curr. Pharm. Design*, *18*, 2048-2057.

[72] Yaku, H.; Murashima, T.; Miyoshi, D.; Sugimoto, N. (2012). Specific binding of anionic porphyrin and phthalocyanine to the G-quadruplex with a variety of in vitro and in vivo applications. *Molecules*, *17*, 10586-10613.

[73] Ruttkay-Nedecky, B.; Kudr, J.; Nejdl, L.; Maskova, D.; Kizek, R.; Adam, V. (2013). G-quadruplexes as sensing probes. *Molecules*, *18*, 14760-14779.

[74] He, H.Z.; Chan, D.S.H.; Leung, C.H.; Ma, D.L. (2013). G-quadruplexes for luminescent sensing and logic gates. *Nucleic Acids Res.*, *41*, 4345-4359.

[75] Xu, H.; Gao, S.; Yang, Q.; Pan, D.; Wang, L.; Fan, C. (2010). Amplified fluorescent recognition of G-quadruplex folding with a cationic conjugated polymer and DNA intercalator. *ACS Appl. Mater. Inter.*, *2*, 3211-3216.

[76] Zhou, X.M.; Xing, D.; Zhu, D.B.; Jia, L. (2009). Magnetic bead and nanoparticle based electrochemiluminescence amplification assay for direct and sensitive measuring of telomerase activity. *Anal. Chem.*, *81*, 255-261.

[77] Wu, L.; Wang, J.; Feng, L.; Ren, J.; Wei, W.; Qu, X. (2012). Label-free ultrasensitive detection of human telomerase activity using porphyrin-functionalized graphene and electrochemiluminescence technique. *Adv. Mater.*, *24*, 2447-2452.

[78] Seydack, M. (2005). Nanoparticle labels in immunosensing using optical detection methods. *Biosens. Bioelectron.*, *20*, 2454-2469.

[79] Jans, H.; Huo, Q. (2012). Gold nanoparticle-enabled biological and chemical detection and analysis. *Chem. Soc. Rev.*, *41*, 2849-2866.

[80] Wang, J.; Wu, L.; Ren, J.; Qu, X. (2012). Visualizing human telomerase activity with primer-modified Au nanoparticles. *Small*, *8*, 259-264.

[81] Duan, R.; Wang, B.; Zhang, T.; Zhang, Z.; Xu, S.; Chen, Z.; Lou, X.; Xia, F. (2014). Sensitive and bidirectional detection of urine telomerase based on the four detection-color states of difunctional gold nanoparticle probe. *Anal. Chem.*, *86*, 9781-9785.

[82] Sato, S.; Kondo, H.; Nojima, T.; Takenaka, S. (2005). Electrochemical telomerase assay with ferrocenyl naphthalene diimide as a tetraplex DNA-specific binder. *Anal. Chem.*, *77*, 7304-7309.

[83] Li, Y.; Liu, B.W.; Li, X.; Wei, Q.L. (2010). Highly sensitive electrochemical detection of human telomerase activity based on bio-barcode method. *Biosens. Bioelectron.*, *25*, 2543-2547.

[84] Yi, Z.; Wang, H.B.; Chen, K.; Gao, Q.; Tang, H.; Yu, R.Q.; Chu, X. (2014). A novel electrochemical biosensor for sensitive detection of telomerase activity based on structure-switching DNA. *Biosens. Bioelectron.*, *53*, 310-315.

[85] Homola, J. (2008). Surface plasmon resonance sensors for detection of chemical and biological species. *Chem. Rev.*, *108*, 462-493.

[86] Buckle, M.; Williams, R.M.; Negroni, M.; Buc, H. (1996). Real time measurements of elongation by a reverse transcriptase using surface plasmon resonance. *Proc. Natl. Acad. Sci. USA*, *93*, 889-894.

[87] Pemberton, I.K.; Buckle, M. (1999). Real time in vitro analysis of transcription by RNA polymerase on immobilized DNA fibres. *J. Mol. Recognit.*, *12*, 322-327.

[88] Li, Y.; Yan, Y.; Lei, Y.; Zhao, D.; Yuan, T.; Zhang, D.; Cheng, W.; Ding, S. (2014). Surface plasmon resonance biosensor for label-free and highly sensitive detection of point mutation using polymerization extension reaction. *Colloids Surf. B Biointerfaces.*, *120*, 15-20.

[89] Maesawa, C.; Inaba, T.; Sato, H.; Iijima, S.; Ishida, K.; Terashima, M.; Sato, R.; Suzuki, M.; Yashima, A.; Ogasawara, S., et al. (2003). A rapid biosensor chip assay for measuring of telomerase activity using surface plasmon resonance. *Nucleic Acids Res.*, *31, e4*.

[90] Nie, S.; Emory, S.R. (1997). Probing single molecules and single nanoparticles by surface-enhanced Raman scattering. *Science*, *275*, 1102-1106.

[91] Sharma, B.; Frontiera, R.R.; Henry, A.I.; Ringe, E.; Van Duyne, R.P. (2012). SERS: materials, applications, and the future. *Mater. Today*, *15*, 16-25.

[92] Orendorff, C.J.; Gole, A.; Sau, T.K.; Murphy, C.J. (2005). Surface-enhanced Raman spectroscopy of self-assembled monolayers: sandwich architecture and nanoparticle shape dependence. *Anal. Chem.*, *77*, 3261-3266.

[93] Le Ru, E.C.; Blackie, E.; Meyer, M.; Etchegoint, P.G. (2007). Surface enhanced Raman scattering enhancement factors: a comprehensive study. *J. Phys. Chem. C*, *111*, 13794-13803.

[94] Zong, S.; Wang, Z.; Chen, H.; Cui, Y. (2013). Ultrasensitive telomerase activity detection by telomeric elongation controlled surface enhanced Raman scattering. *Small*, *9*, 4215-4220.

[95] Zong, S.; Wang, Z.; Chen, H.; Hu, G.; Liu, M.; Chen, P.; Cui, Y. (2014). Colorimetry and SERS dual-mode detection of telomerase activity: combining rapid screening with high sensitivity. *Nanoscale*, *6*, 1808-1816.

[96] Ronaghi, M.; Uhlén, M.; Nyrén, P. (1998). A sequencing method based on real-time pyrophosphate. *Science*, *281*, 363-365.

[97] Ronaghi, M.; Karamohamed, S.; Pettersson, B.; Uhlén, M.; Nyrén, P. (1996). Real-time DNA sequencing using detection of pyrophosphate release. *Anal. Biochem.*, *242*, 84-89.

[98] Xu, S.Q.; He, M.; Yu, H.P.; Wang, X.Y.; Tan, X.L.; Lu, B.; Sun, X.; Zhou, Y.K.; Yao, Q.F.; Xu, Y.J., et al. (2002). Bioluminescent method for detecting telomerase activity. *Clin. Chem.*, *48*, 1016-1020.

[99] Ma, C.B.; Xia, K.; Chen, H.C.; Zeng, W.M.; Han, R.; Tang, J.H. (2012). Label-free highly sensitive detection of telomerase activity in cancer cell by chemiluminescence imaging. *Mol. Cell. Probe.*, *26*, 212-214.

[100] Lee, H.W.; Chinnapen, D.J.F.; Sen, D. (2004). Structure-function investigation of a deoxyribozyme with dual chelatase and peroxidase activities. *Pure Appl. Chem.*, *76*, 1537-1545.

[101] Majhi, P.R.; Shafer, R.H. (2006). Characterization of an unusual folding pattern in a catalytically active guanine quadruplex structure. *Biopolymers*, *82*, 558-569.

[102] Kong, D.M.; Cai, L.L.; Guo, J.H.; Wu, J.; Shen, H.X. (2009). Characterization of the G-quadruplex structure of a catalytic DNA with peroxidase activity. *Biopolymers*, *91*, 331-339.

[103] Cheng, X.H.; Liu, X.J.; Bing, T.; Cao, Z.H.; Shangguan, D.H. (2009). General peroxidase activity of G-quadruplex-hemin complexes and its application in ligand

screening. *Biochemistry*, *48*, 7817-7823.
[104] Kong, D.M.; Yang, W.; Wu, J.; Li, C.X.; Shen, H.X. (2010). Structure-function study of peroxidase-like G-quadruplex-hemin complexes. *Analyst*, *135*, 321-326.
[105] Xiao, Y.; Pavlov, V.; Niazov, T.; Dishon, A.; Kotler, M.; Willner, I. (2004). Catalytic beacons for the detection of DNA and telomerase activity. *J. Am. Chem. Soc.*, *126*, 7430-7431.
[106] Li, H.; Fu, H.W.; Zhao, T.; Kong, D.M. (2015). Simple, PCR-free telomerase activity detection using G-quadruplex-hemin DNAzyme. *RSC Adv.*, *5*, 6475-6480.
[107] Tian, T.; Peng, S.; Xiao, H.; Zhang, X.E.; Guo, S.; Wang, S.R.; Zhou, X.; Liu, S.M. (2013). Highly sensitive detection of telomerase based on a DNAzyme strategy. *Chem. Commun.*, *49*, 2652-2654.
[108] Wang, M.; Fu, Z.; Li, B.; Zhou, Y.; Yin, H.; Ai, S. (2014). One-step, ultrasensitive, and electrochemical assay of microRNAs based on T7 exonuclease assisted cyclic enzymatic amplification. *Anal. Chem.*, *86*, 5606-5610.
[109] Cui, L.; Zhu, Z.; Lin, N.; Zhang, H.; Guan, Z.; Yang, C.J. (2014). A T7 exonuclease-assisted cyclic enzymatic amplification method coupled with rolling circle amplification: a dual-amplification strategy for sensitive and selective microRNA detection. *Chem. Commun.*, *50*, 1576-1578.
[110] Wu, Z.K.; Zhou, D.M.; Wu, Z.; Chu, X.; Yu, R.Q.; Jiang, J.H. (2015). Single-base mismatch discrimination by T7 exonuclease with target cyclic amplification detection. *Chem. Commun.*, *51*, 2954-2956.
[111] Wang, H.B.; Ou, L.J.; Huang, K.J.; Wen, X.G.; Wang, L.L.; Liu, Y.M. (2013). A sensitive biosensing strategy for DNA detection based on graphene oxide and T7 exonuclease assisted target recycling amplification. *Can. J. Chem.*, *91*, 1266-1271.
[112] Yan, M.; Bai, W.; Zhu, C.; Huang, Y.; Yan, J.; Chen, A. (2016). Design of nuclease-based target recycling signal amplification in aptasensors. *Biosens. Bioelectron.*, *77*, 613-623.
[113] Wang, H.B.; Wu, S.; Chu, X.; Yu, R.Q. (2012). A sensitive fluorescence strategy for telomerase detection in cancer cells based on T7 exonuclease-assisted target recycling amplification. *Chem. Commun.*, *48*, 5916-5918.
[114] Liu, X.; Li, W.; Hou, T.; Dong, S.; Yu, G.; Li, F. (2015). Homogeneous electrochemical strategy for human telomerase activity assay at single-cell level based on T7 exonuclease-aided target recycling amplification. *Anal. Chem.*, *87*, 4030-4036.
[115] Van Ness, J.; Van Ness, L.K.; Galas, D.J. (2003). Isothermal reactions for the amplification of oligonucleotides. *Proc. Natl. Acad. Sci. USA*, *100*, 4504-4509.
[116] Tan, E.; Wong, J.; Nguyen, D.; Zhang, Y.; Erwin, B.; Van Ness, L.K.; Baker, S.M.; Galas, D.J.; Niemz, A. (2005). Isothermal DNA amplification coupled with DNA nanosphere-based colorimetric detection. *Anal. Chem.*, *77*, 7984-7992.
[117] Wang, L.J.; Zhang, Y.; Zhang, C.Y. (2013). Ultrasensitive detection of telomerase activity at the single-cell level. *Anal. Chem.*, *85*, 11509-11517.
[118] Zhang, Y.; Wang, L.J.; Zhang, C.Y. (2014). Highly sensitive detection of telomerase using a telomere-triggered isothermal exponential amplification-based DNAzyme biosensor. *Chem. Commun.*, *50*, 1909-1911.
[119] Tian, L.; Weizmann, Y. (2013). Real-time detection of telomerase activity using the exponential isothermal amplification of telomere repeat assay. *J. Am. Chem. Soc.*, *135*, 1661-1664.

[120] Gollahon, L.S.; Holt, S.E. (2000). Alternative methods of extracting telomerase activity from human tumor samples. *Cancer Lett.*, *159*, 141-149.

[121] Patel, S.D.; Isalan, M.; Gavory, G.; Ladame, S.; Choo, Y.; Balasubramanian, S. (2004). Inhibition of human telomerase activity by an engineered zinc finger protein that binds G-quadruplexes. *Biochemistry*, *43*, 13452-13458.

[122] Reed, J.; Gunaratnam, M.; Beltran, M.; Reszka, A.P.; Vilar, R.; Neidle, S. (2008). Trap-lig, a modified telomere repeat amplification protocol assay to quantitate telomerase inhibition by small molecules. *Anal. Biochem.*, *380*, 99-105.

[123] Chen, Y.C.; Huang, F.C.; Lin, J.J. (2015). The addition of a spin column step in the telomeric repeat application protocol removes telomerase inhibitors. *Anal. Biochem.*, *478*, 49-51.

[124] Yaku, H.; Murashima, T.; Miyoshi, D.; Sugimoto, N. (2010). Anionic phthalocyanines targeting G-quadruplexes and inhibiting telomerase activity in the presence of excessive DNA duplexes. *Chem. Commun.*, *46*, 5740-5742.

[125] Kha, H.; Zhou, W.; Chen, K.; Karan-Tamir, B.; Miguel, T.S.; Zeni, L.; Kearns, K.; Mladenovic, A.; Rasnow, B.; Robinson, M., et al. (2004). A telomerase enzymatic assay that does not use polymerase chain reaction, radioactivity, or electrophoresis. *Anal. Biochem.*, *331*, 230-234.

[126] Pavlov, V.; Willner, I.; Dishon, A.; Kotler, M. (2004). Amplified detection of telomerase activity using electrochemical and quartz crystal microbalance measurements. *Biosens. Bioelectron.*, *20*, 1011-1021.

[127] Li, Y.; Wen, Y.; Wang, L.; Liang, W.; Xu, L.; Ren, S.; Zou, Z.; Zuo, X.; Fan, C.; Huang, Q., et al. (2015). Analysis of telomerase activity based on a spired DNA tetrahedron TS primer. *Biosens. Bioelectron.*, *67*, 364-369.

[128] Pavlov, V.; Xiao, Y.; Gill, R.; Dishon, A.; Kotler, M.; Willner, I. (2004). Amplified chemiluminescence surface detection of DNA and telomerase activity using catalytic nucleic acid labels. *Anal. Chem.*, *76*, 2152-2156.

[129] Niazov, T.; Pavlov, V.; Xiao, Y.; Gill, R.; Willner, I. (2004). DNAzyme-functionalized au nanoparticles for the amplified detection of DNA or telomerase activity. *Nano Lett.*, *4*, 1683-1687.

[130] Zhang, Z.; Wu, L.; Wang, J.; Ren, J.; Qu, X. (2013). A Pt-nanoparticle electrocatalytic assay used for PCR-free sensitive telomerase detection. *Chem. Commun.*, *49*, 9986-9988.

[131] Wu, L.; Wang, J.; Ren, J.; Qu, X. (2014). Ultrasensitive telomerase activity detection in circulating tumor cells based on DNA metallization and sharp solid-state electrochemical techniques. *Adv. Funct. Mater.*, *24*, 2727-2733.

[132] Bekkaoui, F.; Poisson, I.; Crosby, W.; Cloney, L.; Duck, P. (1996). Cycling probe technology with RNase H attached to an oligonucleotide. *Biotechniques*, *20*, 240-248.

[133] Yaku, H.; Murashima, T.; Miyoshi, D.; Sugimoto, N. (2013). A highly sensitive telomerase activity assay that eliminates false-negative results caused by PCR inhibitors. *Molecules*, *18*, 11751-11767.

[134] Tian, L.; Cronin, T.M.; Weizmann, Y. (2014). Enhancing-effect of gold nanoparticles on DNA strand displacement amplifications and their application to an isothermal telomerase assay. *Chem. Sci.*, *5*, 4153-4162.

[135] Lodish, H.; Berk, A.; Zipursky, S.L.; Matsudaira, P.; Baltimore, D.; Darnell, J. *Molecular cell biology*. 4th edition. New York: W. H. Freeman and Company: 2000.

[136] Mori, K.; Sato, S.; Kodama, M.; Habu, M.; Takahashi, O.; Nishihara, T.; Tominaga, K.; Takenaka, S. (2013). Oral cancer diagnosis via a ferrocenylnaphthalene diimide-based electrochemical telomerase assay. *Clin. Chem.*, *59*, 289-295.

[137] Vilkner, T.; Janasek, D.; Manz, A. (2004). Micro total analysis systems. Recent developments. *Anal. Chem.*, *76*, 3373-3386.

[138] Dittrich, P.S.; Tachikawa, K.; Manz, A. (2006). Micro total analysis systems. Latest advancements and trends. *Anal. Chem.*, *78*, 3887-3907.

[139] Patabadige, D.E.W.; Jia, S.; Sibbitts, J.; Sadeghi, J.; Sellens, K.; Culbertson, C.T. (2016). Micro total analysis systems: fundamental advances and applications. *Anal. Chem.*, *88*, 320-338.

[140] Tsuchiya, H.; Okochi, M.; Nagao, N.; Shikida, M.; Honda, H. (2008). On-chip polymerase chain reaction microdevice employing a magnetic droplet-manipulation system. *Sens. Actuators B Chem.*, *130*, 583-588.

[141] Shikida, M.; Takayanagi, K.; Inouchi, K.; Honda, H.; Sato, K. (2006). Using wettability and interfacial tension to handle droplets of magnetic beads in a micro-chemical-analysis system. *Sens. Actuators B Chem.*, *113*, 563-569.

[142] Shikida, M.; Takayanagi, K.; Honda, H.; Ito, H.; Sato, K. (2006). Development of an enzymatic reaction device using magnetic bead-cluster handling. *J. Micromech. Microeng.*, *16*, 1875-1883.

[143] Lehmann, U.; Vandevyver, C.; Parashar, V.K.; Gijs, M.A.M. (2006). Droplet-based DNA purification in a magnetic lab-on-a-chip. *Angew. Chem., Int. Ed.*, *45*, 3062-3067.

[144] Yeung, S.W.; Hsing, I.M. (2006). Manipulation and extraction of genomic DNA from cell lysate by functionalized magnetic particles for lab on a chip applications. *Biosens. Bioelectron.*, *21*, 989-997.

[145] Bucak, S.; Jones, D.A.; Laibinis, P.E.; Hatton, T.A. (2003). Protein separations using colloidal magnetic nanoparticles. *Biotechnol. Progr.*, *19*, 477-484.

[146] Hubbuch, J.J.; Thomas, O.R.T. (2002). High-gradient magnetic affinity separation of trypsin from porcine pancreatin. *Biotechnol. Bioeng.*, *79*, 301-313.

[147] Cooper, R.M.; Leslie, D.C.; Domansky, K.; Jain, A.; Yung, C.; Cho, M.; Workman, S.; Super, M.; Ingber, D.E. (2014). A microdevice for rapid optical detection of magnetically captured rare blood pathogens. *Lab Chip*, *14*, 182-188.

[148] Pivetal, J.; Toru, S.; Frenea-Robin, M.; Haddour, N.; Cecillon, S.; Dempsey, N.M.; Dumas-Bouchiat, F.; Simonet, P. (2014). Selective isolation of bacterial cells within a microfluidic device using magnetic probe-based cell fishing. *Sens. Actuators B Chem.*, *195*, 581-589.

[149] Otieno, B.A.; Krause, C.E.; Latus, A.; Chikkaveeraiah, B.V.; Faria, R.C.; Rusling, J.F. (2014). On-line protein capture on magnetic beads for ultrasensitive microfluidic immunoassays of cancer biomarkers. *Biosens. Bioelectron.*, *53*, 268-274.

[150] Glynn, M.; Kirby, D.; Chung, D.; Kinahan, D.J.; Kijanka, G.; Ducrée, J. (2014). Centrifugo-magnetophoretic purification of CD4+ cells from whole blood toward future HIV/AIDS point-of-care applications. *J. Lab. Autom.*, *19*, 285-296.

[151] Leïchlé, T.; Bourrier, D. (2015). Integration of lateral porous silicon membranes into planar microfluidics. *Lab Chip*, *15*, 833-838.

[152] Kulasingam, S.L.; Hughes, J.P.; Kiviat, N.B.; Mao, C.; Weiss, N.S.; Kuypers, J.M.; Koutsky, L.A. (2002). Evaluation of human papillomavirus testing in primary screening for cervical abnormalities: comparison of sensitivity, specificity, and frequency of

referral. *J. Amer. Med. Assoc.*, *288*, 1749-1757.

[153] Solomon, D. (2003). Results of a randomized trial on the management of cytology interpretations of atypical squamous cells of undetermined significance. *Am. J. Obstet. Gynecol.*, *188*, 1383-1392.

[154] McMurray, H.R.; McCance, D.J. Human papillomavirus type 16 E6 activates TERT gene transcription through induction of c-Myc and release of USF-mediated repression. *J. Virol.* 2003, *77*, 9852-9861.

[155] Cuzick, J.; Clavel, C.; Petry, K.U.; Meijer, C.J.L.M.; Hoyer, H.; Ratnam, S.; Szarewski, A.; Birembaut, P.; Kulasingam, S.; Sasieni, P., et al. (2006). Overview of the european and north american studies on HPV testing in primary cervical cancer screening. *Int. J. Cancer*, *119*, 1095-1101.

[156] Schiffman, M.; Castle, P.E.; Jeronimo, J.; Rodriguez, A.C.; Wacholder, S. (2007). Human papillomavirus and cervical cancer. *Lancet*, *370*, 890-907.

[157] Cuzick, J.; Arbyn, M.; Sankaranarayanan, R.; Tsu, V.; Ronco, G.; Mayrand, M.H.; Dillner, J.; Meijer, C.J.L.M. (2008). Overview of human papillomavirus-based and other novel options for cervical cancer screening in developed and developing countries. *Vaccine*, *26*, K29-K41.

[158] Dillner, J.; Rebolj, M.; Birembaut, P.; Petry, K.U.; Szarewski, A.; Munk, C.; de Sanjose, S.; Naucler, P.; Lloveras, B.; Kjaer, S., et al. (2008). Long term predictive values of cytology and human papillomavirus testing in cervical cancer screening: joint european cohort study. *Br. Med. J. (Clin. Res. Ed.)*, *337*, 969-972.

[159] Howie, H.L.; Katzenellenbogen, R.A.; Galloway, D.A. (2009). Papillomavirus E6 proteins. *Virology*, *384*, 324-334.

[160] Wang, J.; Xie, L.Y.; Allan, S.; Beach, D.; Hannon, G.J. (1998). Myc activates telomerase. *Gene. Dev.*, *12*, 1769-1774.

[161] Wu, K.J.; Grandori, C.; Amacker, M.; Simon-Vermot, N.; Polack, A.; Lingner, J.; Dalla-Favera, R. (1999). Direct activation of TERT transcription by c-MYC. *Nat. Genet.*, *21*, 220-224.

[162] Trimble, C.L.; Piantadosi, S.; Gravitt, P.; Ronnett, B.; Pizer, E.; Elko, A.; Wilgus, B.; Yutzy, W.; Daniel, R.; Shah, K., et al. (2005). Spontaneous regression of high-grade cervical dysplasia: effects of human papillomavirus type and HLA phenotype. *Clin. Cancer Res.*, *11*, 4717-4723.

[163] Castle, P.E.; Stoler, M.H.; Solomon, D.; Schiffman, M. (2007). The relationship of community biopsy-diagnosed cervical intraepithelial neoplasia grade 2 to the quality control pathology-reviewed diagnoses: an ALTS report. *Am. J. Clin. Pathol.*, *127*, 805-815.

[164] Castle, P.E.; Schiffman, M.; Wheeler, C.M.; Solomon, D. (2009). Evidence for frequent regression of cervical intraepithelial neoplasia-grade 2. *Obstet. Gynecol.*, *113*, 18-25.

[165] Ronco, G.; Giorgi-Rossi, P.; Carozzi, F.; Confortini, M.; Palma, P.D.; Del Mistro, A.; Ghiringhello, B.; Girlando, S.; Gillio-Tos, A.; De Marco, L., et al. (2010). Efficacy of human papillomavirus testing for the detection of invasive cervical cancers and cervical intraepithelial neoplasia: a randomised controlled trial. *Lancet Oncol.*, *11*, 249-257.

[166] Ronco, G.; Giorgi-Rossi, P.; Carozzi, F.; Confortini, M.; Palma, P.D.; Del Mistro, A.; Gillio-Tos, A.; Minucci, D.; Naldoni, C.; Rizzolo, R., et al. (2008). Results at recruitment from a randomized controlled trial comparing human papillomavirus testing alone with conventional cytology as the primary cervical cancer screening test. *J. Natl.*

Cancer Inst., *100*, 492-501.
[167] Okita, K.; Yamanaka, S. (2011). Induced pluripotent stem cells: opportunities and challenges. *Phil. Trans. R. Soc. B*, *366*, 2198-2207.

BIOGRAPHICAL SKETCHES

Hidenobu Yaku
Advanced Research Division,
Panasonic Corporation
Japan

Education: PhD in 2014 at Konan University

Research and Professional Experience: DNA/Protein detection, Cancer drug design

Professional Appointments: Chief Researcher, Advanced Research Division, Panasonic Corporation

Selected Publications Last 3 Years:

Sensors, 15, 9388-9403 (2015)
J. Phys. Chem. B (Cover), 118, 2605-2614 (2014)
Org. Biomol. Chem., 12, 936-941 (2014)
Jpn. J. Appl. Phys., 53, 05FS03-1-05FS03-5 (2014)
Methods, 64, 19-27 (2013)
Molecules, 18, 11751-11767 (2013)
Biochemistry, 52, 5620-5628 (2013)

Daisuke Miyoshi
Faculty of Frontiers of Innovative Research in Science and Technology (FIRST);
Konan University, Japan

Education: PhD in 2003 at Konan University

Research and Professional Experience: DNA chemistry, Biophysical chemistry

Professional Appointments: 2014 Professor, Faculty of Frontiers of Innovative Research in Science and Technology (FIRST), Konan University

Selected Publications Last 3 Years:

Biochimie, 121, 204-208 (2016)
J. Biol. Inorg. Chem., 20, 1049-1058 (2015)
J. Phys. Chem. B (Cover), 119, 11969-11977 (2015)

ChemBioChem, 16, 1803-1810 (2015)
Sensors, 15, 9388-9403 (2015)
Chem. Commun., 51, 1479-1482 (2015)
ChemMedChem, 9, 2150-2155 (2014)
FEBS Open Bio., 4, 643-650 (2014)
ChemMedChem (Back cover), 9, 2156-2163 (2014)
Org. Biomol. Chem., 12, 936-941 (2014)
Chem. Rev., 114, 2733-2758 (2014)
J. Phys. Chem. B (Cover), 118, 2605-2614 (2014)
Small, 10, 330-336 (2014)
Quadruplex Nucleic Acids (Springer), 330, 87-110 (2013)
Molecules, 18, 11751-11767 (2013)
J. Phys. Chem. B (Cover), 117, 12391-12401 (2013)
Biochemistry, 52, 5620-5628 (2013)
Methods, 64, 19-27 (2013)
J. Phys. Chem. B (Cover), 117, 963-972 (2013)

INDEX

K

L

M

P

Q

R

S

T

X

Y

Z

μ